8
Plant Cell Monographs

Series Editor: David G. Robinson

Plant Cell Monographs

Recently Published Titles

Viral Transport in Plants
Volume Editors: Waigmann, E., Heinlein, M.
Vol. 7, 2007

Nitric Oxide in Plant Growth, Development and Stress Physiology
Volume Editors: Lamattina, L., Polacco, J.
Vol. 6, 2007

The Expanding Cell
Volume Editors: Verbelen, J.-P., Vissenberg, K.
Vol. 5, 2007

The Plant Endoplasmic Reticulum
Volume Editor: Robinson, D. G.
Vol. 4, 2006

The Pollen Tube
A Cellular and Molecular Perspective
Volume Editor: Malhó, R.
Vol. 3, 2006

Somatic Embryogenesis
Volume Editors: Mujib, A., Šamaj, J.
Vol. 2, 2006

Plant Endocytosis
Volume Editors:
Šamaj, J., Baluška, F., Menzel, D.
Vol. 1, 2005

Editor

Odd-Arne Olsen studied genetics and cell biology (M.Sc.) at the University of Oslo (Norway). He received a doctorate in agriculture from the Agricultural University of Norway. Olsen has held positions as research director and professor of genetics at the Agricultural University of Norway. He has made sabbatical research visits to UC Davis, Cornell and Stanford. From 2000–2005 he worked at Pioneer hi-Bred International, a DuPont Company, in Johnston, Iowa. Since 2006 Olsen has been working at Monsanto Company in St. Louis, Missouri. His major research interests include endosperm developmental biology and biotechnology.

Endosperm

Developmental and Molecular Biology

Volume Editor: Odd-Arne Olsen

With 23 Figures and 5 Tables

Springer

Volume Editor:

Dr. Odd-Arne Olsen
Monsanto Company
700 Chesterfield Pkwy West
Chesterfield, MO 63017
USA
odd-arne.olsen@monsanto.com

Series Editor:

Professor Dr. David G. Robinson
Ruprecht-Karls-University of Heidelberg
Heidelberger Institute for Plant Sciences (HIP)
Department Cell Biology
Im Neuenheimer Feld 230
D-69120 Heidelberg
Germany

Library of Congress Control Number: 2007924314

ISSN 1861-1370
ISBN 978-3-540-71234-3 Springer Berlin Heidelberg New York
DOI 10.1007/978-3-540-71235-0

Springer is a part of Springer Science+Business Media

springer.com

Editor: Dr. Christina Eckey, Heidelberg
Desk Editor: Anette Lindqvist, Heidelberg
Cover design: WMXDesign GmbH, Heidelberg
Typesetting and Production: LE-TEX Jelonek, Schmidt & Vöckler GbR, Leipzig

Printed on acid-free paper 149/3180 YL – 5 4 3 2 1 0

Preface

The biological function of endosperm is to support the developing and germinating embryo, and as such is believed to have played an important role for the evolutionary success of angiosperms. The ontogeny of the endosperm remained an enigma until the discovery of double fertilization a little more than a hundred years ago. Soon thereafter, plant anatomists revealed an unusual pattern of development: many types of endosperm go through a syncytial phase before becoming cellular. The mechanism underlying the cellularization process of these so-called nuclear endosperms was discovered only recently, and most of the molecular and developmental biological insight into endosperms has been gained only in the last two decades. To my knowledge, this book is the first comprehensive treaty dedicated in its entirety to endosperm developmental and molecular biology.

Cellularization of nuclear endosperm occurs through an unusual series of events in which all microtubular structures known to exist in plants are involved. This book details the structure, cell biology and molecular aspects of this process in cereals as well as in *Arabidopsis thaliana*. After cellularization, the cereal endosperm consists of four distinct cell types: the starchy endosperm, the aleurone layer, the basal transfer cell layer and the cells of the embryo surrounding region. The developmental and molecular biology of each of these cell types are described in separate chapters. Genetic studies of endosperm traits revealed unusual patterns of inheritance that also occur in animal systems known as maternal imprinting, which is described in a chapter summarizing recent findings in maize as well as in *Arabidopsis thaliana*. Due to its importance as a source for food, feed and as an industrial raw material, much knowledge has accumulated about the synthesis of starch and storage proteins in the cereals, topics that are covered in separate chapters in this book. Anthocyanins give rise to the colorful seeds of maize and played an essential role in the discovery of transposons, the first plant transcription factors and gene structure–function relationships. One chapter is therefore dedicated to anthocyanin synthesis. Efforts to cultivate maize endosperm in vitro are desirable for several reasons and efforts have been ongoing for more than 60 years. One chapter in the book summarizes past and recent efforts in this area.

Plant biotechnology is now established commercially, and is likely to have an impact on efforts to enhance the value of cereal grains. It is expected that the combination of advances in endosperm cell and molecular biology as well as the biotechnology reviewed in this book will lead to further advances in improving cereals for food, feed and industrial applications.

January 2007 Odd-Arne Olsen

Contents

Plant Cell Monogr (8)
O.-A. Olsen: Endosperm
DOI 10.1007/7089_2007_106/Published online: 6 March 2007

The Developmental Biology of Cereal Endosperm

R. C. Brown (✉) · B. E. Lemmon

Department of Biology, University of Louisiana at Lafayette, Lafayette, LA 70504-2451, USA
rcb@louisiana.edu

Abstract This chapter summarizes the current understanding of structural aspects of cereal endosperm development with emphasis on the cytoskeleton and its role in morphogenesis. The developmental pathway of nuclear-type endosperm includes unusual cell and microtubule cycles and methods of wall placement. After double fertilization, the primary endosperm nucleus divides repeatedly without cytokinesis resulting in a large syncytium that lines the periphery and surrounds the central vacuole. During interphase this "liquid" endosperm is elegantly organized into functional units termed nuclear cytoplasmic domains (NCDs) by radial microtubule systems (RMSs) emanating from the nuclei. Cellularization is driven by the RMS microtubule cycle in which hoop-like cortical microtubules and preprophase microtubule bands (PPBs) are absent. It comprises alternation of two stages: deposition of anticlinal walls without mitosis leading to formation of alveoli, and periclinal walls following mitosis in the alveoli. Following centripetal cellularization, three principal tissues are differentiated: the central endosperm in which food reserves are stored, a transfer region specialized for uptake of metabolites, and the peripheral aleurone that releases enzymes to mobilize reserves at germination. The PPB microtubule cycle typical of meristematic growth replaces the RMS cycle when cell divisions associated with aleurone expansion and grain enlargement occur in the peripheral layers.

1 Introduction

The nutritive endosperm of seeds is a unique feature of angiosperms that is initiated by double fertilization. Although double fertilization does occur in gymnosperms, it results in competing embryos rather than in a new nutritive tissue. The nutritive tissue in seeds of gymnosperms is female gametophyte, which develops without fertilization. However, as in ginkgo, the nutritive gametophyte does not develop unless pollination occurs. This limits investment of maternal resources to infertile seeds. In angiosperms the much reduced female gametophyte does not serve a nutritive function and development of endosperm is delayed until fertilization. Recent studies have suggested that the female gametophytes of stem angiosperms consists of a module of four cells (Williams and Friedman 2002), whereas the majority of angiosperms have an eight nucleate embryo sac that can be interpreted as two such modules with opposite polarity. Two polar nuclei, one from each of the two modules, occupy a single central cell.

Development of the nutritive endosperm is launched by fertilization of the single polar nucleus in the four nucleate embryo sac resulting in diploid endosperm or the fused polar nuclei in the eight nucleate embryo sac resulting in a triploid endosperm. About 70% of flowering plants have triploid endosperm, suggesting that the higher level of ploidy is advantageous. Endosperm develops rapidly compared to the embryo. It is often partially or fully resorbed by the embryo early in development resulting in seeds that are exalbuminous, or it may persist until germination resulting in albuminous seeds. Both types of seeds may contain reserves of agricultural importance.

The endosperm, which is essential for reproduction but which leaves no descendants, plays the vital role of gatekeeper in the angiosperm life cycle by mediating nutrient transfer from mother to offspring; serving as a major site of gene imprinting; and detecting inappropriate hybridizations/polyploidy, thus preventing a futile investment in unfit seeds. As will be seen (Penterman et al., in this volume), the transcriptional control of endosperm development is complex and can be subjected to parental conflict, which modulates the investment in endosperm and ultimately determines the perpetuation of the successful genome in the next generation.

Endosperm has long been considered to develop in three general patterns: (1) ab initio cellular, in which cytokinesis follows the first division of the primary endosperm nucleus; (2) nuclear, in which a multicellular cytoplasm (syncytium) occurs before cellularization; and (3) helobial, in which the first cytokinesis is transverse and results in micropylar and chalazal chambers that develop independently. Recent studies have recognized many variants of these patterns based on the type of founder gametophyte, modality, apical/basal (micropylar/chalazal) polarity, extent of development, storage products, and other criteria (Floyd and Friedman 2000). Because the model plants, cereals and crucifers, both have nuclear type endosperm development, the chapters in this book will emphasize this pattern and will serve as an introduction to the biology of this fascinating plant generation.

Even though the endosperm is determinate and short-lived, the developmental processes are varied and complex. Endosperm development can be visualized as a hierarchy of domains: nuclear-cytoplasmic domains in the syncytial stage, early micropylar and chalazal developmental domains, and later developmental domains manifested in differentiation of endosperm into embryo surrounding region (ESR), basal endosperm transfer layer (BETL), aleurone, and the central storage tissue consisting of cells with nutrient reserves. Chapters in this book review developmental processes and their controls, details of the structure and function of the specialized tissue types, and the synthesis and utilization of storage compounds in the central endosperm.

The cereal grain is one of evolution's most intriguing and ultimately important accomplishments. Members of the grass family (Poaceae) produce a unique fruit, the grain, in which a single seed is tightly encased within a persistent pericarp. The seed contains a highly structured embryo surrounded by a copious nutritive endosperm that will be utilized at germination. These traits, which adapt grains for dispersal, storage, and seedling establishment, make cereals the principal source of food upon which civilization depends. Recognition of the agricultural preeminence and the potential for bioengineering of this unique and renewable source of energy and material has lead to a resurgence of interest in the biology of the endosperm (Becraft et al. 2001; Brown et al. 2002a; Costa et al. 2004; Olsen 2004).

Endosperm undergoes a brief, determinant, and highly specialized pattern of development involving differences in cell and microtubule cycles, and methods of wall placement. In cereals, as in most plants, endosperm development is of the nuclear type. This is an unusual type of development characterized by a long period of nuclear divisions without cytokinesis, resulting in a large multinucleate cytoplasm (syncytium) before cellularization occurs. Importantly for plant research, nuclear endosperm development also occurs in the model plant *Arabidopsis* and other members of the Brassicaceae (Berger 1999; Brown et al. 1999; Nguyen et al. 2001). Although there are differences in the later stages of differentiation and maturation, the early syncytial stage and the process of cellularization appear virtually identical in cereals and mustards. The emphasis of this review is on developmental chronology and mechanisms involving the cytoskeleton. The cytoskeleton plays a fundamental role in plant morphogenesis at the cellular level, including cell division and the control of placement and subsequent expansion of walls.

Endosperm development is initiated by double fertilization. In cereals, one sperm fertilizes the egg to produce the diploid zygote and the other fertilizes the diploid central cell nucleus, or joins with the two polar nuclei in triple fusion, to produce the triploid primary endosperm nucleus. Although the two fusion products are alike genetically (except for ploidy) and in a like environment, they develop into totally different structures. The zygote enters the developmental pattern typical of meristematic plant growth and gives rise to the embryo. The primary endosperm nucleus enters the pathway leading to the highly specialized nutritive endosperm. From the first divisions of the zygote and primary endosperm nucleus, the microtubule cycles that drive the developmental pathways are different (Brown and Lemmon 2001). The interphase microtubules of endosperm are nuclear-based RMSs rather than hoop-like cortical arrays and preprophase bands of microtubules (PPBs). Whereas the embryo quickly organizes meristems, the endosperm is essentially without meristems and matures into a short-lived, terminal structure specialized for food storage.

2
Nuclear Endosperm Development

Genes involved in the initiation of endosperm were discovered as gain-of-function mutations in which the unfertilized central cell nucleus begins syncytial development (Kiyosue et al. 1999; Ohad et al. 1999; Sorensen et al. 2001). This suggests that the wild-type genes function in the normal suppression of central cell proliferation until release by fertilization. It is possible that this suppression was a critical step in the evolution of endosperm, which develops after fertilization, as distinct from the development of gametophytic food storage tissues in gymnosperms, which occurs before fertilization. Details of these genes and their role(s) in the regulation of endosperm development is discussed by Penterman et al., in this volume.

Four stages occur in cereal endosperm development (Bosnes et al. 1992): (1) syncytial, a period of nuclear divisions without cytokinesis resulting in a large syncytium that lines the periphery of the central cell; (2) cellularization, a period during which cytokinesis results in discrete cells filling the central cell; (3) growth and differentiation, which results in distinct tissues; and (4) maturation, an important process characterized by engorgement of tissues with storage reserves and the development of desiccation tolerance and dormancy.

2.1
Syncytial

The behavior of nuclei during early development of the endosperm in maize has been deduced from analyses of color sectors in mature kernels (for summary see Walbot 1994). Anthocyanin pigmentation in the aleurone and iodine staining of starchy endosperm in the interior of the grain provide clonal evidence that sectors stem from the first three mitoses. When viewed from the top, kernels with two half-sector hemispheres result from the first mitosis, which places a nucleus to the right and left of the zygote. The next mitosis results in quarter-sectors and the third mitosis results in eight sectors. At the eight nucleate stage, the nuclei are in the peripheral layer of syncytial cytoplasm surrounding the central vacuole. It is generally assumed that nuclei migrate to the specific positions to initiate the characteristic sectors in the mature kernel, but the pattern could as well result from precise spindle orientations. The nuclei continue to proliferate synchronously until there are about 256–512 nuclei in the syncytial cytoplasm before the onset of cellularization in maize.

The initial rounds of nuclear division in endosperm of *Arabidopsis* (Webb and Gunning 1991; Brown et al. 2003) provide evidence that the precise placement of syncytial nuclei can be accomplished by spindle orientation and vacuolation. Following double fertilization, the primary endosperm nucleus begins to divide almost immediately and there are at least 16 endosperm nuclei before the first division of the zygote. Each of the nuclei resulting from the first mitosis develops a radial microtubule system (RMS) and the sister nuclei are displaced toward the poles with a large vacuole between them. Spindles for the next round of division are oriented oblique to the long axis of the central cell and one spindle places a nucleus near the zygote at the micropylar pole while the other places one near the chalazal pole, the other two nuclei occupying the central portion of the central cell. Following the third round of nuclear division, two nuclei reside in the common chalazal cytoplasm. Subsequently, the syncytium is subtly differentiated into micropylar and chalazal poles, a general feature of nuclear endosperm development (Maheshwari 1950; Floyd and Friedman 2000). Bipolarity leading to different developmental domains establishes conditions for differentiation of the endosperm for nutrient uptake, storage, and release. In cereals, the micropylar region is usually termed the embryo-surrounding region (ESR) and the opposite pole develops into the basal endosperm transfer layer (BETL), a highly specialized region of nutrient uptake.

Following the first few rounds of synchronous divisions, the syncytium containing a single layer of nuclei is restricted to the periphery of the central cell by a large central vacuole. The syncytium is organized into nuclear cytoplasmic domains (NCDs) by radial microtubule systems (RMSs) emanating from the nuclei (Figs. 1 and 2). This fundamental organization is similar in cereals (Olsen 2004), crucifers (Brown et al. 1999; Nguyen et al. 2001; Berger et al., in this volume) and the megagametophytes of gymnosperms, e.g., ginkgo (Brown et al. 2002b). The RMSs serve to position the single layer of NCDs equidistantly into a regular polygonal (usually hexagonal) pattern. Thus, the syncytium, once thought to be simply a "liquid" containing nuclei, is now known to be elegantly organized by the cytoskeleton. The syncytial endosperm can be viewed as an unusual plant system in which structural units comparable to cells are organized by the cytoskeleton into a "tissue" that remains wall-less.

The NCD model supposes that the radial microtubules are nucleated with − ends at the γ-tubulin microtubule organizing centers (MTOCs) surrounding nuclei (Fig. 2). The free + ends of the dynamically unstable radial microtubules contact and sense + ends of opposing microtubules radiating from adjacent nuclei. Dynamic interaction continues until there is a balance among nuclei as reflected in more or less uniform RMSs with zones of + end interaction marking boundaries of NCDs. The biochemistry of this interaction is not known but an ever increasing number of molecules are being discovered at the + ends of microtubules (e.g., Yang et al. 2003; Dhonukshe et al. 2005).

During nuclear proliferation by rapid rounds of mitosis, the RMSs undergo disassembly in prophase and are absent in metaphase/anaphase. After mitosis, the interzonal arrays between sister nuclei disappear without organizing a functional phragmoplast. RMSs reassemble around new nuclei and

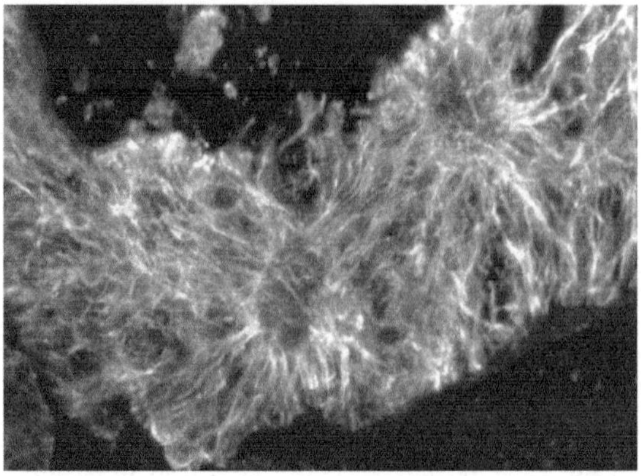

Fig. 1 Portion of the syncytial cytoplasm of maize viewed from the top showing organization of the common cytoplasm into NCDs by RMSs

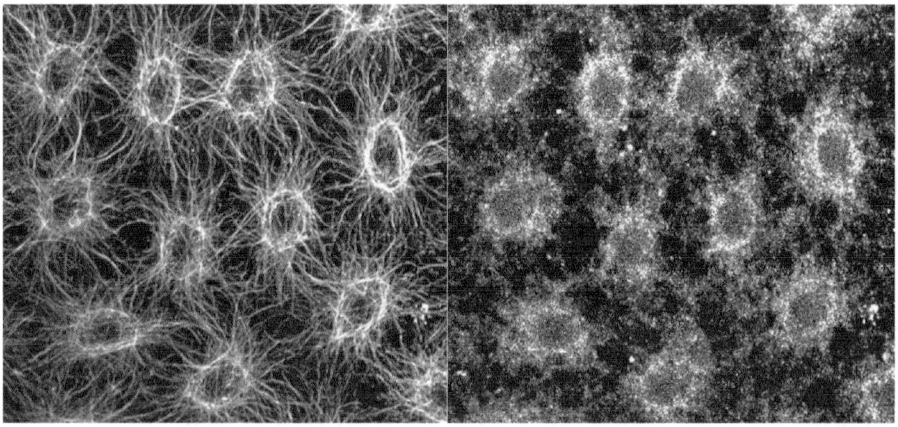

Fig. 2 RMSs that evenly space nuclei in the syncytium (seen here in face view) emanate from γ-tubulin surrounding the nuclear envelopes. *Left panel* microtubules. *Right panel* γ-tubulin

re-establish the network of interacting RMSs that define NCDs. This microtubule cycle is repeated throughout the syncytial phase.

An interesting feature of syncytial development in the endosperm of both cereals and mustards is that the nuclear divisions tend to occur synchronously early on and as coordinated waves of mitosis that move through the syncytium during the later stages of nuclear proliferation. In the small grains barley (Brown et al. 1994) and rice (Brown et al. 1996a), where the developing seed has vascular tissue adjacent to the long central chamber, developmental waves start at the vascular crease and move in both directions around the central chamber. These observations would seem to argue for factors in addition to the embryo as a source of signals in endosperm development.

The coordination of syncytial development appears to be a function of its organization into NCDs. Developmental cues coordinating mitosis and eventually cellularization are transmitted from one to another in an orderly manner. Thus, it appears that the interphase RMSs serve several interrelated functions in cytoplasmic organization: (1) they define and space the NCDs into a hexagonal pattern, (2) they establish a fabric providing integrity to the "liquid" endosperm, and (3) they somehow function in communicating developmental signals.

3 Cellularization

The filling of the central cell with walled cells is a multistage process comprising the initial deposition of walls among nuclei in the common cytoplasm at the periphery, and centripetal addition of new layers of cells until closure in the center (Fig. 3).

3.1 Initial Anticlinal Walls

The initial development of cell plates among non-sister nuclei in the absence of mitosis was one of the more intractable problems in understanding the developmental biology of endosperm. However, when the results of intensive studies on cereal grains (barley, rice, wheat, and maize) and mustards are combined, a clear picture emerges. After a certain number of nuclear divisions has populated the cytoplasm with nuclei, nuclear divisions cease and the syncytial cytoplasm again becomes organized into NCDs by nuclear-based RMSs (Brown et al. 1994; Olsen et al. 1995). At this time, there is an abrupt change to cellularization of the syncytium that is initiated in the absence of mitosis.

Anticlinal walls (perpendicular to the central cell wall) are initiated at multiple sites where RMSs interact along boundaries of the monolayer of NCDs (Fig. 3). The cell plate patches consolidate and spread to form the initial walls between adjacent NCDs. The outer edges join to the central cell wall in a fashion similar to somatic cell plate fusion with a parental wall, but the interior edges remain free in the cytoplasm adjacent to the central vacuole. Unfortunately, these unusual anticlinal walls have been termed "free-growing" in the literature (reviewed by Olsen et al. 1995). This term has led to much confusion, particularly since some workers (e.g., Morrison and O'Brien 1976; Mansfield and Briarty 1990) interpreted the origin of these walls as ingrowths of the central cell wall. Other investigators (Brown et al. 1994, 1999; Olsen

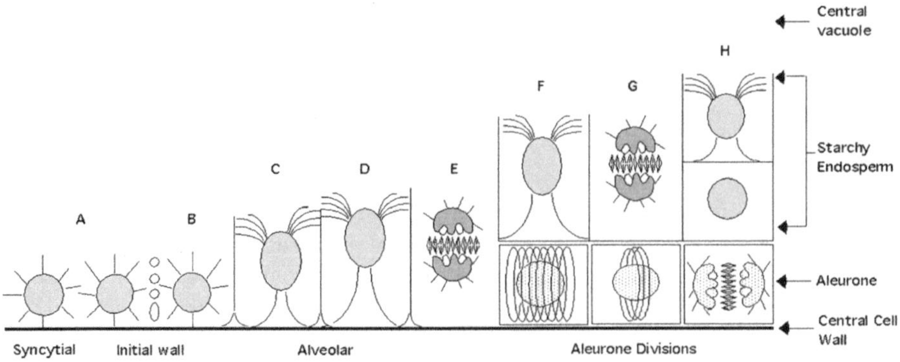

Fig. 3 Major events in endosperm cellularization. **A** During the syncytial stage, nuclei are equidistantly positioned in the peripheral cytoplasm by RMSs. **B** Cellularization is initiated by deposition of cell plate vesicles between adjacent NCDs. Vesicle coalescence results in anticlinal walls that join with the surrounding central cell wall at the periphery and continue to grow into the interior resulting in open-ended compartments (alveoli). **C–D.** Adventitious phragmoplasts mediate continued alveolar wall growth. **E** Nuclei in alveoli undergo periclinal mitosis followed by cytokinesis mediated by interzonal phragmoplasts. **F** The interior nuclei polarize and re-enter the alveolar growth pattern. The peripheral daughter cells, which will become aleurone, enter a different pathway with cortical hoop-like microtubules. **G** Periclinal divisions again occur in alveoli. PPBs mark the future division plane only in peripheral cells. **H** Periclinal divisions cut off the first complete cells of the starchy endosperm and a second interior layer of alveoli. Anticlinal wall growth is resumed in the alveoli and this two-step method of cellularization continues until closure in the center. Anticlinal divisions occur in the periphery to accommodate enlargement of the developing grain. In cereals with multiple aleurone layers (e.g., barley) divisions occur in both planes

et al. 1995; Otegui and Staehelin 2000) found no evidence whatsoever that the initial anticlinal walls begin as ingrowths. Another explanation, that the anticlinal walls are deposited in interzonal phragmoplasts after mitosis (Fineran et al. 1982), was questioned (van Lammeren 1988) and later proven to be incorrect as there is a hiatus between the cessation of mitosis and the onset of cellularization. Even if a mitosis followed by cytokinesis does occur, it cannot account for walls deposited between non-sister nuclei, which would be as numerous as sister nuclei. The origin of anticlinal walls either as ingrowths or as occurring in interzonal phragmoplasts has been discounted in favor of the NCD model.

The NCD model (Brown and Lemmon 1992, reviewed by Pickett-Heaps et al. 1999) provides the most plausible explanation for the unusual phenomenon of cell plate initiation in the absence of mitosis and prominent interzonal phragmoplasts. This model holds that wall placement occurs at the boundaries of NCDs and is a function of + end interaction of the opposing RMSs. A summary diagram (Fig. 2A in Olsen et al. 1995) illustrated the anticlinal wall starting at multiple sites along shared boundaries of NCDs and not yet fused with each other or with the central cell wall. As fusion occurs, the new wall joins with the nearby central cell wall producing wall flanges that can easily be mistaken for ingrowths.

Ultrastructural details of wall initiation in rice (Brown and Lemmon 1996a) and *Arabidopsis* (Otegui and Staehelin 2000) substantiate the NCD model. In the micropylar syncytium of *Arabidopsis*, cell plates are initiated in individual phragmoplast units of 4–12 microtubules ("mini-phragmoplasts") at the perimeter of NCDs (Otegui and Staehelin 2000; Otegui, in this volume). Thus, the initial anticlinal walls originate in a patchwork fashion among the interacting RMSs defining adjacent NCDs and spread to merge with each other, forming an interconnected network of walls that completes cellularization in this narrow chamber (Otegui and Staehelin 2000; Nguyen et al. 2001). Similar events may account for the precocious cellularization in the ESR of cereals.

In the large central chamber, junction of the initial anticlinal walls with the central cell wall is immediately followed by unidirectional growth into the interior of the central cell (Fig. 3). Phragmoplasts that form adventitiously at the interfaces of RMSs mediate continued growth of the anticlinal walls leading to the phenomenon of alveolation, which partially compartmentalizes the cytoplasm leaving a thin layer of syncytium adjacent to the tonoplast. The NCDs become polarized in axes perpendicular to the central cell wall and continued anticlinal wall formation results in elongated alveoli. Numerous works (e.g., Olsen et al. 1995; XuHan 1995, Brown et al. 1996a,b, 1999; Nguyen et al. 2001) have demonstrated that alveolation is the typical mechanism for cellularization of a syncytium peripheral to a large central vacuole. In the elongate cereal grains (e.g., rice, barley, wheat) this is a very large region and the process of alveolation dominates cellularization. It is even more prominent in the large megagametophytes of gymnosperms, where in some cases the entire process of initial cellularization is attributable to a single peripheral layer of alveoli that elongate centripetally until meeting in the center to effect closure (Singh 1978).

Two phenomena contribute to alveolation, polarization and vacuolation. Each NCD becomes polarized in an axis perpendicular to the central cell wall and vacuoles accumulate around the nucleate column of cytoplasm (Brown and Lemmon 1996a,b; Nguyen et al. 2001). The vacuoles serve to isolate a thin layer of shared cytoplasm between adjacent NCDs in which the anticlinal walls are formed. Little is known of either process but it seems likely that vacuolation is important in driving the elongation of NCDs. Certainly vacuolation is conspicuous as anticlinal walls continue to grow centripetally. A mutant (*mangled*) of *Arabidopsis* that causes defects in endosperm development shows sequence and structural similarity to a yeast protein required for vacuole protein sorting and vacuole biogenesis (Rojo et al. 2000).

Concomitant with elongation of alveoli is a dramatic reorganization of the nuclear-based radial microtubules into highly polar systems. As seen from the side, the microtubule system in an alveolus resembles a tree with axially aligned microtubules in the trunk of cytoplasm containing the nucleus, root-like processes at the base adjacent to the central cell wall, and a canopy extending into the syncytial cytoplasm adjacent to the central vacuole and overtopping the anticlinal walls (Brown et al. 1994; Brown and Lemmon 2001). Phragmoplasts form adventitiously (not in interzones between telophase nuclei) at the interfaces of opposing microtubule systems emanating from adjacent NCDs. In face view, as seen from the central vacuole, the alveoli are arranged in a honeycomb-like pattern. The arboreal configuration, which is a consistent and conspicuous feature of alveoli in both cereals and mustards, is unlike any other cytoskeletal configuration known in plant cells.

The adventitious phragmoplasts guiding unidirectional wall deposition in the syncytial cytoplasm remain tethered to adjacent nuclei which themselves advance centripetally in the vacuolate cytoplasm within the alveoli. The complex of merged alveolar phragmoplasts and the syncytial front of cytoplasm in which they are formed is continuously elevated as the six walls surrounding each alveolar NCD grow into the center of the central cell.

Descriptions of centripetally growing alveolar walls based on TEM are remarkably consistent (Brown et al. 1996a, 1997; Olsen et al. 1995; Wilson et al. 2006). The phragmoplast microtubules are at right angles to the forming wall at its leading edge. The leading edge, which ends blindly in the cytoplasm, is often enlarged and bulb-shaped, and associated with ER cisternae, dictyosomes, and vesicles. Flattened ER cisternae are appressed to the sides of the developing walls. The contribution of this distinctive ER system to the process of alveolar wall construction remains to be resolved. The abundant ER cisternae have been suggested to function in trapping of dictyosome vesicles (Gori 1987) or formation of plasmodesmata (Dute and Peterson 1992). An alternative hypothesis is that the entire secretory apparatus is transported to the region of wall advancement in the polarized alveoli. The forming walls exhibit a distinct gradation with terminus of fusing vesicles, a lamellar zone, and a continuous zone (Brown et al. 1997). (1 : 3)-β-D-Glucan (callose) is first detected when walls become lamellar and the walls remain callosic for a prolonged period (Brown et al. 1994, 1997; Otegui and Staehelin 2000).

The mature cell walls of the aleurone and starchy endosperm in cereal grains contain significant quantities of (1 : 3,1 : 4)-β-D-glucans and low levels of cellulose (Bacic and Stone 1981; Shibuya and Iwasaki 1985; Otegui, in this volume). Although the composition of walls in the mature grains of various cereals is well known from biochemical studies, the pattern of deposition of cell wall polysaccharides during the cellularization phase of endosperm development has only recently been studied (Brown et al. 1997; Wilson et al. 2006). Callose and cellulose appear in the first formed cell walls between 3 and 4 DAP. By 6 DAP, callose remains only in collars surrounding plasmodesmata. (1 : 3,1 : 4)-β-D-glucan is deposited in the developing cell walls at approximately 5 DAP followed by hetero-(1 : 4)-β-D-mannans at 6 DAP. Deposition of arabinogalactan-proteins and arabinoxylan in the wall begins at 7 and 8 DAP, respectively.

An important aspect of nuclear endosperm development is the temporal relationship of alveolar wall formation to the mitotic cycle. A long interval, up to 48 h in barley, may occur between cessation of mitosis and the wave of anticlinal wall deposition that initiates alveoli (Brown et al. 1994). However, it is more usual for the hiatus to be shorter as it is in rice (Brown et al. 1996a) and maize (Randolph 1936; Walbot 1994).

3.2 Periclinal Division

Each alveolus, starting with only one wall (the central cell wall) and growing usually six merged anticlinal walls, remains open-ended until it is divided periclinally by cytokinesis following mitosis. Prior to the wave of periclinal divisions, the alveolar phragmoplasts are disassembled and nuclei become more centrally located in the alveoli. This has been observed in both cereals (e.g., Brown et al. 1994, 1996a) and mustards (e.g., Nguyen et al. 2001). The factors involved in positioning nuclei are not known. Alveoli are of nearly uniform length and the prophase nuclei are suspended in rafts of cytoplasm (phragmosomes) nearly equidistant from the central cell wall. Phragmosomes predict the plane of the future division as is typical of vacuolate cells (Sinnott and Bloch 1941; Lloyd et al. 1992) but no PPBs are formed in alveoli (Brown et al. 1994, 1999) and no specific arrangement of actin appears to indicate division site selection (Nguyen et al. 2001). After mitosis, interzonal phragmoplasts/cell plates expand to junction with the anticlinal walls of the alveoli (Fig. 3). In this manner, the peripheral portion of each alveolus receives its final wall and becomes a cell while the inner portion remains an alveolus. Following the wave of periclinal divisions, microtubules again emanate from the tips of interphase nuclei in the inner alveolar layer, interact at their interfaces and organize adventitious phragmoplasts that direct renewed growth of the anticlinal walls between the non-sister nuclei.

The second layer of alveolar growth in endosperm is identical to the first and is again followed by periclinal division. This repeated cycle of anticlinal wall formation between non-sister nuclei followed by periclinal wall formation between sister nuclei completes centripetal cellularization of the endosperm. The alternating cycles are conspicuous in rice where orderly files of several cells are produced (Brown et al. 1996a). Nuclei of any given file are sister nuclei and the periclinal walls between them are the result of interzonal phragmoplast activity following mitosis. The anticlinal walls between adjacent cell files are all formed in association with adventitious phragmoplasts between non-sister nuclei. This regular pattern of centripetal cellularization in cereals may be obscured by later divisions that expand the starchy endosperm.

4
Growth and Differentiation

The fully differentiated endosperm of grasses is structurally simple, consisting of three principal cell types: (1) an inner mass of starchy endosperm cells with distinct embryo surrounding region (2) a basal endosperm transfer region (BETL) adjacent to the main maternal vasculature, and (3) a specialized epidermis (aleurone) of one to three cell layers. The following text describes the developmental basis for each cell type. For more in-depth descriptions, please see individual chapters in this volume describing each cell type.

4.1
Central or Starchy Endosperm

After closure, additional cell divisions occur in the endosperm. These later divisions depart from the RMS-driven NCD developmental pathway of initial cellularization. Cells of the central endosperm develop cortical microtubules but no PPBs (Brown et al. 1994; Clore et al. 1996) and no set pattern of division plane has been discerned. These cells in the interior become multifaceted and greatly enlarged, differentiating into cells specialized for the storage of food. Whereas most plants store reserves as proteins and oils, the cereals store large quantities of starch in amyloplasts (see Hannah, in this volume). This unique tissue, known as the starchy endosperm, comprises the bulk of the endosperm in cereal grains.

The starchy endosperm of cereals also contains storage proteins (Lopes and Larkins 1993; Halford and Shewry, in this volume), which are deposited either in protein bodies sequestered from the ER or in protein storage vacuoles (Müntz 1998; Kumamuru et al., in this volume). A small amount of oil bodies similar to those present in high concentration in the aleurone layer are found in the starchy endosperm of wheat (Hargin and Morrison 1980), and probably other cereals as well.

Endoreduplication resulting in extremely high levels of DNA appears to play an important role in endosperm development (Lopes and Larkins 1993; Nguyen, Sabelli and Larkins, in this volume). In the starchy endosperm of maize, the DNA content may reach 96C (Grafi and Larkins 1995). Endoreduplication begins during the differentiation stage (10–12 DAP) after cellularization has been completed. It requires the induction of S-phase kinase and the inhibition of M-phase promoting factors.

The starchy endosperm of cereals is generally thought to be dead at maturity and recent studies have shown that programmed cell death is linked to the ethylene pathway. In the maize mutant *shrunken2* that overproduces ethylene, cell death is accelerated (Young et al. 1997; Nguyen et al., in this volume). During seed germination in cereals, a rush of enzymes produced by the aleurone degrades the starchy endosperm. In albuminous seeds with living endosperm such as celery (Jacobsen et al. 1976) and tomato (Nonogaki et al. 2000), the breakdown of endosperm is due to hydrolases produced in the endosperm itself as a response to embryo-generated gibberellin.

Rapidly accumulating evidence indicates that the embryo-surrounding region (ESR), a pocket within the starchy endosperm in which the embryo develops, is a distinct developmental domain. In maize, the ESR is at first a restricted region of dense endosperm that surrounds the entire embryo at 5 DAP, the suspensor at 7 DAP, and only the lower part of the suspensor at 9 DAP (Schel et al. 1984; Opsahl-Ferstad et al. 1997; Cosségal et al., in this volume). Several genes are specifically expressed in the ESR (Bonello et al. 2000). The absence of gene expression in spontaneously occurring embryoless endosperms suggests a dependence on signaling from the embryo for ESR gene expression (Opsahl-Ferstad et al. 1997). The gene products may play a role in embryo nutrition and/or in establishing an interface between the embryo and the endosperm, leading to creation of an embryonic cavity.

4.2
Basal Endosperm Transfer Layer

The transfer cells of endosperm (see Royo et al., in this volume), which are characterized by heavy secondary wall ingrowths, are located adjacent to the maternal vasculature. All metabolites must enter the endosperm by loading from the maternal tissue as there are no direct connections to the maternal vascular tissue. Wall protuberances and labyrinthine structures are generally accepted as evidence of transfer cells in which amplification of the plasma membrane is associated with an increased capacity for uptake of assimilate into the cell (Gunning and Pate 1969). Transfer cells at the interface between gametophyte and sporophyte constitute a placental region recently recognized as one of the key anatomical features defining land plants (Graham et al. 2000). It is very interesting that the interface between the maternal sporophyte generation and the intercalated endosperm generation found only in angiosperms has placental characteristics.

BETL develops at the base of large grains such as maize and along the ventral side of small grains. In maize, aleurone transfer cells are located at the broad region overlying the hilar pad (Davis et al. 1990). In addition to the basal-most cell layer, two or three adjacent layers of endosperm cells possess wall ingrowths in a gradient decreasing towards the interior of the endosperm (Schel et al. 1984). In the small cereal grains the vascular tissue extends the length of the ventral surface of the elongated grain. Nucellar cells in the ventral crease develop into the nucellar projection, a specialized region of elongate cells with extensive wall infoldings between the vascular bundle and the endosperm. The adjacent endosperm exhibits transfer cell type walls grading from the aleurone to the subaleurone (Wang et al. 1994; Linnestad et al. 1998).

Genetic imbalance provides evidence for an essential role for the endosperm transfer layer in nutrient loading in cereals. Ultrastructural studies of seeds with an atypical genomic ratio of two maternal to two paternal revealed that the transfer layer was almost completely suppressed (Charlton et al. 1995). Another factor promoting differentiation of the transfer layer is proper uptake of nutrients. An invertase-deficient mutant (*miniature-1*) of maize results in aberrant endosperm development. Development is normal until 9 DAP but soon thereafter a gap occurs between the endosperm and the maternal pedicel tissues. The hilar pad disintegrates and aleurone transfer cells do not differentiate (Miller and Chourey 1992). Thus, the wild-type allele is required for normal development of both endosperm and maternal cells, emphasizing the complex interplay of various tissues in seed development (Cheng et al. 1996). This is further demonstrated by the fact that endosperm grown in vitro never develops transfer cells although the differentiation of the starchy and aleurone tissues appears to follow a normal time course (Gruis et al. 2006; Gruis et al., in this volume).

4.3 Aleurone

Following cellularization, the peripheral layer enters a specialized pathway that leads to the differentiation of aleurone. In the cereal grains, the aleurone layer is the only layer of the endosperm that may be pigmented, as in maize and rice. The aleurone consists of a single layer of cells in maize (Walbot 1994) and wheat (Morrison et al. 1975), from one to several layers in rice depending upon anatomical position and growing conditions (Bechtel and Pomeranz 1980), and three layers in barley (Jones 1969). Mutants of maize with altered numbers of aleurone layers are known (Wolf et al. 1972; Shen et al. 2003).

The persistent aleurone, which has evolved important functions in regulation of dormancy, is a central component in the biology of cereal grains. Agriculture is dependent upon a balance between dormancy at harvest (preventing preharvest sprouting) and high rates of germination upon planting. In general, abscisic acid and gibberellin act antagonistically to control dormancy breakage and germination (Kermode 2005). Abscisic acid promotes maturation including protein synthesis and suppresses precocious germination. Gibberellin acts as a transcriptional activator of genes encoding amylases and proteases in the aleurone. Following gene expression, the aleurone cells undergo programmed cell death involving DNA fragmentation and wall degradation (Wang et al. 1998) releasing the hydrolytic enzymes. Starchy endosperm is degraded into sugars and other simple molecules that support embryo growth during germination. Mature aleurone cells are characterized by thick, highly autofluorescent walls, prominent nuclei, and aleurone grains (protein storage bodies). The aleurone grains produced in protein storing vacuoles (PSVs) are very complex, consisting of a protein matrix in which globoid inclusions (phytin, niacin, and phenolics) or carbohydrates are embedded (Bethke et al. 1998). Both types may develop in the same PSV (Jacobsen et al. 1971). Lipid droplets surround the aleurone grains (Jones 1969; Morrison et al. 1975; Kyle and Styles 1977). In addition to the PSV, a second kind of vacuole, the lytic vacuole, occurs in the aleurone cells of barley (Swanson et al. 1998). Additionally, the aleurone may develop pigmentation caused by accumulation of anthocyanins (see Kone, in this volume).

The regulation and mechanism of aleurone cell differentiation has long been a key topic in endosperm biology because of its importance and the obvious mosacism described in early genetic studies. Molecular studies have opened the door to in-depth analysis of the controls of this key pathway and are discussed by Becraft, in this volume.

The first sign of aleurone formation is a preponderance of anticlinal cell divisions in the peripheral layer, resulting in tabular cells. After the establishment of aleurone cell shape, small vacuoles accumulate and the cytoplasm becomes dense. Differentiation begins at 4 DAP in rice (Juliano and Bechtel 1985; Brown et al. 1996a), 6–8 DAP in wheat (Morrison et al. 1975), 8 DAP in barley (Bosnes et al. 1992; Brown et al. 1994), and 10–15 DAP in maize (Kyle and Styles 1977). This variation may be due in part to different criteria for determination of the onset of differentiation (e.g., anticlinal divisions vs. onset of aleurone grain accumulation). In the multilayered aleurone of barley, the peripheral cells are the first to differentiate with the second and third layers differentiating sequentially afterward. The earliest aleurone-specific transcript known in barley is Ltp2, which is detectable at about 8 DAP, i.e., simultaneously with the first morphological signs of aleurone cell differentiation (Kalla et al. 1994). The aleurone specificity of this gene has been confirmed in both transgenic rice (Kalla et al. 1994) and maize (Shen et al. 2003) expressing the Ltp2 promotor::GUS construct.

The morphogenetic pathway of aleurone has been shown to include hooplike cortical microtubules and PPBs in barley (Brown et al. 1994) and maize (unpublished observations). It is significant that the switch to the PPB microtubule cycle typical of vegetative growth occurs only in the later stages of endosperm development when cells are added in an orderly fashion to the growing aleurone. In maize, PPBs occur in division of derivatives of the peripheral cambium-like zone as well as in the aleurone proper. Randolph (1936) recognized that meristematic activity at the periphery after cellularization is a developmental feature peculiar to the large grains that gives rise to files of cells that contribute significantly to the increase in grain size. The genetic/epigenetic factors that control switching between RMS and PPB microtubule cycles are unknown. Early nuclear endosperm development is unique in that it is driven in large part by the RMS cycle, typically found only in cells of the reproductive lineage (Brown and Lemmon 2001). The existence of a genetic mechanisms for division plane control in the aleurone is supported by mutants in maize such as *xcl1* (extra cell layer1), in which the aleurone layer possesses one extra cell layer as a result of aberrant periclinal divisions (Kessler and Sinha 2000), and *dal1* and *dal2* (disorganized aleurone layer 1 and 2), which have relaxed control over aleurone division plane determination (Lid et al. 2004).

5 Conclusions

Life history of the endosperm is distinctly unplantlike. Growth is without meristems and is determinate. The unique developmental pathway of cereal endosperm includes a specific program of changes in the cell and microtubule cycles leading to an assemblage of plant tissues specialized for uptake of metabolites (transfer cells), storage of food reserves (starchy endosperm), and enzyme synthesis (aleurone). Progress has been made in discovering the genetic controls of endosperm initiation and aleurone differentiation. Little is known of the signals involved in regulation of the underlying unique cell and microtubule cycles nor the factors responsible for terminating growth. The basic pattern of nuclear endosperm development is driven by a different microtubule cycle than that of the typical vegetative growth plan via organization of meristems. Yet, in the final developmental stages of complex aleurone, the PPB method of wall placement replaces the RMS method and there is even some evidence of meristematic activity at the periphery in large grains. These fundamental aspects of structure and development provide a framework for understanding the genetic/epigenetic controls of this unusual and important plant system.

Acknowledgements The author's research on endosperm has been made possible through the resources of the University of Louisiana, especially the Center for Microscopy, The Louisiana State University Rice Research Station, and fruitful collaborations with investigators at the Norwegian University of Life Sciences and Pioneer Hi-Bred International, a DuPont Company.

References

Bacic A, Stone BA (1981) Chemistry and organization of aleurone cell wall components from wheat and barley. Aust J Plant Physiol 8:475–495

Bechtel DB, Pomeranz Y (1980) The rice kernel. In: Pomeranz Y (ed) Advances in cereal science and techology, vol 3. Am Assoc Cereal Chem, St Paul, Minnesota, pp 73–113

Becraft PW, Brown RC, Lemmon BE, Olsen O-A, Opsahl-Ferstad H-G (2001) Endosperm development. In: Bhojwani SS (ed) Current trends in the embryology of angiosperms. Kluwer Academic, Dordrecht, the Netherlands, pp 353–374

Bethke PC, Swanson SJ, Hillmer S, Jones RL (1998) From storage compartment to lytic organelle: The metamorphosis of the aleurone storage vacuole. Ann Bot 82:399–412

Berger F (1999) Endosperm development. Curr Opin Plant Biol 2:28–32

Bonello J-F, Opsahl-Ferstad H-G, Perez P, Dumas C, Rogowsky PM (2000) Esr genes show different levels of expression in the same region of maize endosperm. Gene 246:219–227

Bosnes M, Weideman F, Olsen O-A (1992) Endosperm differentiation in barley wild-type and sex mutants. Plant J 2:661–674

Brown RC, Lemmon BE (1992) Cytoplasmic domain: A model for spatial control of cytokinesis in reproductive cells of plants. EMSA Bull 22:48–53

Brown RC, Lemmon BE (2001) The cytoskeleton and spatial control of cytokinesis in the plant life cycle. Protoplasma 215:35–49

Brown RC, Lemmon BE, Olsen O-A (1994) Endosperm development in barley: microtubule involvement in the morphogenetic pathway. Plant Cell 6:1241–1252

Brown RC, Lemmon BE, Nguyen H (2002a) Endosperm development. In: O'Neill SD, Roberts AJ (eds) Plant reproduction: annual plant reviews, vol 6. Sheffield Academic, Sheffield, UK, pp 193–220

Brown RC, Lemmon B, Nguyen H (2002b) The microtubule cycle during successive mitotic waves in the syncytial female gametophyte of ginkgo. J Plant Res 115:491–494

Brown RC, Lemmon BE, Nguyen H (2003) Events during the first four rounds of mitosis establish three developmental domains in the syncytial endosperm of *Arabidopsis*. Protoplasma 222:167–174

Brown RC, Lemmon BE, Nguyen H, Olsen O-A (1999) Development of endosperm in *Arabidopsis thaliana*. Sex Plant Reprod 12:32–42

Brown RC, Lemmon BE, Olsen O-A (1996a) Development of the endosperm in rice (*Oryza sativa* L.): cellularization. J Plant Res 109:301–13

Brown RC, Lemmon BE, Olsen O-A (1996b) Polarization predicts the pattern of cellularization in cereal endosperm. Protoplasma 192:168–177

Brown RC, Lemmon BE, Stone BA, Olsen O-A (1997) Cell wall (1->3)- and (1->3, 1->4)-β-glucans during early grain development in rice (*Oryza sativa* L.). Planta 202:414–426

Charlton WL, Keen CL, Merriman C, Lynch P, Greenland AJ, Dickinson HG (1995) Endosperm development in: *Zea mays*: implication of gametic imprinting and paternal excess in regulation of transfer layer development. Development 121:3089–3097

Cheng W-H, Taliercio EW, Chourey PS (1996) The minature1 seed locus of maize encodes a cell wall invertase required for normal development of endosperm and maternal cells in the pedicel. Plant Cell 8:971–983

Clore AM, Dannenhoffer JM, Larkins BA (1996) EF-1alpha is associated with a cytoskeletal network surrounding protein bodies in maize endosperm cells. Plant Cell 8:2003–2014

Costa L, Gutierrez-Marcos J, Dickinson HG (2004) No longer a yolk: the short life and complex times of the plant endosperm. Trends Plant Sci 9:507–514

Davis RW, Smith JD, Cobb BG (1990) A light and electron microscope investigation of the transfer cell region of maize caryopses. Canad J Bot 68:471–479

Dhonukshe P, Mathur J, Hülskamp M, Gadella Jr TWJ (2005) Microtubule plus-ends reveal essential links between intracellular polarization and localized modulation of endocytosis during divison-plane establishment in plant cells. BMC Biology 3:11

Dute RR, Peterson CM (1992) Early endosperm development in ovules of soybean, *Glycine max* (L.) Merr. (Fabaceae). Ann Bot 69:263–271

Fineran BA, Wild DJC, Ingerfeld M (1982) Initial wall formation in the endosperm of wheat *Triticum aestivum*: a reevaluation. Canad J Bot 60:1776–1795

Floyd SK, Friedman WE (2000) Evolution of endosperm developmental patterns among basal flowering plants. Int J Plant Sci 161:57–81
Gori P (1987) The fine structure of the developing *Euphorbia dulcis* endosperm. Ann Bot 60:563–569
Grafi G, Larkins BA (1995) Endoreduplication in maize endosperm: involvement of M phase-promoting factor inhibition and induction of S phase-related kinases. Science 269:1262–1264
Graham LE, Cook ME, Busse JS (2000) The origin of plants: body plan changes contributing to a major evolutionary radiation. PNAS 97:4535–4540
Gruis DF, Guo H, Selinger D, Tian Q, Olsen O-A (2006) Surface position, not signaling from surrounding maternal tissues, specify aleurone epidermal cell fate in maize endosperm organ cultures. Plant Physiol 141:898–909
Gunning BES, Pate JS (1969) Transfer cells: plant cells with wall ingrowths, specialized in relation to short distance transport of solutes-their occurrence, structure, and development. Protoplasma 68:107–133
Hargin KD, Morrison WR (1980) The distribution of acyl lipids in the germ, aleurone, starch and nonstarch endosperm of four wheat varieties. J Sci Food Agric 31:877–888
Jacobsen JV, Knox RB, Pyliotis NA (1971) The structure and composition of aleurone grains in the barley aleurone layer. Planta 101:189–209
Jacobsen JV, Pressman E, Pyliotis NA (1976) Gibberellin-induced separation of cells in isolated endosperm of celery seed. Planta 129:113–122
Jones RL (1969) The fine structure of barley aleurone cells. Planta 85:359–375
Juliano BO, Bechtel DB (1985) The rice grain and its gross composition. In: Juliano BO (ed) Rice: chemistry and technology, 2nd edn. Am Assoc Cereal Chem, St Paul, MN, pp 17–57
Kalla R, Shimamoto K, Potter R, Nielsen PS, Linnestad C, Olsen O-A (1994) The promoter of the barley aleurone-specific gene encoding a putative 7 kDa lipid transfer protein confers aleurone cell-specific expression in transgenic rice. Plant J 6:849–860
Kermode AR (2005) Role of abscisic acid in seed dormancy. J Plant Growth Regul 24:319–344
Kessler SA, Sinha NR (2000) Characterization of xcl (extra cell layers), a mutation affecting plane of cell division during maize development. In: Abstracts of Plant Biology 2000, San Diego, CA, 15–19 July 2000. American Society Plant Physiologists, Rockville, MD, p 48
Kiyosue T, Ohad N, Yadegari R, Hannon M, Dinneny J, Wells D, Katz A, Margossian L, Harada J, Goldberg RB, Fischer R (1999) Control of fertilization-independent endosperm development by the MEDEA polycomb gene in *Arabidopsis*. Proc Natl Acad Sci USA 96:4186–4191
Kyle DJ, Styles ED (1977) Development of aleurone and sub-aleurone layers in maize. Planta 137:185–193
Lid SE, Al RH, Krekling T, Meeley RB, Ranch J, Opsahl-Ferstad H-G, Olsen O-A (2004) The maize disorganized aleurone layer 1 and 2 (dil1, dil2) mutants lack control of mitotic division plane in the aleurone layer of developing endosperm. Planta 218:370–378
Lid SE, Gruis D, Jung R, Lorentzen JA, Ananiev E, Chamberlin M, Niu X, Meeley R, Nichols S, Olsen O-A (2002) The defective kernel 1 (dek1) gene required for aleurone cell development in the endosperm of maize grains encodes a membrane protein of the calpain gene superfamily. Proc Natl Acad Sci USA 99:5460–5465
Linnestad C, Doan DNP, Brown RC, Lemmon BE, Meyer DJ, Jung R, Olsen O-A (1998) Nucellain, a barley homolog of the dicot vacuolar-processing protease, is localized in nucellar cell walls. Plant Physiol 118:1169–1180

Lloyd CW, Venverloo CJ, Goodbody KC, Shaw PJ (1992) Confocal laser microscopy and three-dimensional reconstruction of the nucleus-associated micotubules in the division plane of vacuolated plant cells. J Microsc 166:99–109

Lopes MA, Larkins BA (1993) Endosperm origin, development, and function. Plant Cell 5:1383–1399

Maheshwari P (1950) An introduction to the embryology of angiosperms. McGraw-Hill, New York

Mansfield SG, Briarty LG (1990) Endosperm cellularization in *Arabidopsis thaliana* (L). Ara Info Serv 27:65–72

Miller ME, Chourey PS (1992) The maize invertase-deficient minature-1 seed mutation is associated with aberrant pedicel and endosperm development. Plant Cell 4:297–305

Morrison IN, Kuo J, O'Brien TP (1975) Histochemistry and fine structure of developing wheat aleurone cells. Planta 123:105–116

Morrison IN, O'Brien TP (1976) Cytokinesis in the developing wheat grain; division with and without a phragmoplast. Planta 130:57–67

Müntz K (1998) Deposition of storage proteins. Plant Mol Biol 38:77–99

Nguyen H, Brown RC, Lemmon BE (2001) Patterns of cytoskeletal organization reflect distinct developmental domains in endosperm of *Coronopus didymus* (Brassicaceae). Int J Plant Sci 162:1–14

Nonogaki H, Gee OH, Bradford KJ (2000) A germination-specific endo β-mannanase gene is expressed in the micropylar endosperm cap of tomato seeds. Plant Physiol 123:1235–1245

Ohad N, Yadegari R, Margossian L, Hannon M, Michaeli D, Harada JJ, Goldberg RB, Fischer RL (1999) Mutations in FIE, a WD polycomb group gene, allow endosperm development without fertilization. Plant Cell 11:407–415

Olsen O-A (2004) Nuclear endosperm development in cereals and *Arabidopsis thaliana*. Plant Cell 16:S214–S227

Olsen O-A, Brown RC, Lemmon BE (1995) Pattern and process of wall formation in developing endosperm. BioEssays 17:803–812

Opsahl-Ferstad H-G, Le Deunff E, Dumas C, Rogowsky PM (1997) ZmEsr, a novel endosperm-specific gene expressed in a restricted region around the maize embryo. Plant J 12:235–246

Otegui M, Staehelin LA (2000) Syncytial-type cell plates: a novel kind of cell plate involved in endosperm cellularization of *Arabidopsis*. Plant Cell 12:933–947

Pickett-Heaps JD, Gunning BES, Brown RC, Lemmon BE, Cleary AL (1999) The cytoplast concept in dividing plant cells: cytoplasmic domains and the evolution of spatially organized cell division. Am J Bot 86:153–172

Randolph LF (1936) Developmental morphology of the caryopsis in maize. J Agric Res 53:881–961

Rojo E, Gilmore CS, Somerville CR, Raikel NK (2000) Characterization of mangled, a gene required for normal mitosis and cell growth in *Arabidopsis* ovules. Am Soc Cell Biol Annual Meeting, San Francisco, Dec 2000, late abstract L51

Schel JHN, Kieft H, van Lammeren AAM (1984) Interactions between embryo and endosperm during early developmental stages of maize caryopses (*Zea mays*). Canad J Bot 62:2842–2853

Shen B, Li C, Min Z, Meeley RB, Tarczynski MC, Olsen O-A (2003) Sal1 determines the number of aleurone cell layers in maize endosperm and encodes a class E vacuolar sorting protein. Proc Natl Acad Sci USA 100:6552–6557

Shibuya N, Iwasaki T (1985) Structural features of rice bran hemicellulose. Phytochem 24:285–289

Singh H (1978) Embryology of gymnosperms. In: Handbuch der Pflanzenanatomie 10. Borntraeger, Berlin, pp 110–287
Sinnot EW, Bloch R (1941) Division in vacuolate plant cells. Am J Bot 28:225–232
Sorensen MB, Chaudhury AM, Robert H, Bancharel E, Berger F (2001) Polycomb group genes control pattern formation in plant seed. Curr Biol 11:277–281
Swanson SJ, Bethke PC, Jones RL (1998) Barley aleurone cells contain two types of vacuoles: characterization of lytic organelles by use of fluorescent probes. Plant Cell 10:685–698
Van Lammeren AAM (1988) Structure and function of the microtubular cytoskeleton during endosperm development in wheat: an immunofluorescence study. Protoplasma 146:18–27
Walbot V (1994) Overview of key steps in aleurone development. In: Freeling M, Walbot V (eds) The maize handbook. Springer, Berlin Heidelberg New York, pp 78–80
Wang HL, Offler CE, Patrick JW (1994) Nucellar projection transfer cells in the developing wheat grain. Protoplasma 182:39–52
Wang M, Oppedijk BJ, Caspers MPM, Lamers GEM, Boot MJ, Geerlings DNG, Bakhuizen B, Meijer AH, van Duijn B (1998) Spatial and temporal regulation of DNA fragmentation in the aleurone of germinating barley. J Exp Bot 49:1293–1301
Webb MC, Gunning BES (1991) The microtubular cytoskeleton during development of the zygote, proembryo and free-nuclear endosperm in *Arabidopsis thaliana* (L.) Heynh. Planta 184:187–195
Williams JH, Friedman WE (2002) Identification of diploid endosperm in an early angiosperm lineage. Nature 415:522–526
Wilson SM, Burton RA, Doblin MS, Stone BA, Newbigin EJ, Fincher GB, Bacic A (2006) Temporal and spatial appearance of wall polysaccharides during cellularization of barley (*Hordeum vulgare*) endosperm. Planta 224:655–667
Wolf MJ, Cutler HC, Zuber MS, Khoo U (1972) Maize with multilayer aleurone of high protein content. Crop Sci 12:440–442
XuHan X (1995) Seed development in *Phaseolus vulgaris* L., *Populus nigra* L., and *Ranunculus scleratus* L. with special reference to the microtubular cytoskeleton. CIP-Gegevens Koninklijke Bibliotheek, Den Haag, Netherlands
Yang C-Y, Spielman M, Coles JP, Li Y, Ghelani S, Bourdon V, Brown RC, Lemmon BE, Scott RJ, Dickinson HG (2003) TETRASPORE encodes a kinesin required for male meiotic cytokinesis in *Arabidopsis*. Plant J 34:229–240
Young TE, Gallie DR, DeMason DA (1997) Ethylene-mediated programmed cell death during maize endosperm development for wild-type and *shrunken2* genotypes. Plant Physiol 115:737–751

Plant Cell Monogr (8)
O.-A. Olsen: Endosperm
DOI 10.1007/7089_2007_107/Published online: 31 March 2007

Endoreduplication and Programmed Cell Death in the Cereal Endosperm

H. N. Nguyen · P. A. Sabelli · B. A. Larkins (✉)

Department of Plant Sciences, University of Arizona, Forbes 303, 210036,
Tucson, AZ 85721, USA
larkins@Ag.arizona.edu

Abstract Endoreduplication and programmed cell death are important processes during the development and maturation of cereal endosperm. Endoreduplication results in the formation of multiple copies of each chromosome in the nucleus, and it leads to cells with ploidies of 24C to 96C, and sometimes even higher. Although it has generally been thought that this process is associated with cell enlargement and increased levels of gene expression, recent studies of maize endosperm development provided data that do not support this hypothesis. Programmed cell death commences in starchy endosperm cells around mid-development, and ultimately affects nearly all of the endosperm cells, except for those in the aleurone layer. The progression of programmed cell death, which appears to be mediated by increasing levels of ethylene and gibberellic acid, is mediated by cysteine proteases that appear to be similar to vacuolar processing enzymes. Presumably, the loss of viability of starchy endosperm cells facilitates hydrolysis of stored nutrients and their uptake by the embryo during seed germination.

1 Introduction

The cereal endosperm is important both economically and scientifically. It is a major source of food and feed for human and animal nutrition, and therefore is of great economic importance. Scientifically, the cereal endosperm makes a good model system in several respects: (1) its genetics is well known; (2) it has a determinente development with a short period of growth and maturation, so it is ideal for developmental and cell biology studies; and (3) it is a simple tissue composed of a few specialized cell types that produce large quantities of storage metabolites, making the cereal endosperm a good model for biochemical and molecular studies as well. Cereal endosperm undergoes the "nuclear" pattern of development, which is the most common and best studied type of endosperm formation (Olsen 2004; chapter by Brown and Lemmon, in this volume). This developmental pathway consists of three distinct stages: syncytial, mitotic, and differentiation. Following fertilization, the primary endosperm nucleus goes through multiple rounds of division without cytokinesis, resulting in a syncytium or multinucleate cytoplasm. Cell walls then form around individual nuclei, so that the endosperm becomes multicellular. A mitotic stage follows and creates the basic cell number for the body of the endosperm.

Differentiation of the cereal endosperm results in four distinct cell types: aleurone cells, starchy endosperm cells, transfer cells, and cells of the embryo surrounding region (Becraft 2000; Olsen 2001; chapters by Becraft, Royo, and Cossegal, in this volume). The bulk of the endosperm is composed of starchy endosperm cells and their primary function is the synthesis of starch and storage proteins. Concurrent with the accumulation of these storage metabolites, starchy endosperm cells undergo several rounds of endoreduplication, a modified cell cycle that produces polyploid nuclei. The final stage of starchy endosperm cell maturation involves programmed cell death (PCD), which is the ordered disassembly of a cell through a genetically regulated program (Buckner et al. 2000). At maturity, all endosperm cells, except those of the aleurone, undergo PCD. Endoreduplication and PCD are not unique to cereal grains, but they are substantial processes in cereal endosperm development.

In this review, we focus on two aspects of starchy endosperm development: endoreduplication and PCD, and their regulation. Although these processes have long been known, it is only in recent years that there have been significant advances in our understanding of their mechanisms and functions in plant cells in general, and in endosperm in particular. Endoreduplication has generally been thought to be important for cell enlargement and increasing the transcription of genes involved in the synthesis of stored metabolites; however, recent data seem to be in contradiction with these hypotheses. PCD presumably facilitates the hydrolysis and absorption of nutrients by the germinating embryo. The pathway of PCD is well understood in animal cells, but it is only beginning to be elucidated in plant cells. While PCD in plants has many similarities to that of animals, there are also some differences between the two.

2
Maize Endosperm Development

We primarily consider studies of maize endosperm in this review, as endoreduplication and PCD have been most extensively studied in this species. Cellularization of the syncytial endosperm is complete by around 4 days after pollination (DAP), and a period of active mitosis follows, peaking at 8–10 DAP (Kowles and Phillips 1985). This phase determines the number of cells in the endosperm and may be important for grain yield. During the mitotic phase, endosperm cells manifest a standard cell cycle, and the newly formed daughter cells maintain their initial triploid DNA content. The most rapid period of growth in maize endosperm is between 8 and 12 DAP, due to cell division and cell enlargement. Thereafter, cell division is limited to peripheral cell layers, i.e., aleurone and subaleurone, as cells in the central region cease to divide (Kowles and Phillips 1985).

Endoreduplication begins around 8 DAP, starting with cells in the center of the endosperm and slowly spreading outward, so that at the peak of endoreduplication, around 15 DAP, most starchy endosperm cells have gone through at least one round of endoreduplication (Fig. 1a). The switch from a mitotic to an endoreduplication cell cycle is gradual, so that by 15 DAP the endosperm contains a heterogeneous population of polyploid nuclei with DNA contents ranging from 3C to 96C, and sometimes higher (Fig. 1b) (Kowles and Phillips 1985; Larkins et al. 2001).

Enlargement of starchy endosperm cells begins in the center of the endosperm around 10–12 DAP, and coincides temporally and spatially with the onset of endoreduplication. Coupled with the increase in cell size is an increase of nuclear size. By measuring the size of endoreduplicated nuclei purified by fluorescence-activated cell sorting (FACS), one can see a positive correlation between DNA ploidy and nuclear size (Fig. 2a). At 18 DAP, central endosperm cells, which have larger nuclei, are more highly endoreduplicated than those in peripheral endosperm cells (Fig. 2b).

Shortly after the onset of endoreduplication and cell enlargement, starchy endosperm cells begin to accumulate large quantities of storage metabolites, such as starch and a variety of storage proteins, including prolamins

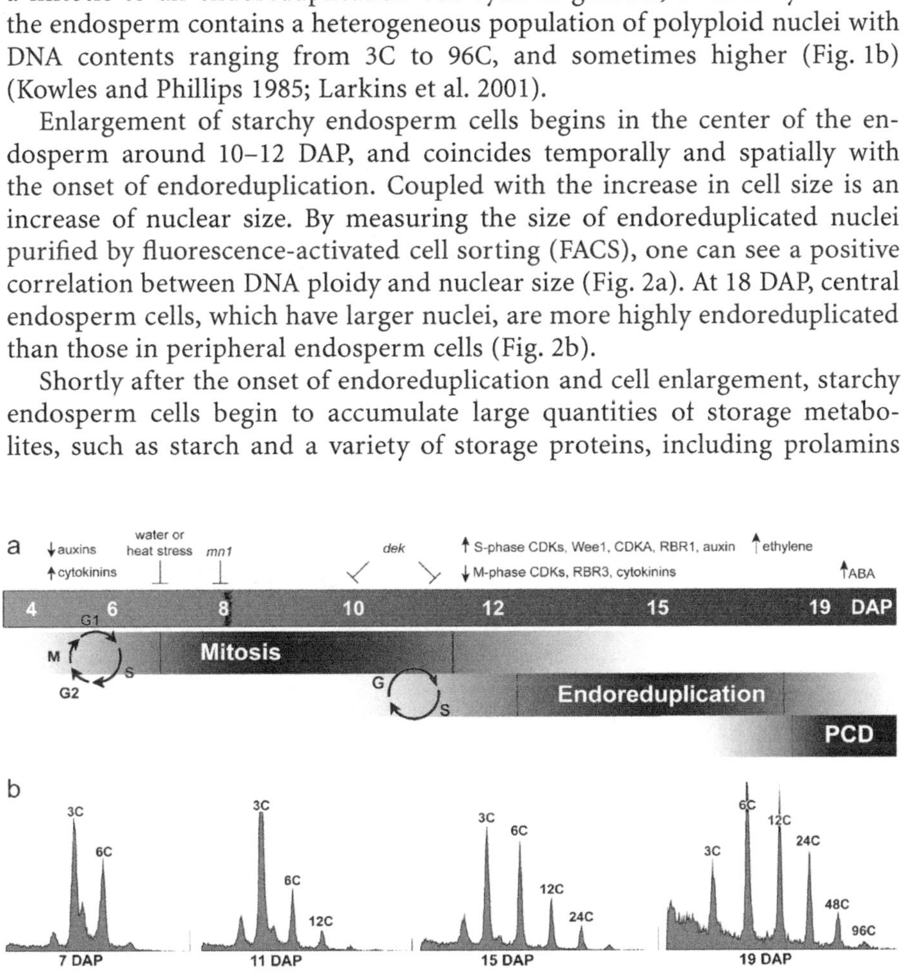

Fig. 1 Physiological and cell cycle factors that influence maize endosperm development. **a** Timeline of maize endosperm development. Endosperm cells begin to engage in mitosis from around 4 DAP and reach a peak between 8 and 10 DAP. Subsequently, starchy endosperm cells begin to engage in endoreduplication, resulting in polyploid nuclei. PCD begins after 16 DAP. At the top are listed factors that affect the different phases of endosperm development. Mutations are shown in italics. **b** Flow cytometric profiles showing DNA content of nuclei at different stages of maize endosperm development. At 7 DAP, endosperm cells are primarily mitotic, with nuclei in either G1 (3C DNA content) or G2 (6C). Later developmental stages are characterized by the presence of endoreduplicating cells, as shown by nuclei with increasing ploidy levels (indicated above the peaks)

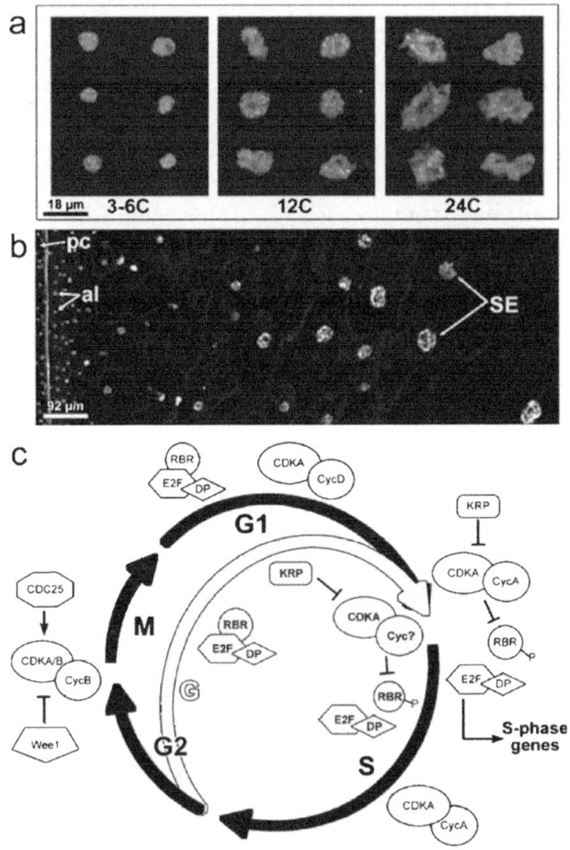

Fig. 2 The endoreduplication cycle creates polyploid nuclei of increasing size. **a** Endosperm nuclei purified by fluorescence-activated cell sorting. There is a positive correlation between nuclear size and the ploidy level. **b** Spatial distribution of endoreduplicated nuclei in 18-DAP endosperm. Nuclei in the central starchy endosperm cells (*SE*) are larger and, therefore, more highly endoreduplicated than those of the aleurone (*al*) cells. Pericarp, *pc*. **c** Diagram showing the key cell cycle regulators active during the mitotic and endoreduplication cycle in maize endosperm. CDKA is constitutively expressed in both types of cell cycle, and it associates with different cyclins during cell cycle progression. During the cell cycle, many S-phase genes are regulated by the E2F/DP family of transcription factors, which are repressed by RBR proteins in early G1. During late G1, the G1/S transition, and throughout S-phase, RBR becomes phosphorylated and inhibited by CycD/ and CycA/CDKA complexes, so that E2F/DP-responsive genes are transcribed. These CDK complexes can be inhibited by the maize CDK inhibitor, KRP. CycB/CDKB activity peaks during G2/M; its kinase component is inhibited by Wee1 and activated by CDC25-related proteins. During endoreduplication (*gray arrow*), cells alternate between G- and S-phases, bypassing M-phase. Endoreduplication is regulated by CDKA bound to an unidentified cyclin. This cyclin/CDKA complex is inhibited by KRP. As in the mitotic cycle, E2F/DP is repressed by RBR during G-phase and activated during S-phase, when RBR is phosphorylated by the kinase complex. Regulators of the endoreduplication cycle are *shaded gray*

(zeins), an 18-kD alpha-globulin, and an 11S legumin (Kodrzycki et al. 1989; Dolfini et al. 1992; Woo et al. 2001; Yamagata et al. 2003; chapter by Halford and Shrewry, in this volume). Transcripts encoding these proteins start to appear at around 10 DAP and peak at 14–18 DAP (Dolfini et al. 1992; Yamagata et al. 2003). The predominant storage proteins, the zein family of prolamins, associate with each other to form protein bodies and account for at least half the protein content at seed maturity (Lending and Larkins 1989; Shrewry and Tatham 1999). Starch also starts to accumulate around 10–12 DAP (Ingle et al. 1965; chapter by Hannah, in this volume), and at maturity the central endosperm cells are packed with starch granules and protein bodies.

PCD begins around 16 DAP in the starchy endosperm (Young and Gallie 2000a). Like endoreduplication, the onset of PCD is gradual and starts with cells in the center of the endosperm. It then slowly spreads to neighboring cells and progresses until all the starchy endosperm cells are dead. At seed maturity, only the aleurone cells are viable in the endosperm.

3 Endoreduplication

Endoreduplication is a widespread phenomenon in the plant kingdom that has been found in many differentiated and specialized cells, such as those of the hypocotyl (Gendreau et al. 1997), epidermis (Kudo and Kimura 2002), and trichomes (Schnittger et al. 2002), and storage cells such as those of the endosperm (Lopes and Larkins 1993; Larkins et al. 2001), cotyledons, and mesocarp of fruits (Sugimoto-Shirasu and Roberts 2003). Endoreduplication is correlated with an increased metabolic rate, high levels of gene transcription, and terminal differentiation (Kowles and Phillips 1985; Brunori et al. 1993), but its precise role is unknown. Transcriptional and translational activity are thought to increase proportionately with each doubling of the genome, so the metabolic activity of a highly polyploid cell could be functionally equivalent to that of many diploid cells (D'Amato 1984).

In cereal endosperm, endoreduplication is usually restricted to starchy endosperm cells, with the exception of barley, which shows polyploid aleurone cells (Keown et al. 1977). The extent of endoreduplication in the endosperm varies depending on the cereal species. Wheat endosperm goes through only about three rounds of endoreduplication, resulting in a ploidy of 24C (Chojecki et al. 1986), whereas in rice, ploidy levels can be as high as 74C (Ramachandran and Raghavan 1989), and in sorghum, 96C (Kladnik et al. 2006). Maize endosperm is by far the best studied system, and cells can achieve ploidies of 96C and higher (Kowles and Phillips 1985; Larkins et al. 2001; Dilkes et al. 2002; Leiva-Neto et al. 2004).

3.1 Cell Cycle Overview

3.1.1 The Mitotic Cell Cycle

The standard mitotic cell cycle consists of periods of DNA synthesis (S-phase) and chromosome separation (M-phase), preceded by gaps G1 and G2, respectively (Fig. 2c). During this cycle, the orderly progression of events that cause chromosomes to duplicate and separate is largely controlled by cyclin-dependent protein kinases (CDKs), which consist of two components: a serine/threonine protein kinase and a cyclin regulatory subunit. The activity of CDKs is controlled by changes in cyclins, the phosphorylation state of the CDK complex, and its association with inhibitors (Morgan 1997; Sherr and Roberts 1995). Protein kinases related to Wee1 (Parker et al. 1992; Sun et al. 1999b; Sorrell et al. 2002) inactivate CDKs by phosphorylating tyrosine-15 and threonine-14 on the kinase component. Conversely, CDC25-related phosphatases activate CDKs by dephosphorylating these residues (Kumagai and Dunphy 1991; Landrieu et al. 2004). Cyclin-dependent kinase inhibitors (CKIs) inhibit CDKs by binding to the cyclin/CDK complex, or to the kinase component, and cause allosteric changes, thereby preventing CDK activity. Multiple CKIs have been isolated from yeast (Mendenhall 1998), mammals, and more recently from plants (Morgan 1997; Verkest et al. 2005; Coelho et al. 2005; De Clercq and Inze 2006). Two CKIs from maize endosperm, Zeama;KRP;1 and Zeama;KRP;2, have been characterized (Coelho et al. 2005), and they will be discussed in the next section.

As in mammals and yeast, specific cyclins change in concentration during the different phases of the plant cell cycle, thereby causing an oscillation in CDK activity (reviewed by DeWitte and Murray 2003 and de Jager et al. 2005). G1 progression and transition into S-phase are controlled by cyclin D/CDKs, while S-phase progression primarily involves cyclin A/CDKs. In animals, cyclin E/CDKs regulate the G1/S transition (Sauer et al. 1995; Coverly et al. 2002), but no E-type cyclins have been identified thus far in plants. It is hypothesized that their function is replaced by cyclin A (Yu et al. 2003). Cyclin B accumulates at the beginning of G2 and controls the transition into M-phase. During anaphase, both cyclins A and B are degraded (Criqui et al. 2001), thus causing a downregulation in CDK activity in late M-phase, which resets the cell cycle.

Flowering plants, like mammals, utilize different CDKs during different phases of the cell cycle. This contrasts with yeasts, which have a single CDK called Cdc2/Cdc28. Plants contain five classes of CDKs, of which only two, CDKA and CDKB, have been shown to regulate cell cycle progression (Joubes et al. 2000; Porceddu et al. 2001; Sorrell et al. 2001; Menges and Murray 2002). CDKA, which contains a PSTAIRE motif and is capable of complementing yeast cdc28/cdc2 mutants (Joubes et al. 2000), is expressed throughout the cell cycle. It combines with D- and A-type cyclins during cell cycle progression and its kinase activity increases from late S- through M-phase (Fig. 2c). In CDKB, which is unique to plants, the cyclin-interacting PSTAIRE motif is replaced by PPTALRE or PPTTLRE. Its expression and activity peak at the G2/M-phase transition (Sorrell et al. 2001; Menges and Murray 2002).

According to the current model for eukaryotic cell cycle progression, which is mainly derived from research on yeast and mammals, low levels of CDK activity after M-phase allow the orderly assembly of prereplication complexes (pre-RCs) during early G1-phase (Gerbi et al. 2002). These complexes are composed of ORC, CDC6, CDT1, and MCM proteins, and they bind DNA replication origins and license the chromatin for replication (Fung and Poon 2005). As CDK activity rises near the end of G1, the pre-RCs are converted to initiation complexes (Gerbi et al. 2002) and eventually replication enzymes are recruited to begin DNA synthesis.

One major point of cell cycle regulation is at the transition into S-phase, which involves the activation of the E2F family of transcription factors. The E2F pathway regulates the expression of many S-phase genes (De Veylder et al. 2002; Sozzani et al. 2006), but during early G1, it is repressed by the binding of retinoblastoma-related (RBR) proteins (Fung and Poon 2005). During late G1, RBR proteins are phosphorylated by cyclin D and cyclin A/CDKs, respectively, causing them to disassociate from E2F. The E2F pathway is then able to transactivate a battery of genes that leads to entry into S-phase. The sustained high CDK activity throughout S-phase results in phosphorylation of licensing factors and their disassociation from chromatin (Findeisen et al. 1999; Li et al. 2004), which prevents refiring of the replication origins and ensures that only one round of replication occurs per cell cycle. During G2, a second rise in CDK activity drives the cell into M-phase.

3.1.2
The Endoreduplication Cell Cycle

Endoreduplication results from a modified cell cycle that bypasses M-phase and simply manifests alternating G- and S-phases (Fig. 2c) (Edgar and Orr-Weaver 2001; Larkins et al. 2001). As a consequence, there are multiple rounds of DNA replication without mitosis, resulting in polyploid cells. Recent studies suggest that endoreduplication is achieved by a suppression of M-phase-associated CDK activity, along with an oscillation of S-phase CDK activity (Grafi and Larkins 1995). Mis- or overexpression of cyclin B in *Arabidopsis* trichomes induced mitosis in cells that normally undergo four rounds of endoreduplication, suggesting that the absence of M-phase CDKs is sufficient for endoreduplication to occur (Walker et al. 2000; Schnittger et al. 2002). Research on *Drosophila* salivary glands and rat trophoblast cells also showed that in endoreduplicating cells, there is a downregulation of M-phase CDKs, while the activity of S-phase-associated CDKs is maintained, and perhaps oscillates (MacAuley et al. 1998; Weiss et al. 1998).

3.2
The Endoreduplication Cell Cycle in Maize Endosperm

Grafi and Larkins (1995) showed that in maize endosperm, kinase activity associated with a CDK complex, the M-phase promoting factor (MPF), peaked at 10 DAP, when mitosis was most active, and declined as endoreduplication progressed. In fact, when 10-DAP maize endosperm extract was mixed with that from 15 DAP, kinase activity decreased in a concentration-dependent manner, suggesting an inhibition of M-phase CDKs during endoreduplication. When 8- and 16-DAP endosperm extracts were incubated with proteins known to bind CDKs, greater kinase activity was recovered from 16-DAP than 8-DAP extract, suggesting that there is an increase in S-phase-associated CDK activity during the endoreduplication phase compared to the mitotic phase. So, maize endosperm shows a reduction of M-phase-associated CDK activity along with maintenance of S-phase-associated CDK activity during the period of endoreduplication, similar to what was observed in animal systems.

Maize CDKA and CDKB were recently characterized and found to be differentially expressed in developing endosperm (Dante 2005). CDKA protein levels remain relatively constant throughout endosperm development, while CDKB decreases after the mitotic stage, suggesting it does not contribute to the CDK activity associated with endoreduplication. CDKA, on the other hand, plays an important role in endoreduplication. Transgenic maize plants were generated that overexpressed a dominant-negative mutant form of CDKA (CDKA-DN), in which a change of Asp-146 to Asn-146 abolished the kinase activity, but did not interfere with cyclin binding (Leiva-Neto et al. 2004). The transgene was driven by the 27-kD gamma-zein promoter, so that CDKA-DN expression was endosperm-specific and was produced throughout the starchy endosperm. Because this is a strong promoter, the high level of CDKA-DN expression out-competed the endogenous CDKA for cyclin binding, thereby effectively reducing S-phase kinase activity and interfering with the DNA replication process. As a result, there was a 50% reduction in endoreduplication during transgenic endosperm development, with no cells exceeding a DNA ploidy of 24C at 18 DAP.

The cyclin subunit(s) that interacts with CDKA during endoreduplication was not identified, but a number of maize cyclins, including cyclins B1;3 (Sun et al. 1999a), A1;3, D5;1, and D2;1 (Dante 2005), have been characterized and found to be expressed in maize endosperm. Cyclin A1;3-associated kinase activity was greatest during the mitotic stage and then declined as the endosperm progressed into the endoreduplication stage, suggesting that cyclin A1;3 is more likely involved in the mitotic than the endoreduplication cell cycle. Cyclins B1;3, D5;1, and D2;1, on the other hand, showed an increase in kinase activity at 11 DAP, but then decreased at 15 DAP. At 11 DAP, the endosperm is active in both the mitotic and endoreduplication cell cycles. The sustained (although reduced) kinase activity associated with these three cyclins throughout endosperm development suggested that they may be involved in the regulation of both cell cycles. As cyclin D is often associated with the transition into S-phase, it would not be surprising that cyclins D2;1 and D5;1 are involved in endoreduplication. But the fact that cyclin B1;3, a mitotic cyclin, maintained kinase activity after mitosis ceased was unexpected and suggested that cyclin B1;3/CDK may be involved in the G/S transition or progression through S-phase. To determine if any of these cyclins interacted with CDKA, affinity-purified CDKA-DN complexes were immunoblotted with antibodies against cyclins A1;3, B1;3, and D2;1. However, this assay failed to detect a cyclin protein, suggesting an unidentified cyclin interacts with CDKA to regulate endoreduplication.

Only a few CDK regulatory factors have been identified in maize that could be involved in controlling the endoreduplication cell cycle by inhibiting CDK activity. These include a Wee1 homolog (Sun et al. 1999b) and two CKIs, Zeama;KRP;1 and Zeama;KRP;2 (Coelho et al. 2005). When ZmWee1 was overexpressed in fission yeast (*Schizosaccharomyces pombe*), cell division was inhibited. In maize, ZmWee1 is expressed at low levels during early endosperm development and peaks during endoreduplication, suggesting that its expression might inhibit cell division at this stage. Zeama;KRP;1 and Zeama;KRP;2 are expressed in the endosperm and were found to inhibit cyclin A1;3- and D5;1-associated kinase activities, both of which are considered to be G1/S CDKs. However, KRP1 and KRP2 protein accumulation differed as the endosperm developed. While KRP1 protein was constant throughout development, that of KRP2 decreased after the onset of endoreduplication (Coelho et al. 2005), suggesting that KRP2 may only be involved in inhibition of G1/S CDKs during the mitotic stage. In *Drosophila*, it has been demonstrated that for endoreduplication to continue, an oscillation of S-phase CDKs is required. S-phase CDK activity is lost after DNA replication initiation, and it begins to increase again prior to a new S-phase (Weiss et al. 1998). Perhaps KRP1 inhibits S-phase CDKs after replication initiation, creating the oscillation required to sustain endoreduplication in maize endosperm.

The RBR/E2F pathway appears to play an important role in cell cycle regulation during endosperm development. Three different RBR genes have been isolated from maize: RBR1 (Grafi et al. 1996), RBR2 (Ach et al. 1997), and RBR3 (Sabelli et al. 2005; Sabelli and Larkins 2006). RBR1 and RBR2 share a high degree of sequence identity and are likely paralogs with redundant functions. Studies on RBR1 (Grafi et al. 1996; Desvoyes et al. 2006) have shown that it is able to interact with cell cycle progression-inducing proteins, such as SV40 T-antigen, papillomavirus E7, wheat dwarf virus RepA, and plant cyclin Ds, indicating that, similar to oncogenic transformation in mammals, relief from an RBR block induces cell cycle entry and progression in plants. RBR3, however, shares only 50% sequence identity with RBR1 and appears to be an E2F target that is repressed by RBR1 (Sabelli et al. 2005; Sabelli and Larkins 2006). In maize, both RBR1 and RBR3 are expressed in developing endosperm, but RBR3 is preferentially expressed during the mitotic phase and is drastically downregulated after the onset of endoreduplication, suggesting that it may be more involved in the G1/S regulation of mitotic cells than endoreduplicating cells. Interestingly, RBR1 expression seems upregulated during endoreduplication (Sabelli et al. 2005; Sabelli and Larkins 2006), which is contradictory to the expectation that negative regulators of the G/S transition would be downregulated. It could be that, like CKIs, RBR1 is required in order to maintain cyclic levels of S-phase CDKs by repressing E2F-regulated expression of DNA replication licensing factors during G-phase. When RBR1 is phosphorylated, E2F repression is relieved, allowing S-phase to proceed.

3.3
Factors that Affect Endoreduplication

Endoreduplication can be affected by a number of physiological and environmental factors. In a cross between two inbred lines, the endoreduplication pattern of the endosperm in the F_1 generation generally resembles that of the maternal parent (Kowles et al. 1997; Dilkes et al. 2002). This "maternal effect" could be due to gene dosage, in that two copies of the maternal and one copy of the paternal genome contribute to the genetic makeup of the endosperm. Alterations to this 2 : 1 ratio cause deleterious changes during development of the endosperm (LeBlanc et al. 2002). A maternal excess results in precocious endoreduplication and a shortened mitotic phase, although it does not alter the degree of endoreduplication. Excess of the paternal genome, on the other hand, delays entry into endoreduplication, not by prolonging the mitosis phase, but by causing an extended mitotic hiatus prior to the switch to endoreduplication. Also, DNA amplification is severely reduced in all cells that are committed to endoreduplication (LeBlanc et al. 2002).

A possible explanation for the maternal effect on endosperm development in general, and endoreduplication in particular, could be due to imprinting, in which specific paternal genes are silenced and only their maternal counterparts are expressed. Genomic imprinting is a phenomenon observed in plants and animals, often during the reproductive phase (Guitton and Berger 2005; chapter by Penterman, in this volume). In *Arabidopsis*, a group of transcriptional regulators belonging to the Polycomb group complex, including *FIS*, *FIE*, and *MEA*, are imprinted during early seed development (Sørensen et al. 2001). After fertilization, only the maternal alleles in the Polycomb group genes are expressed in the endosperm, while the paternal alleles are silenced, at least during the early stages of development. A homolog to *Arabidopsis* *FIE* gene has been found in maize, where it occurs as two copies, *FIE1* and *FIE2* (Danilevskaya et al. 2003). Presumably, the function of these genes in maize endosperm is similar to that in *Arabidopsis*, in that the paternal *FIE1* allele is not expressed, while that of *FIE2* is initially silenced but is expressed 5 DAP. Undoubtedly, there are genes involved in endoreduplication that are imprinted, thereby contributing to the described maternal effect.

Hormones appear to play an important role in the induction of endoreduplication in the endosperm. The levels of several different phytohormones change during its development. It is known that cytokinins stimulate cell division, and indeed during the mitotic stage of maize endosperm development, the levels of cytokinins, such as zeatin and zeatin-riboside, increase sharply and peak at 9 DAP, and then abruptly decline between 9 and 15 DAP (Lur and Setter 1993). Conversely, auxin levels are low when cytokinins peak, and then start to increase at 10 DAP, when the cytokinins begin to decline. Thus, a high auxin/cytokinin ratio coincides with the onset of endoreduplication (Lur and Setter 1993; Cheikh and Jones 1994). Apparently, auxin is involved in the induction of endoreduplication, but not its continuation. Application of exogenous 2,4-dichlorophenoxyacetic acid (2,4-D), a synthetic auxin, during the mitotic phase caused precocious DNA amplification, but when applied during the peak of endoreduplication (around 12–15 DAP), it did not increase DNA content compared to the control (Lur and Setter 1993). When the anti-auxin, *p*-chlorophenoxyisobutyric acid (PCIB), was applied to the endosperm during the period of endogenous auxin increase, i.e., the onset of endoreduplication, there was no difference in endosperm DNA content compared to the control. But when PCIB was applied during the peak of endoreduplication, DNA accumulation was reduced by 20–25% (Lur and Setter 1993). PCIB has been shown to inhibit auxin-stimulated events rather than auxin production itself, indicating that it is auxin-induced events that cause endoreduplication and not auxin directly.

Endoreduplication in maize endosperm is inhibited by environmental stresses, such as water deficit and heat. However, it appears that the reduction in endoreduplication is more a result of perturbation during early events of development, rather than a direct consequence of the inhibition of the endoreduplication cell cycle per se. When the kernel is treated with high heat (Engelen-Eigles et al. 2000, 2001) or deprived of water (Artlip et al. 1995; Setter and Flannigan 2001) during the mitotic phase, there is a marked reduction of both cell number and DNA content; however, when either form of stress is applied during active endoreduplication, the reduction in DNA amplification is not as dramatic.

The reduced level of endoreduplication as a consequence of physiological and genetic perturbations during the mitotic phase shows that there is a relationship between the mitotic and endoreduplication phases of endosperm development. Of 35 maize mutants categorized as defective kernel (*dek*) mutants, all but one had endosperm with both a reduced cell number and a lower level of endoreduplication (Kowles et al. 1992). This suggests a regulatory mechanism that coordinates both the mitotic and endoreduplication phases of development. However, the mitotic cell cycle is not necessarily coupled with that of endoreduplication, as one *dek* mutant showed a 70% reduction in cell number, but little or no reduction in DNA content when compared to the wild type. Another mutation that shows uncoupling of the mitotic and endoreduplication phases of endosperm development is *miniature1* (*mn1*) (Vilhar et al. 2002). Maize plants with this mutation produce small kernels with endosperm that is only 20% the volume of wild type. *Mn1* encodes a cell wall invertase, INCW2, that provides hexose sugars to the developing endosperm. The *mn1* mutation limits the amount of energy available for endosperm development, causing a reduction in mitotic activity and inhibition of cell expansion, but it does not severely affect endoreduplication. *mn1* starchy endosperm cells go through at least six rounds of endoreduplication by 16 DAP and have the same ploidy distribution as wild-type endosperm.

3.4 Role of Endoreduplication

The presumed role of endoreduplication is to promote cell enlargement and high levels of gene expression to facilitate the rapid growth and maturation of the tissue. However, recent work has shown there is not always a tight relationship between cell size and DNA ploidy levels. In *Arabidopsis* roots, there was no correlation between DNA content and the length of cortical cells (Beemster et al. 2002). Also, work with transgenic maize endosperm suggested that ploidy is not tightly coupled with cell size or gene expression. In the CDKA-DN maize mutant previously described (Leiva-Neto et al. 2004), the defective CDKA resulted in a 50% reduction in mean nuclear ploidy. However, there was no detectable effect on the size of endosperm cells, even in the central starchy endosperm where one might expect to detect a difference. Likewise, there was no marked reduction in gene expression when compared to wild type. Conversely, in the small kernels of the *mn1* mutant, endoreduplication is not reduced, even though cell enlargement is affected (Vilhar et al. 2002). While there is a positive correlation between cell size and ploidy within the kernel, when an *mn1* cell was compared with a wild-type cell of the same ploidy class, the *mn1* cell tended to be smaller than the wild type. Therefore, while there is often a correlation between cell size and nuclear ploidy, endoreduplication is not necessarily coupled to the control of cell size or high levels of gene expression, although it could be associated with processes that are coincident with these events.

3.5 Conclusions

Currently, the role of endoreduplication during maize endosperm development is unclear. In the CDKA-DN mutant, the transgene is active in the endosperm at around 10 DAP, by which time it was possible to measure the progression of endoreduplication. Using this promoter, it was only possible to reduce endoreduplication and not completely inhibit it, suggesting that the mutant form of CDKA did not completely abolish wild-type CDKA activity. Perhaps only a few endoreduplication cycles are required to influence cell enlargement and transcription rates. It would be necessary to completely abolish endoreduplication in the endosperm in order to test this hypothesis. Alternatively, perhaps the role of endoreduplication is not to influence cell enlargement or gene expression in the endosperm, but rather to contribute to seed germination. The endosperm acts as a source of stored metabolites for the seedling at germination. It is possible that the polyploid nuclei of the starchy endosperm are a source of nucleotides or nucleosides for the embryo, and this allows for rapid cell division and protein biosynthesis during the early stages of seedling development, as de novo synthesis of nucleotides might be an energy-expensive and time-consuming process.

4 Programmed Cell Death

The final phase of starchy endosperm development in cereal grains involves cell death. Indeed, cell death is an important part of the development and maintenance of many tissues and organs in plants and animals (Buckner et al. 2000; Krishnamurthy et al. 2000). PCD can be triggered in cells that have served their function, such as transmitting cells of the postpollination stigma and the antipodal and synergid cells in the postfertilization megagametophyte, or cells that die as part of differentiation, such as xylem tracheary elements and cells that recognize pathogens (Krishnamurthy et al. 2000). The regulation of PCD has been well studied in animal cells, but the process is not equally well understood in plants. The trigger for PCD can be from signals generated outside or inside the cell, and they lead to signaling events that result in a cascade of hydrolytic activity (Krishnamurthy et al. 2000). Some of the signals generated in animal cells include calcium fluxes, production of free radicals, release of cytochrome *c* from mitochondria, and increased levels of reduced glutathione (Ellis et al. 1991). Upon perception of the signal, a cascade of biochemical reactions results in the activation of the effectors of PCD. The key effectors in animal PCD are cysteine-containing aspartate-specific proteases known as caspases. The activity of caspases leads to cytoplasmic shrinkage, plasma and nuclear membrane blebbing, chromatin condensation, DNA internucleosomal fragmentation, and, finally, phagocytosis by neighboring cells (Ellis et al. 1991). Similarly, PCD in plants also manifests DNA fragmentation, chromatin condensation, and nuclear membrane blebbing (Krishnamurthy et al. 2000; Dominguez et al. 2001). However, as plant cells are immobile and surrounded by walls, they are not phagocytosed but remain in place; the cell remnants may or may not be crushed.

In plants, genes encoding caspases have not been identified (Dominguez and Cejudo 2006; Lam 2005), but there is indirect evidence that caspase-like activity takes place in dying cells. For example, extracts from plant cells undergoing PCD can induce PCD-like activity in cellular extracts (Dominguez and Cejudo 2006) as well as cleave a variety of caspase substrates (Woltering 2004). Recent research in plants has identified cysteine proteases that are vacuolar processing enzymes (VPEs), which have caspase-like activity and induce PCD (Hatsugai et al. 2004; Lam 2004; Greenwood et al. 2005). Also, plant subtilisin-like proteases, which are serine proteases, were shown to have aspartate-specific protease activity and be involved in cell death progression (Coffeen and Wolpert 2004; Roberts et al. 2006).

One consequence of PCD is the disassembly of the nucleus, which starts with fragmentation of chromatin by endonucleases. DNA breaks occur at nucleosome-linker sites, resulting in oligosomes of 180 bp. This reaction, which is shown by gel electrophoresis as a DNA ladder with multimers of 180-bp fragments, is often used as an indicator of PCD (Danon and Gallois 1998; Enari et al. 1998; Young and Gallie 2000a). DNA fragmentation can also be used to identify individual cells undergoing PCD by using a terminal deoxynucleotidyl transferase-mediated dUTP nick end-labeling (TUNEL) assay (Gavrieli et al. 1992). This assay uses the binding of terminal deoxynucleotidyl transferase to the 3′ end of DNA to incorporate labeled deoxyuridine at break sites, which can then be visualized by light microscopy.

In cereal grains, PCD is important for the proper maturation of the endosperm, and it occurs according to a developmentally regulated process in several of tissues in the developing caryopsis. It is responsible for the destruction of the nucellus following fertilization, and maturation of the seed coat and starchy endosperm.

4.1 PCD in the Nucellus

The mature cereal caryopsis is surrounded by remnants of the pericarp and nucellus. After fertilization, both these tissues rapidly undergo PCD. While there is no direct evidence showing the involvement of caspase or caspase-like proteases in these tissues, transcript analysis of developing barley pericarp showed an upregulation of proteases, such as serine proteases and VPE-type cysteine proteases (Sreenivasulu et al. 2006). Furthermore, morphological indicators reveal that PCD takes place in these tissues. In wheat kernels, fragmentation of DNA is well advanced by 5 days after flowering (DAF), and continues until at least 15 DAF (Dominguez et al. 2001); TUNEL assays localized the DNA breaks to cells in the nucellus. These cells had some morphological features similar to those animal cells undergoing PCD, including cytoplasmic degradation, increased vacuolization, and the loss of plasma membrane integrity (Dominguez et al. 2001). Eventually, these cells lysed and their volume was replaced by the endosperm. It has long been thought that degradation of the nucellus is important to provide nutrients that are absorbed by the growing endosperm during its early stages of development (Norstog 1978).

By mid-development (13–15 DAP) of wheat endosperm, PCD occurs in the nucellar projection, which is the interface between the endosperm and the vascular strand in the pericarp (Dominguez et al. 2001). Cells of the nucellar projection differentiate into transfer cells and are important for the movement of nutrients into the endosperm. A similar process occurs in the maize kernel. As shown by TUNEL assay, at around 12 DAP cells in the placental–chalazal (P-C) layer, a patch of tissue between the endosperm and vascular tissue, also undergo PCD (Kladnik et al. 2004, 2005). But unlike the nucellus, the remnants of the cells in the nucellar projection and P-C layer are not absorbed by the endosperm, but rather become conducting tissue for nutrient transport.

4.2 PCD in the Endosperm

The onset of PCD in the endosperm is a gradual process that starts in a few central starchy endosperm cells and spreads until it reaches the aleurone. PCD begins around 16 DAP in two separate regions: cells in the center of the starchy endosperm and cells at the crown of the kernel, adjacent to the silk. Cell death slowly spreads from these two areas until they eventually merge, so that by 28 DAP most of the upper portion of the starchy endosperm is no longer viable. Wheat endosperm also goes through PCD beginning around mid-development (Young and Gallie 1999). However, unlike maize, in which PCD progresses in an orderly fashion starting in specific regions, that in wheat is more stochastic, with PCD affecting cells scattered throughout the endosperm. This process continues until 30 DAP, by which time the entire wheat starchy endosperm has gone through PCD. As is true of maize, in wheat endosperm all but the aleurone cells are dead at maturity.

The effectors of PCD in cereal endosperm have not been identified, but transcript profiling of developing barley endosperm has shown that there is increased expression of a number of cysteine proteases, including gamma- and beta-VPE (Sreenivasulu et al. 2006). This suggests that these, or related enzymes, might bring about PCD.

Although the aleurone cells of the endosperm remain viable at seed maturation, after germination they eventually undergo PCD. This process begins with the cells closest to the embryo and spreads to those farther away (Dominguez et al. 2004). PCD does not begin immediately after seed imbibition, because the aleurone initially functions as a secretory tissue, producing hydrolytic enzymes (proteases, nucleases, and amylases) that are secreted into the starchy endosperm to break down the stored metabolites. However, after these enzymes are produced, PCD becomes evident through rapid vacuolation of the cytoplasm, loss of plasma membrane integrity, nuclear breakdown, and DNA fragmentation (Bethke et al. 1999; Fath et al. 2000; Dominguez et al. 2004). In wheat, DNA laddering is detected as soon as 4 days after imbibition, while in maize it is not observed until 12 days after imbibition (Dominguez et al. 2004). DNA laddering was not observed in barley aleurone cells, but as in wheat, they stained positively in TUNEL assays, indicating that DNA fragmentation had occurred.

4.3
PCD in the Endosperm is Hormonally Regulated

While the proteolytic effectors of PCD in cereal grains are unclear, it is known that PCD in starchy endosperm and aleurone cells is associated with changes in hormone levels. In starchy endosperm, PCD accompanies a peak in ethylene production. In maize kernels, two ethylene peaks are detected: the initial one coincides with the first wave of PCD in the center of the endosperm, around 16 DAP, and the second one at around 20 DAP, when cells in the proximal region begin PCD. Application of exogenous ethylene at 16 DAP led to more extensive PCD and DNA fragmentation than in untreated control kernels (Young et al. 1997). Furthermore, inhibition of endogenous ethylene biosynthesis by aminoethoxyvinylglycine (AVG) resulted in less DNA fragmentation than in control kernels, suggesting that ethylene either initiates or stimulates the induction of PCD.

In addition to ethylene, abscisic acid (ABA) levels are also elevated during the period of PCD. ABA influences events during late kernel development, such as the acquisition of desiccation tolerance and dormancy by the embryo. While ABA was not found to be directly involved in the onset of PCD, alterations of its biosynthetic pathway and perception affect the onset of PCD by causing changes in ethylene biosynthesis. In maize viviparous mutants (*vp1* and *vp9*) that perturb ABA production or perception, there is premature DNA fragmentation as a result of a two- to fourfold elevation in ethylene (Young and Gallie 2000b). While ABA may not be directly involved in PCD, it appears to inhibit ethylene production and consequently delays the onset of this process. Analysis of the barley endosperm transcriptome confirmed the involvement of ethylene and ABA, as transcripts for ABA- and ethylene-receptive genes increase during the latter stages of endosperm development (Sreenivasulu et al. 2006).

PCD in aleurone cells is also hormonally regulated, but by giberellic acid (GA) rather than ethylene. Upon germination, gibberellins are synthesized by the embryo and diffuse into the endosperm, where they are perceived by aleurone cells. The initial response to exogenous GA application is the production of hydrolytic enzymes, and eventually PCD. When barley aleurone protoplasts were incubated with GA, most of the cells died after showing cytoplasmic vacuolation and shrinkage, and DNA degradation (Bethke et al. 1999). Degermed wheat kernels treated with GA had higher levels of endonuclease activity, leading to higher levels of DNA fragmentation when compared to untreated kernels (Dominguez et al. 2004). Also, treatment of wheat kernels with paclobutrazol, a GA synthesis inhibitor, resulted in no DNA fragmentation in the aleurone compare to untreated kernels (Dominguez et al. 2004). As in the starchy endosperm, ABA also plays a role in inhibiting or delaying PCD in aleurone cells. Aleurone protoplasts can be kept alive for up to 6 months when cultured in a medium containing ABA, whereas cell death progresses when they are exposed to GA (Bethke et al. 1999; Fath et al. 2000).

4.4
Conclusions

The function of programmed cell death in cereal endosperm is presumed to be for the facilitation of rapid hydrolysis and absorption of nutrients by the germinating embryo, since cell membranes are no longer maintained and movement of metabolites into and out of cells no longer requires energy. However, to determine if germination is affected, further research is required, perhaps through the discovery of mutants or the generation of transgenic plants lacking PCD in the endosperm. There is emerging evidence that PCD in cereal endosperm may be mediated by caspase-like VPEs, but the specific enzymes are not known. Also, the substrates for these VPEs and downstream targets have yet to be identified. The emerging technology of system-wide profiling of transcripts and proteins may help to unravel the PCD pathway in the cereal endosperm.

References

Ach RA, Durfee T, Miller AB, Taranto P, Hanley-Bowdoin L, Zambryski P, Gruissem W (1997) *RRB1* and *RRB2* encode maize retinoblastoma-related proteins that interact with a plant D-type cyclin and geminivirus replication protein. Mol Cell Biol 17:5077–5086

Artlip TS, Madison JT, Setter TL (1995) Water deficit in developing endosperm of maize:cell division and nuclear DNA endoreduplication. Plant Cell Environ 18:1034–1040

Becraft PW (2007) Aleurone Cell Development (in this volume). Springer, Heidelberg

Becraft PW, Asuncion-Crabb Y (2000) Positional cues specify and maintain aleurone cell fate in maize endosperm development. Development 127:4039–4048

Beemster GTS, De Vusser K, De Tavernier E, De Bock K, Inze D (2002) Variation in growth rate between *Arabidopsis* ecotypes is correlated with cell division and A-type cyclin-dependent kinase activity. Plant Physiol 129:854–864

Bethke PC, Lonsdale JE, Fath A, Jones RL (1999) Hormonally regulated programmed cell death in barley aleurone cells. Plant Cell 11:1033–1046

Brown, Lemmon (2007) The Developmental Biology of Cereal Endosperm (in this volume). Springer, Heidelberg

Brunori A, Forino LMC, Frediani M, Ruberti F (1993) Cell number and polyploidy in the starchy endosperm of *Triticum*. J Genet Breed 47:217–220

Buckner B, Janick-Buckner D, Gray J, Johal GS (2000) Cell death in maize. Physiol Plant 108:231–239

Cheikh N, Jones RJ (1994) Disruption of maize kernel growth and development by heat stress. Plant Physiol 106:45–51

Chojecki AJS, Bayliss MW, Gale MD (1986) Cell production and DNA accumulation in the wheat endosperm, and their association with grain weight. Ann Bot 58:809–817

Coelho CM, Dante RA, Sabelli PA, Sun Y, Dilkes BP, Gordon-Kamm WJ, Larkins BA (2005) Cyclin-dependent kinase inhibitors in maize endosperm and their potential role in endoreduplication. Plant Physiol 138:2323–2336

Coffeen WC, Wolpert TJ (2004) Purification and characterization of serine proteases that exhibit caspase-like activity and are associated with programmed cell death in *Avena sativa*. Plant Cell 16:857–873

Cossegal M (2007) The Embryo Surrounding Region (in this volume). Springer, Heidelberg

Coverley D, Laman H, Laskey RA (2002) Distinct roles for cyclins E and A during DNA replication complex assembly and activation. Nat Cell Biol 4:523–528

Criqui MC, Weingartner M, Capron A, Parmentier Y, Shen W-H, Heberle-Bors E, Laszlo B, Genschik P (2001) Sub-cellular localization of GFP-tagged tobacco mitotic cyclins during the cell cycle and after spindle checkpoint activation. Plant J 28:569–581

D'Amato F (1984) Role of polyploidy in reproductive organs and tissues. In: Johri BM (ed) Embryology of angiosperms. Springer, Berlin, pp 523–566

Danilevskaya ON, Hermon P, Hantke S, Muszynski MG, Kollipara K, Ananiev EV (2003) Duplicated *fie* genes in maize:expression pattern and imprinting suggest distinct functions. Plant Cell 15:425–438

Danon A, Gallois P (1998) UV-C radiation induces apoptotic-like changes in *Arabidopsis thaliana*. FEBS Lett 437:131–136

Dante RA (2005) Characterization of cyclin-dependent kinases and their expression in developing maize endosperm. PhD dissertation, University of Arizona, Tucson, AZ

De Clercq A, Inze D (2006) Cyclin-dependent kinase inhibitors in yeast, animals, and plants: a functional comparison. Crit Rev Biochem Mol Biol 41:293–313

de Jager SM, Maughan S, Dewitte W, Scofield S, Murray JAH (2005) The developmental context of cell-cycle control in plants. Semin Cell Dev Biol 16:385–396

De Veylder L, Beeckman T, Beemster GTS, de Almeida Engler J, Ormenese S, Maes S, Naudts M, Van Der Schueren E, Jacqmard A, Engler G, Inze D (2002) Control of proliferation, endoreduplication and differentiation by the *Arabidopsis* E2Fa–DPa transcription factor. EMBO J 21:1360–1368

Desvoyes B, Ramirez-Parra E, Xie Q, Chua N-H, Gutierrez C (2006) Cell type-specific role of the retinoblastoma/E2F pathway during *Arabidopsis* leaf development. Plant Physiol 140:67–80

Dewitte W, Murray JAH (2003) The plant cell cycle. Annu Rev Plant Biol 54:235–264

Dilkes BP, Dante RA, Coelho CM, Larkins BA (2002) Genetic variation in endoreduplication in maize endosperm is subject to parent-of-origin specific effects. Genetics 160:1163–1177

Dolfini S, Landoni M, Tonelli C, Bernard L, Viotti A (1992) Spatial regulation in the expression of structural and regulatory storage-protein genes in *Zea mays* endosperm. Dev Genet 13:264–276

Dominguez F, Cejudo FJ (2006) Identification of a nuclear-localized nuclease from wheat cells undergoing programmed cell death that is able to trigger DNA fragmentation and apoptotic morphology on nuclei from human cells. Biochem J 397:529–536

Dominguez F, Moreno J, Cejudo FJ (2001) The nucellus degenerates by a process of programmed cell death during the early stages of wheat grain development. Planta 213:352–360

Dominguez F, Moreno J, Cejudo FJ (2004) A gibberellin-induced nuclease is localized in the nucleus of wheat aleurone cells undergoing programmed cell death. J Biol Chem 279:11530–11536

Edgar BA, Orr-Weaver TL (2001) Endoreplication cell cycles: more for less. Cell 105:297–306

Ellis RE, Yuan J, Horvitz HR (1991) Mechanisms and functions of cell death. Annu Rev Cell Biol 7:663–698

Enari M, Sakahira H, Yokoyama H, Okawa K, Iwamatsu A, Nagata S (1998) A caspase-activated DNase that degrades DNA during apoptosis, and its inhibitor ICAD. Nature 391:43–50

Engelen-Eigles G, Jones RJ, Phillips RL (2000) DNA endoreduplication in maize endosperm cells: the effect of exposure to short-term high temperature. Plant Cell Environ 23:657–663

Engelen-Eigles G, Jones RJ, Phillips RL (2001) DNA endoreduplication in maize endosperm cells is reduced by high temperature during the mitotic phase. Crop Sci 41:1114–1121

Fath A, Bethke P, Lonsdale J, Meza-Romero R, Jones R (2000) Programmed cell death in cereal aleurone. Plant Mol Biol 44:255–266

Findeisen M, El-Denary M, Kapitza T, Graf R, Strausfeld U (1999) Cyclin A-dependent kinase activity affects chromatin binding of ORC, Cdc6, and MCM in egg extracts of *Xenopus laevis*. Eur J Biochem 264:415–426

Fung TK, Poon RY (2005) A roller-coaster ride with the mitotic cyclins. Semin Cell Dev Biol 16:335–342

Gavrieli Y, Sherman Y, Ben Sasson SA (1992) Identification of programmed cell death in situ via specific labeling of nuclear DNA fragmentation. J Cell Biol 119:493–501

Gendreau E, Traas J, Desnos T, Grandjean O, Caboche M, Höfte H (1997) Cellular basis of hypocotyl growth in *Arabidopsis thaliana*. Plant Physiol 114:295–305

Gerbi SA, Strezoska Z, Waggener JM (2002) Initiation of DNA replication in multicellular eukaryotes. J Struct Biol 140:17–30

Grafi G, Larkins BA (1995) Endoreduplication in maize endosperm: involvement of M phase-promoting factor inhibition and induction of S phase-related kinases. Science 269:1262–1264

Grafi G, Burnett RJ, Helentjaris T, Larkins BA, DeCaprio JA, Sellers WR, Kaelin WG Jr (1996) A maize cDNA encoding a member of the retinoblastoma protein family: involvement in endoreduplication. Proc Natl Acad Sci USA 93:8962–8967

Greenwood JS, Helm M, Gietl C (2005) Ricinosomes and endosperm transfer cell structure in programmed cell death of the nucellus during *Ricinus* seed development. Proc Natl Acad Sci USA 102:2238–2243

Guitton AE, Berger F (2005) Control of reproduction by *Polycomb* Group complexes in animals and plants. Int J Dev Biol 49:707–716

Halford NG, Shewry P (2007) The Structure and Expression of Cereal Storage Protein Genes (in this volume). Springer, Heidelberg

Hannah C (2007) Starch Formation in the Cereal Endosperm (in this volume). Springer, Heidelberg

Hatsugai N, Kuroyanagi M, Yamada K, Meshi T, Tsuda S, Kondo M, Nishimura M, Hara-Nishimura I (2004) A plant vacuolar protease, VPE, mediates virus-induced hypersensitive cell death. Science 305:855–858

Ingle J, Beitz D, Hageman RH (1965) Changes in composition during development and maturation of maize seeds. Plant Physiol 40:835–839

Joubes J, Chevalier C, Dudits D, Heberle-Bors E, Inzé D, Umeda M, Renaudin JP (2000) CDK-related protein kinases in plants. Plant Mol Biol 43:607–620

Keown AC, Taiz L, Jones RL (1977) The nuclear content of developing barley aleurone cells. Am J Bot 64:1248–1253

Kladnik A, Chamusco K, Dermastia M, Chourey P (2004) Evidence of programmed cell death in post-phloem transport cell of the maternal pedicel tissue in developing caryopsis of maize. Plant Physiol 136:3572–3581

Kladnik A, Chamusco K, Chourey PS, Dermastia M (2005) In situ detection of programmed cell death in the maize caryopsis. Period Biol 107:11–16

Kladnik A, Chourey PS, Pring DR, Dermastia M (2006) Development of the endosperm of *Sorghum bicolor* during the endoreduplication-associated growth phase. J Cereal Sci 43:209–215

Kodrzycki R, Boston RS, Larkins BA (1989) The *opaque-2* mutation of maize differentially reduces zein gene transcription. Plant Cell 1:105–114

Kowles RV, Phillips RL (1985) DNA amplification patterns in maize endosperm nuclei during kernel development. Proc Natl Acad Sci USA 82:7010–7014

Kowles RV, McMullen MD, Yerk G, Phillip RL, Kraemer S, Srienc F (1992) Endosperm mitotic activity and endoreduplication in maize affected by defective kernel mutations. Genome 35:68–77

Kowles RV, Yerk GL, Haas KM, Phillips RL (1997) Maternal effects influencing DNA endoreduplication in developing endosperm of *Zea mays*. Genome 40:798–805

Krishnamurthy KV, Krishnaraj R, Chozhavendan R, Christopher FS (2000) The programme of cell death in plants and animals: a comparison. Curr Sci 79:1169–1181

Kudo N, Kimura Y (2002) Nuclear DNA endoreduplication during petal development in cabbage: relationship between ploidy levels and cell size. J Exp Bot 53:1017–1023

Kumagai A, Dunphy WG (1991) The CDC25 protein controls tyrosine dephosphorylation of the CDC2 protein in a cell-free system. Cell 64:903–914

Lam E (2005) Vacuolar proteases livening up programmed cell death. Trends Cell Biol 15:124–127

Landrieu I, da Costa M, De Veylder L, Dewitte F, Vandepoele K, Hassan S, Wieruszeski J-M, Faure J-D, Van Montague M, Inze D, Lippens G (2004) A small CDC25 dual-specificity tyrosine-phosphatase isoform in *Arabidopsis thaliana*. Proc Natl Acad Sci USA 101:13380–13385

Larkins BA, Dilkes BP, Dante RA, Coelho CM, Woo Y-M, Liu Y (2001) Investigating the hows and whys of DNA endoreduplication. J Exp Bot 52:183–192

LeBlanc O, Pointe C, Hernandez M (2002) Cell cycle progression during endosperm development in *Zea mays* depends on parental dosage effects. Plant J 32:1057–1066

Leiva-Neto JT, Grafi G, Sabelli PA, Dante RA, Woo Y-M, Maddock S, Gordon-Kamm WJ, Larkins BA (2004) A dominant negative mutant of cyclin-dependent kinase A reduces endoreduplication but not cell size or gene expression in maize endosperm. Plant Cell 16:1854–1869

Lending CR, Larkins BA (1989) Changes in the zein composition of protein bodies during maize endosperm development. Plant Cell 1:1011–1023

Li C-J, Vassilev A, DePamphilis ML (2004) Role for Cdk1 (Cdc2)/cyclin A in preventing the mammalian origin recognition complex's largest subunit (Orc1) from binding to chromatin during mitosis. Mol Cell Biol 24:5875–5886

Lopes MA, Larkins BA (1993) Endosperm origin, development, and function. Plant Cell 5:1383–1399

Lur H-S, Setter TL (1993) Role of auxin in maize endosperm development: timing of nuclear DNA endoreduplication, zein expression, and cytokinin. Plant Physiol 103:273–280

MacAuley A, Cross JC, Werb Z (1998) Reprogramming the cell cycle for endoreduplication in rodent trophoblast cells. Mol Biol Cell 9:795–807

Mendenhall MD (1998) Cyclin-dependent kinase inhibitors of *Saccharomyces cerevisiae* and *Schizosaccharomyces pombe*. Curr Top Microbiol Immunol 227:1–24

Menges M, Murray JAH (2002) Synchronous *Arabidopsis* suspension cultures for analysis of cell cycle gene activity. Plant J 30:203–212

Morgan DO (1997) Cyclin-dependent kinases: engines, clocks, and microprocessors. Annu Rev Cell Dev Biol 13:261–291

Norstog K (1974) Nucellus during early embryogeny in barley: fine structure. Bot Gaz 135:97–103

Olsen O-A (2001) Endosperm development: cellularization and cell fate specification. Annu Rev Plant Physiol Plant Mol Biol 52:233–267

Olsen O-A (2004) Nuclear endosperm development in cereals and *Arabidopsis thaliana*. Plant Cell 16:S214–S227

Parker LL, Piwnica-Worms H (1992) Inactivation of the $p34^{cdc2}$-cyclin B complex by the human WEE1 tyrosine kinase. Science 257:1955–1957

Penterman J (2007) Genomic Imprinting in *Arabidopsis thaliana* and *Zeamays* (in this volume). Springer, Heidelberg

Porceddu A, Stals H, Reichheld JP, Segers G, De Veylder L, Barroco RP, Casteels P, Van Montagu M, Inze D, Mironov V (2001) A plant-specific cyclin-dependent kinase is involved in the control of G2/M progression in plants. J Biol Chem 276:36354–36360

Ramachandran C, Raghavan V (1989) Changes in nuclear DNA content of endosperm cells during grain development in rice (*Oryza sativa*). Ann Bot 64:459–468

Roberts IN, Passeron S, Barneix AJ (2006) The two main endoproteases present in dark-induced senescent wheat leaves are distinct subtilisin-like proteases. Planta 224:1437–1447

Royo J (2007) Transfer Cells (in this volume). Springer, Heidelberg

Sabelli PA, Larkins BA (2006) Grasses like mammals? Redundancy and compensatory regulation within the retinoblastoma protein family. Cell Cycle 5:352–355

Sabelli PA, Dante RA, Leiva-Neto JT, Jung R, Gordon-Kamm WJ, Larkins BA (2005) RBR3, a member of the retinoblastoma-related family from maize, is regulated by the RBR1/E2F pathway. Proc Natl Acad Sci USA 102:13005–13012

Sauer K, Knoblich JA, Richardson H, Lehner CF (1995) Distinct modes of cyclin E/cdc2c kinase regulation and S-phase control in mitotic and endoreduplication cycle of *Drosophila* embryogenesis. Genes Dev 9:1327–1339

Schnittger A, Schobinger U, Stierhof Y-D, Hulskamp M (2002) Ectopic B-type cyclin expression induces mitotic cycles in endoreduplicating *Arabidopsis* trichomes. Curr Biol 12:415–420

Setter TL, Flannigan BA (2001) Water deficit inhibits cell division and expression of transcripts involved in cell proliferation and endoreduplication in maize endosperm. J Exp Bot 52:1401–1408

Sherr CJ, Roberts JM (1995) Inhibitors of mammalian G1 cyclin-dependent kinases. Genes Dev 9:1149–1163

Shrewry PR, Tatham AS (1999) The characteristics, structures, and evolutionary relationships of prolamins. In: Shrewry PR, Casey R (eds) Seed proteins. Kluwer, Dordrecht, pp 11–33

Sørensen MB, Chaudhury AM, Robert H, Bancharel E, Berger F (2001) Polycomb group genes control pattern formation in plant seed. Curr Biol 11:277–281

Sorrell DA, Menges M, Healy JMS, Deveaux Y, Amano C, Su Y, Nakagami H, Shinmyo A, Doonan JH, Sekine M, Murray JAH (2001) Cell cycle regulation of cyclin-dependent kinases in tobacco cultivar bright yellow-2 cells. Plant Physiol 126:1214–1223

Sorrell DA, Marchbank A, McMahon K, Dickinson JR, Rogers HJ, Francis D (2002) A *WEE1* homologue from *Arabidopsis thaliana*. Planta 215:518–522

Sozzani R, Maggio C, Varotto S, Canova S, Bergounioux C, Albani D, Cella R (2006) Interplay between *Arabidopsis* activating factors E2FB and E2FA in cell cycle progression and development. Plant Physiol 140:1355–1366

Sreenivasulu N, Radchuk V, Strickert M, Miersch O, Weschke W, Wobus U (2006) Gene expression patterns reveal tissue-specific signaling networks controlling programmed cell death and ABA-regulated maturation in developing barley seeds. Plant J 47:310–327

Sugimoto-Shirasu K, Roberts K (2003) Big it up: endoreduplication and cell-size control in plants. Curr Opin Plant Biol 6:544–553

Sun Y, Dilkes BP, Zhang C, Dante RA, Carneiro NP, Lowe KS, Jung R, Gordon-Kamm WJ, Larkins BA (1999a) Characterization of maize (*Zea mays* L.) Wee1 and its activity in developing endosperm. Proc Natl Acad Sci USA 96:4180–4185

Sun Y, Flannigan BA, Setter TL (1999b) Regulation of endoreduplication in maize (*Zea mays* L.) endosperm. Isolation of a novel B1-type cyclin and its quantitative analysis. Plant Mol Biol 41:245–258

Verkest A, Weinl C, Inze D, De Veylder L, Schnittger A (2005) Switching the cell cycle. Kip-related proteins in plant cell cycle control. Plant Physiol 139:1099–1106

Vilhar B, Kladnik A, Blejec A, Chourey PS, Dermastia M (2002) Cytometrical evidence that the loss of seed weight in *miniature1* seed mutant of maize is associated with reduced mitotic activity in the developing endosperm. Plant Physiol 129:23–30

Walker JD, Oppenheimer DG, Concienne J and Larkin JC (2000) *SIAMESE*, a gene controlling the endoreplication cell cycle in *Arabidopsis*. Development 127:3931–3940

Weiss A, Herzig A, Jacobs H, Lehner CF (1998) Continuous cyclin E expression inhibits progression through endoreduplication cycles in *Drosophila*. Curr Biol 8:239–242

Woltering EJ (2004) Death proteases come alive. Trends Plant Sci 9:469–472

Woo Y-M, Hu DW-N, Larkins BA, Jung R (2001) Genomics analysis of genes expressed in maize endosperm identifies novel seed proteins and clarifies patterns of zein gene expression. Plant Cell 13:2297–2317

Yamagata T, Kato H, Kuroda S, Abe S, Davies E (2003) Uncleaved legumin in developing maize endosperm: identification, accumulation, and putative subcellular localization. J Exp Bot 54:913–922

Young TE, Gallie DR (1999) Analysis of programmed cell death in wheat endosperm reveals differences in endosperm development between cereals. Plant Mol Biol 39:915–926

Young TE, Gallie DR (2000a) Programmed cell death during endosperm development. Plant Mol Biol 44:283–301

Young TE, Gallie DR (2000b) Regulation of programmed cell death in maize endosperm by abscisic acid. Plant Mol Biol 42:397–414

Young TE, Gallie DR, DeMason DA (1997) Ethylene-mediated programmed cell death during maize endosperm development of wild-type and *shrunken2* genotypes. Plant Physiol 115:737–751

Yu Y, Steinmetz A, Meyer D, Brown S, Shen W-H (2003) The tobacco A-type cyclin, *Nicta;CYCA3;2*, at the nexus of cell division and differentiation. Plant Cell 15:2763–2777

Plant Cell Monogr (8)
O.-A. Olsen: Endosperm
DOI 10.1007/7089_2007_108/Published online: 31 March 2007

Aleurone Cell Development

Philip W. Becraft

Genetics, Development & Cell Biology/Agronomy, Iowa State University,
Ames, IA 50011, USA
becraft@iastate.edu

Abstract The periphery of the endosperm of many plant species forms an epidermis-like layer called the aleurone. During germination, the aleurone performs an important digestive function, secreting hydrolases to break down the starch and proteins stored in the starchy endosperm cells. Several features of cereal aleurone cells make them an attractive system for studying fundamental questions of cell fate and differentiation. The system is conceptually simple, with a single fate choice between starchy endosperm or aleurone cell types. The surface location makes the cells accessible for study, and they can be readily isolated by peeling from developing grains. Because of these experimental advantages and its importance to crop utilization, aleurone development has been most intensively studied in cereals. This chapter describes a picture of aleurone cell fate specification and development as a dynamic system displaying unique modifications to the cell cycle and cytoskeletal arrays, and surprising plasticity in cell fate decisions.

1 Introduction

In cereals, and some dicots with persistent endosperms, the aleurone layer has important functions in the accumulation of storage compounds during seed development, and in the mobilization of storage compounds during germination. In other seeds, such as *Arabidopsis*, that have a transitory endosperm, an aleurone layer persists in the mature seed (Brown et al. 1999) and functions in seed dormancy and germination (Bethke et al. 2007).

The storage function of aleurone cells in cereals involves the accumulation of high levels of phytic acid, which chelates several minerals. In barley grains, the aleurone is the major storage site for phosphate, magnesium, potassium, and calcium, accumulating over 70% (97% for magnesium) of the endosperm stores of these minerals (Stewart et al. 1988). During germination, the aleurone performs an important digestive function, secreting hydrolases to break down the starch and proteins stored in the starchy endosperm cells, which undergo programmed cell death during endosperm development (Young and Gallie 2000). The germination response is under hormonal control; gibberellic acid (GA) produced by the germinating embryo induces amylase gene expression and secretion, while abscisic acid (ABA) acts antagonistically to suppress these activities. Barley aleurone has served as an important model system for studying the mechanisms of GA and ABA action.

Until recently it was a matter of debate whether the peripheral cell layer of the *Arabidopsis* endosperm should actually be considered an aleurone, but a recent report shows clearly that the structure and function of *Arabidopsis* aleurone cells are very similar to cereals (Bethke et al. 2007). Secretory activity during germination apparently causes cell wall weakening to facilitate seedling emergence from the seed. This secretory activity was induced by GA and suppressed by ABA, just as in cereals. Furthermore, the aleurone layer was shown to be required for seed dormancy, a previously unknown function.

In addition to the importance of aleurone to the plant, aleurone impacts human well-being in several ways. Aleurone has reported dietary and health benefits; aleurone flour dramatically decreased the incidence of colon adeno-

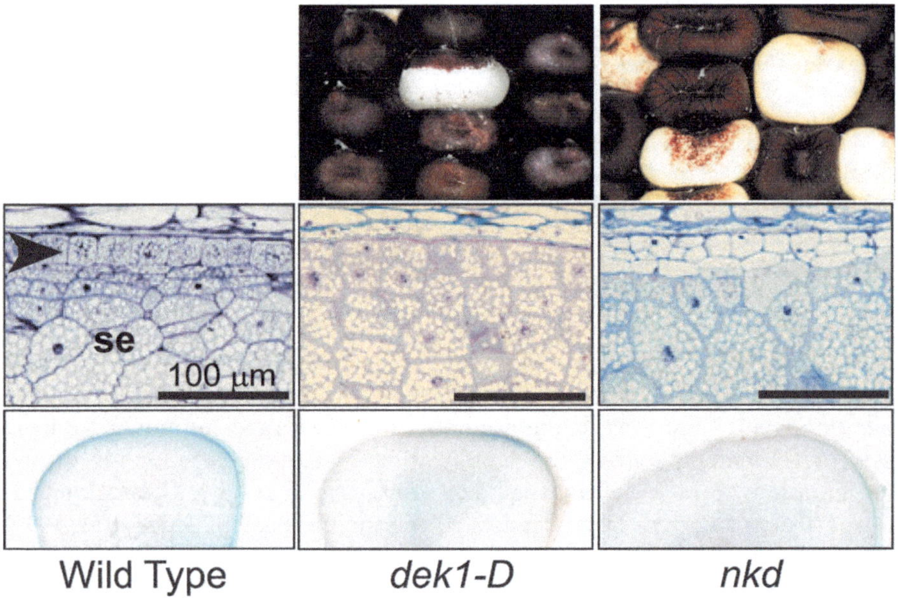

Fig. 1 Examples of mutations affecting maize aleurone development. Wild-type maize contains a single layer of aleurone cells (*arrow*), which accumulates anthocyanin in particular genotypes. Starchy endosperm cells (se) are recognizable by the large unstained starch grains. Aleurone markers, such as the *Vp1::GUS* reporter shown here (*blue*), express around the entire periphery of the endosperm (Cao et al. 2007). *dek1* is an example of a mutation affecting cell fate. Instead of aleurone, the peripheral layer of endosperm cells adapts a starchy endosperm cell fate. The *dek1-D* allele causes partial loss of aleurone, as seen by the partial loss of anthocyanin pigmentation and *Vp1::GUS* expression. In *nkd** mutants, the cell fate decisions appear normal because the peripheral cells are clearly distinct from starchy endosperm. However, they do not acquire the characteristics of normal aleurone cells, including the dense cytoplasm, anthocyanin accumulation, and marker gene expression, indicating that this mutant blocks the differentiation process

mas in rats, while raising red blood cell folate levels and decreasing plasma homocysteine levels in humans (McIntosh et al. 2001; Fenech et al. 2005). Aleurone is linked to environmental concerns; because phytate is not readily digested by livestock, most of the phosphate present in cereal grain-based animal feeds is excreted, posing a serious water pollution problem if runoff from manure is not controlled (Adeola 1999). Finally, aleurone has a major impact on leisure activities; amylases produced during germination convert starch to fermentable sugars, giving aleurone an all-important role in the malting process.

Several features of cereal aleurone cells make them an attractive system for studying fundamental questions of cell fate and differentiation. The system is conceptually simple, with a single fate choice between starchy endosperm or aleurone cell types. The surface location makes the cells accessible for study, and they can be readily isolated by peeling from developing grains. A number of aleurone markers are available, including several genes that have been used to regulate the aleurone-specific expression of GUS or GFP reporters (Gavazzi et al. 1997; Costa et al. 2003; Opsahl-Sorteberg et al. 2004; Wisniewski and Rogowsky 2004; Gruis et al. 2006; Cao et al. 2007). In maize, particular genotypes specifically accumulate anthocyanin pigment in the aleurone cells (see the chapter by Cone, in this volume), which has been used as a convenient marker for performing genetic screens for mutants that affect aleurone development (Gavazzi et al. 1997; Becraft and Asuncion-Crabb 2000). Figure 1 shows examples of anthocyanin pigmentation and expression of the *Vp1::GUS* reporter in wild-type and mutant maize endosperm. Because of these experimental advantages and its importance to crop utilization, aleurone development has been most intensively studied in cereals. The picture that emerges is of a dynamic system displaying unique modifications to the cell cycle and cytoskeletal arrays, and surprising plasticity in cell fate decisions.

2 Overview of Aleurone Ontogeny

Cereals show nuclear endosperm development, where the first several rounds of mitosis occur without cytokinesis to produce a coenocytic endosperm (see the chapter by Brown and Lemmon, in this volume). After a period of free-nuclear division, the nuclei come to occupy a thin layer of cortical cytoplasm surrounding a large central vacuole. Cellularization commences through the process of free cell wall formation whereby phragmoplasts form between both daughter and non-daughter nuclei. These cytoplasmic phragmoplasts direct cell wall deposition as walls grow inward to form a honeycomb-like assembly (see the chapter by Otegui, in this volume). Nuclei are located in separate alveoli, which are open on the inner face to the common syncytial cytoplasm.

Then a periclinal division occurs, accompanied by cytokinesis, to produce a cellular peripheral layer and an alveolar internal layer. This process reiterates in the alveolar layer until the volume of the endosperm is completely cellularized (Kiesselbach 1949; Brown et al. 1994, 1996; reviewed by Olsen 2001, 2004a).

After cellularization, cell division initially occurs throughout the endosperm but becomes localized to a cambium-like region at the periphery. Periclinal divisions produce distinct cell files that radiate from the center of the endosperm (Randolph 1936; Kiesselbach 1949; Morrison et al. 1975; McClintock 1978). As growth continues, cell divisions in the surface layer become strongly biased to the anticlinal plane to allow surface growth to keep pace with the expanding endosperm volume. At this point the aleurone cells are cuboidal in shape, thin-walled, and highly vacuolate. Most lines of maize and wheat contain a single layer of aleurone cells, while rice contains one to three and barley contains three layers.

As differentiation ensues, the cell walls become thick and highly autofluorescent, and the cytoplasm takes on a dense granular appearance (Morrison et al. 1975). Aleurone cells are rich in endoplasmic reticulum, mitochondria, and a variety of membrane bound vesicles (Jones 1969). Lytic vacuoles and protein storage vacuoles are both present (Swanson et al. 1998), and two types of inclusion bodies can be found within the same vacuole (Jakobsen et al. 1971). Globoid bodies, also known as aleurone grains, contain a crystalline matrix of phytin and protein surrounded by lipid droplets, while the second type of inclusion is a protein–carbohydrate body (Jones 1969; Jakobsen et al. 1971; Morrison et al. 1975; Kyle and Styles 1977; Swanson et al. 1998).

During seed maturation, the starchy endosperm cells undergo programmed cell death (Young and Gallie 2000), while the aleurone cells and embryo express a specific maturation program that allows them to survive seed desiccation (reviewed by Vicente-Carbajosa and Carbonero 2005). Maturation is positively regulated by ABA and involves the accumulation of late embryo abundant (LEA) proteins that function as dehydrins to protect cell membranes and proteins in desiccated seeds (Goyal et al. 2005). In particular genotypes of maize, anthocyanins specifically accumulate in the aleurone layer during the maturation phase (reviewed by Dooner et al. 1991).

The final stage in the life of an aleurone cell occurs after seed imbibition when the embryo sends a GA signal that induces the germination function of the aleurone. Hydrolase genes such as α-amylase are induced and their products secreted into the starchy endosperm for the remobilization of stored carbohydrates and amino acids to supply the growing seedling. This is a terminal process in that the aleurone cells expend all their resources and undergo non-apoptotic programmed cell death (Bethke et al. 1999; chapter by Nguyen et al., in this volume).

3
Positional Specification of Aleurone Cell Fate

Cells at the endosperm periphery must somehow interpret their position and assume aleurone cell fate. The first periclinal division is noteworthy because this establishes a peripheral cell layer, which constitutes the initial cells that will ultimately produce the aleurone layer. The propensity of anticlinal divisions led to early models which stated that the aleurone formed a separate lineage from the starchy endosperm (Randolph 1936; Kiesselbach 1949; Coe 1978; Levy and Walbot 1990; Walbot 1994). However, periclinal divisions can be observed in aleurone cells (Morrison et al. 1975) and genetically marked sectors showed that even late in development the aleurone contributed cells internally to the starchy endosperm (Becraft and Asuncion-Crabb 2000). Thus the peripheral cells serve as initials for both the aleurone and starchy endosperm.

Immediately following the first periclinal division the peripheral cells assume a distinct behavior from internal cells. Like most plant cells, the peripheral endosperm cells produce a pre-prophase band (PPB) of microtubules that forms prior to mitosis and predicts the plane of mitosis and cell plate formation (Brown et al. 1994, 1996). Internal cells have the unusual property of lacking a PPB. A phenomenon likely to be related to this is the difference in cell division behavior between peripheral and internal cells; the peripheral cells divide in anticlinal or periclinal planes, while the internal cells divide in random planes (Randolph 1936; Kiesselbach 1949). Thus it would appear that at the time of cellularization, peripheral determinants confer properties to these cells that distinguish them from internal cells.

Despite the early appearance of characters that distinguish aleurone cells from starchy endosperm cells, the identity of peripheral endosperm cells remains surprisingly plastic through late stages of seed development. As shown in Fig. 1, the aleurone is missing in *dek1* mutants and peripheral cells assume starchy endosperm identity (Sheridan 1982; Becraft et al. 2002; Lid et al. 2002). It follows that the *dek1* mutant cells are unable to perceive or respond to the positional cues that normally specify aleurone identity. Induction of *dek1* mutant sectors late in development caused aleurone cells to lose their identity and transdifferentiate into starchy endosperm (Becraft and Asuncion-Crabb 2000). Conversely, reversion of an unstable transposon-induced *dek1* allele allowed starchy endosperm cells in the peripheral layer to transdifferentiate to aleurone cells. In both cases, sectors as small as a single cell could be observed indicating that cell fate remained plastic up through the last cell division in endosperm development. Further evidence of plasticity is seen when maize kernels occasionally become conjoined. At the surfaces that form the junction, the aleurone layers transdifferentiate to starchy endosperm cells (Geisler-Lee and Gallie 2005). The upshot of these studies is that peripheral endosperm cells constantly monitor their position, and that positional cues are required to specify and maintain aleurone cell identity throughout endosperm development.

What is the source of the positional cues? Current evidence suggests that the key determinant for aleurone identity is surface position on the endosperm per se, rather than juxtaposition to nucellus or other maternal tissues. A maize mutant produces an endosperm comprised of spherical masses of cells and each of these spheroids forms its own aleurone layer, even though it is not in direct contact with maternal tissues (Olsen 2004b). Conversely, when two maize kernels become conjoined, cells that had occupied a surface position become internalized and consequently lose their aleurone identity (Geisler-Lee and Gallie 2005). Most compelling is that endosperms grown in culture can establish and maintain an aleurone layer despite the absence of an embryo or any maternal tissues (Gruis et al. 2006). Thus the isolated endosperm appears competent to confer aleurone identity to cells occupying the surface position. An exception is the disorganized endosperm of the *globby* mutant of maize, where isolated aleurone cells can be found internally (Costa et al. 2003).

4 Genes and Molecules

Our understanding of aleurone development at the molecular level is still in the early stages. Genetic studies suggest that a hierarchical system regulates the acquisition of aleurone cell fate and the subsequent differentiation of aleurone characteristics (see Fig. 2). Several maize mutants, including *cr4* and *dek1*, block the formation of aleurone; peripheral cells that would normally form aleurone develop instead as starchy endosperm (Becraft et al. 1996, 2002; Becraft and Asuncion-Crabb 2000; Lid et al. 2002; Wisniewski and Rogowsky 2004). Thus the normal gene products are required for aleurone cell fate specification. Figure 1 shows an example of a *dek1* mutant in maize.

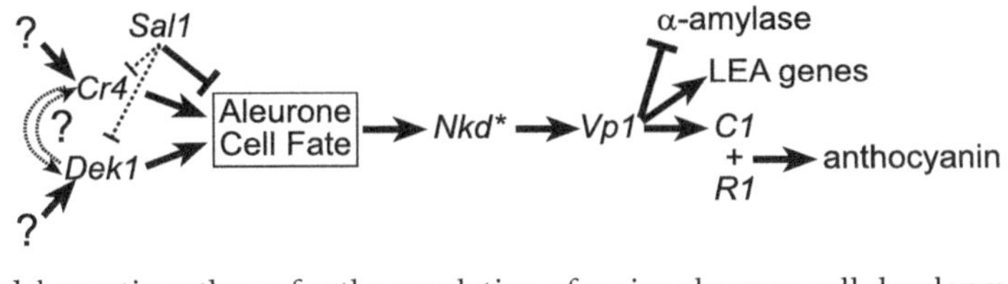

Fig. 2 Model genetic pathway for the regulation of maize aleurone cell development. Aleurone cell fate acquisition is positively regulated by CR4 and DEK1, and inhibited by SAL1. Hypothetical regulatory interactions among CR4, DEK1, and SAL1 are shown with *dotted lines*. As yet unknown factors function upstream of CR4 and/or DEK1. After aleurone cell fate specification, genes such as *Nkd** control cell differentiation, including the expression of *Vp1*. *Vp1*, in the presence of ABA, promotes aleurone maturation, including the accumulation of anthocyanin pigments in the appropriate genotypes

Mutation of *dek1* also blocks aleurone formation in *Arabidopsis*, indicating a conserved function (Lid et al. 2005). The *sal1* mutant has the opposite effect; the fate of peripheral starchy endosperm cells is switched to aleurone, producing a multilayered aleurone (Shen et al. 2003).

Other mutants establish a peripheral layer of cells that is distinct from the starchy endosperm but which lacks some of the characteristics of normal aleurone (Gavazzi et al. 1997; Becraft and Asuncion-Crabb 2000; Lid et al. 2004). These are interpreted as functioning in the aleurone differentiation process, downstream of the cell fate genes. Interestingly, many of the genes that affect the aleurone also affect epidermal development on the plant (Becraft et al. 1996, 1999, 2002; Jin et al. 2000; Kessler et al. 2002; Lid et al. 2004, 2005; Watanabe et al. 2004; Johnson et al. 2005), suggesting that a related mechanism controls the development of both tissues. Hence, it appears appropriate to consider the aleurone as the epidermis of the endosperm.

Several of the mutants that disrupt maize aleurone differentiation, such as *cr4* and the weak *dek1-D* allele, cause mosaic phenotypes (see Fig. 1). The mosaicism shows a predictable pattern with a strong tendency for disruption of aleurone development at the crown region and abgerminal side of the kernel, while the germinal face tends to develop an intact aleurone layer. This is hypothesized to reflect a patterning mechanism that organizes endosperm development (Becraft and Asuncion-Crabb 2000). Interestingly, ectopic cytokinin production under the regulation of a senescence inducible promoter phenocopies these mutants to produce kernels with mosaic aleurones (Geisler-Lee and Gallie 2005). This suggests that a hormone gradient could be involved in establishing endosperm pattern and aleurone fate, or at least that the normal establishment or interpretation of pattern can be disrupted by inappropriate hormone levels.

In addition to the specific genes mentioned, global genome level factors are also important for endosperm development in general, and aleurone development in particular. Aleurone cells in endosperms with a 2 : 2, instead of the normal 2 : 1, balance of maternal and paternal genome contributions lack the large number of spherosomes normally found in aleurone cells (Charlton et al. 1995). It is not known whether this results from imprinting at a specific locus or is a symptom of more general effects of genomic imbalance.

Several of the genes involved in aleurone cell fate specification have been isolated and appear to encode signal transduction components. The *cr4* gene encodes a receptor-like kinase with a tumor necrosis factor receptor (TNFR)-like motif in the extracellular domain (Becraft et al. 1996). This suggests that CR4 could function in the perception of the positional cues that specify aleurone identity. *Dek1* encodes a large protein with 21 predicted transmembrane regions, an extracellular loop, and a cytoplasmic domain containing a calpain protease (Lid et al. 2002; Wang et al. 2003). This configuration also allows the possibility that DEK1 could function as a cell surface receptor.

Genetic studies suggest that *Cr4* and *Dek1* participate in the same biological process, possibly as components of a signal transduction system (Becraft et al. 2002). At this point, the functional relationship between CR4 and DEK1 is unclear at the molecular level. In *Arabidopsis*, active forms of the CR4 ortholog, ACR4, are rapidly proteolyzed in endosomes (Gifford et al. 2005). Thus it is possible that CR4 is a substrate of the DEK1 protease. It is also possible that proteolysis of CR4 represents a processing step necessary for signal transduction. Alternatively, it is possible that CR4 regulates DEK1 activity by phosphorylation of either DEK1 itself or of a DEK1 substrate. The activity of animal calpains can be regulated by the phosphorylation of either the calpain protein itself or of its substrate (Nicolas et al. 2002; Shiraha et al. 2002). Yet another possibility is that both proteins regulate separate signaling events that converge further downstream.

The *Sal1* gene encodes a vacuolar sorting protein related to human CHMP1 (Shen et al. 2003). CHMP1 is involved in the endocytic recycling of receptors from the plasma membrane (Howard et al. 2001). This suggests the intriguing possibility that SAL1 could regulate CR4 and/or DEK1 by internalizing them, either for degradation or as part of the signaling mechanism. The opposite phenotypes of *sal1* versus *cr4* or *dek1* mutants suggests that SAL1 would probably function as a negative regulator of CR4 or DEK1 signaling, making the former possibility more likely. In addition, the aberrant vesicle traffic observed in *cr4* mutants (Jin et al. 2000) suggests the possibility that CR4 might also function to regulate SAL1 activity. This mutual regulation could establish a balance between CR4 and SAL1 activity, which would ultimately determine the number of aleurone cell layers formed.

The maturation phase of aleurone development is regulated in the aleurone in much the same way as it is regulated in the embryo (see Vicente-Carbajosa and Carbonero 2005 for review). ABA is the key hormone that promotes maturation, and maize mutants in ABA biosynthetic genes cause a viviparous phenotype involving precocious germination of the embryo and a concomitant activation of digestive functions in the aleurone. ABA induces the expression of LEA genes and other maturation associated genes in the aleurone (Miyoshi et al. 2002; Furtado and Henry 2005; Bethke et al. 2006). *Viviparous1* encodes a B3 domain transcription factor that is a central regulator of the maturation process. In maize, VP1 also activates anthocyanin production by promoting expression of the *C1* gene (Hattori et al. 1992; chapter by Cone, in this volume). As such, VP1 is the most upstream known factor in the transcriptional regulation of anthocyanin synthesis in the aleurone. *vp1* mutants lack anthocyanin in the aleurone, and express *α-amylase* genes associated with germination (Hoecker et al. 1995). The *Vp1* gene is expressed in the embryo and aleurone (Cao et al. 2007), but how *Vp1* expression is spatially regulated in the endosperm is not known. In addition, the bZIP transcription factor HvABI5 is required for ABA induction of maturation associated genes in barley aleurone (Casaretto and Ho 2003).

5
Conclusions

The aleurone layer is critical to plant survival and several different functions for the aleurone have evolved in different taxa. The most familiar is in cereal grains where aleurone has both storage and digestive functions. The aleurone layer provides a powerful system for studying basic questions of cell fate specification and differentiation. Additionally, it might be possible to manipulate the aleurone layer to engineer altered seed properties, such as increased malting rate or enhanced mineral or lipid content. Several key regulators of aleurone development have been identified. Further work to understand how these factors interact, how they are regulated, and the signal transduction systems they control, will provide insights into the biology of this fascinating system and may provide the knowledge needed to realize the potential for improved seed quality.

References

Adeola O (1999) Nutrient management procedures to enhance environmental conditions: an introduction. J Anim Sci 77:427–429

Becraft PW, Asuncion-Crabb YT (2000) Positional cues specify and maintain aleurone cell fate during maize endosperm development. Development 127:4039–4048

Becraft PW, Stinard PS, McCarty DR (1996) CRINKLY4: a TNFR-like receptor kinase involved in maize epidermal differentiation. Science 273:1406–1409

Becraft PW, Chandok MR, Jin P, Guo T, Asuncion-Crabb Y, Zhang Y (1999) The function of the maize CRINKLY4 receptor-like kinase in a growth factor like signaling system. Symposium of Plant Signal Transduction, New Delhi, India

Becraft PW, Li K, Dey N, Asuncion-Crabb YT (2002) The maize *dek1* gene functions in embryonic pattern formation and in cell fate specification. Development 129:5217–5225

Bethke PC, Lonsdale JE, Fath A, Jones RL (1999) Hormonally regulated programmed cell death in barley aleurone cells. Plant Cell 11:1033–1046

Bethke PC, Hwang Y-S, Zhu T, Jones RL (2006) Global patterns of gene expression in the aleurone of wild-type and dwarf1 mutant rice. Plant Physiol 140:484–498

Bethke PC, Libourel IGL, Aoyama N, Chung Y-Y, Still DW, Jones RL (2007) The *Arabidopsis thaliana* aleurone layer responds to nitric oxide, gibberellin, and abscisic acid and is sufficient and necessary for seed dormancy. Plant Physiol 143:1173-1188

Brown RC, Lemmon BE (2007) The Developmental Biology of Cereal Endosperm (in this volume). Springer, Heidelberg

Brown RC, Lemmon BE, Olsen O-A (1994) Endosperm development in barley: microtubule involvement in the morphogenetic pathway. Plant Cell 6:1241–1252

Brown RC, Lemmon BE, Olsen O-A (1996) Development of the endosperm in rice (*Oryza sativa* L.): cellularization. J Plant Res 109:301–313

Brown RC, Lemmon BE, Nguyen H, Olsen O-A (1999) Development of endosperm in *Arabidopsis thaliana*. Sex Plant Reprod 12:32–42

Cao X, Costa LM, Biderre-Petit C, Kbhaya B, Dey N, Perez P, McCarty DR, Gutierrez-Marcos JF, Becraft PW (2007) Abscisic acid and stress signals induce Viviparous1 (Vp1) expression in seed and vegetative tissues of maize. Plant Physiol p 106.091454

Casaretto J, Ho TD (2003) The transcription factors HvABI5 and HvVP1 are required for the abscisic acid induction of gene expression in barley aleurone cells. Plant Cell 15:271–284

Charlton WL, Keen CL, Merriman C, Lynch P, Greenland AJ, Dickinson HJ (1995) Endosperm development in Zea mays; implication of gametic imprinting and paternal excess in regulation of transfer layer development. Development 121:3089–3097

Coe EH Jr (1978) The aleurone tissue of maize as a genetic tool. In: Walden DB (ed) Maize breeding and genetics. Wiley, New York, pp 447–459

Cone KC (2007) Anthocyanin biosynthesis (in this volume). Springer, Heidelberg

Costa LM, Gutierrez-Marcos JF, Brutnell TP, Greenland AJ, Dickinson HG (2003) The globby1-1 (glo1-1) mutation disrupts nuclear and cell division in the developing maize seed causing alterations in endosperm cell fate and tissue differentiation. Development 130:5009–5017

Dooner HK, Robbins TP, Jorgensen RA (1991) Genetic and developmental control of anthocyanin biosynthesis. Annu Rev Genet 25:173–199

Fenech M, Noakes M, Clifton P, Topping D (2005) Aleurone flour increases red-cell folate and lowers plasma homocyst(e)ine substantially in man. Br J Nutr 93:353–360

Furtado A, Henry RJ (2005) The wheat Em promoter drives reporter gene expression in embryo and aleurone tissue of transgenic barley and rice. Plant Biotechnol J 3:421–434

Gavazzi G, Dolfini S, Allegra D, Castiglioni P, Todesco G, Hoxha M (1997) Dap (defective aleurone pigmentation) mutations affect maize aleurone development. Mol Gen Genet 256:223–230

Geisler-Lee J, Gallie DR (2005) Aleurone cell identity is suppressed following connation in maize kernels. Plant Physiol p 105.064295

Gifford ML, Robertson FC, Soares DC, Ingram GC (2005) ARABIDOPSIS CRINKLY4 function, internalization, and turnover are dependent on the extracellular crinkly repeat domain. Plant Cell 17:1154–1166

Goyal K, Walton LJ, Tunnacliffe A (2005) LEA proteins prevent protein aggregation due to water stress. Biochem J 388:151–157

Gruis D, Guo H, Selinger D, Tian Q, Olsen O-A (2006) Surface position, not signaling from surrounding maternal tissues, specifies aleurone epidermal cell fate in maize. Plant Physiol 141:898–909

Hattori T, Vasil V, Rosenkrans L, Hannah LC, McCarty DR, Vasil IK (1992) The Viviparous-1 gene and abscisic acid activate the C1 regulatory gene for anthocyanin biosynthesis during seed maturation in maize. Genes Dev 6:609–618

Hoecker U, Vasil IK, McCarty DR (1995) Integrated control of seed maturation and germination programs by activator and repressor functions of Viviparous-1 of maize. Genes Dev 9:2459–2469

Howard TL, Stauffer DR, Degnin CR, Hollenberg SM (2001) CHMP1 functions as a member of a newly defined family of vesicle trafficking proteins. J Cell Sci 114:2395–2404

Jakobsen JV, Knox RB, Pyliotis NA (1971) The structure and composition of aleurone grains in the barley aleurone layer. Planta 101:189–209

Jin P, Guo T, Becraft PW (2000) The maize CR4 receptor-like kinase mediates a growth factor-like differentiation response. Genesis 27:104–116

Johnson KL, Degnan KA, Ross Walker J, Ingram GC (2005) AtDEK1 is essential for specification of embryonic epidermal cell fate. Plant J 44:114–127

Jones RL (1969) The fine structure of barley aleurone cells. Planta 85:359–375

Kessler S, Seiki S, Sinha N (2002) Xcl1 causes delayed oblique periclinal cell divisions in developing maize leaves, leading to cellular differentiation by lineage instead of position. Development 129:1859–1869

Kiesselbach TA (1949) The structure and reproduction of corn. Nebraska Agric Exp Stn Res Bull 161:1–96

Kyle DJ, Styles ED (1977) Development of aleurone and sub-aleurone layers in maize. Planta 137:185–193

Levy AA, Walbot V (1990) Regulation of the timing of transposable element excision during maize development. Science 248:1534–1537

Lid SE, Gruis D, Jung R, Lorentzen JA, Ananiev E, Chamberlin M, Niu X, Meeley R, Nichols S, Olsen O-A (2002) The *defective kernel 1* (*dek1*) gene required for aleurone cell development in the endosperm of maize grains encodes a membrane protein of the calpain gene superfamily. Proc Natl Acad Sci USA 99:5460–5465

Lid SE, Al RH, Krekling T, Meeley RB, Ranch J, Opsahl-Ferstad HG, Olsen O-A (2004) The maize disorganized aleurone layer 1 and 2 (dil1, dil2) mutants lack control of the mitotic division plane in the aleurone layer of developing endosperm. Planta 218:370–378

Lid SE, Olsen L, Nestestog R, Aukerman M, Brown RC, Lemmon B, Mucha M, Opsahl-Sorteberg H-G, Olsen O-A (2005) Mutation in the *Arabidopsis thaliana DEK1* calpain gene perturbs endosperm and embryo development while over-expression affects organ development globally. Planta 221:339

McClintock B (1978) Development of the maize endosperm as revealed by clones. In: Subtelny S, Sussex IM (eds) The clonal basis of development. Academic, New York, pp 217–237

McIntosh GH, Royle PJ, Pointing G (2001) Wheat aleurone flour increases cecal beta-glucuronidase activity and butyrate concentration and reduces colon adenoma burden in azoxymethane-treated rats. J Nutr 131:127–131

Miyoshi K, Kagaya Y, Ogawa Y, Nagato Y, Hattori T (2002) Temporal and spatial expression pattern of the OSVP1 and OSEM genes during seed development in rice. Plant Cell Physiol 43:307–313

Morrison IN, Kuo J, O'Brien TP (1975) Histochemistry and fine structure of developing wheat aleurone cells. Planta 123:105–116

Nguyen HN, Sabelli PA, Larkins BA (2007) Endoreduplication and programmed cell death in the cereal endosperm (in this volume). Springer, Heidelberg

Nicolas G, Fournier CM, Galand C, Malbert-Colas L, Bournier O, Kroviarski Y, Bourgeois M, Camonis JH, Dhermy D, Grandchamp B, Lecomte MC (2002) Tyrosine phosphorylation regulates alpha II spectrin cleavage by calpain. Mol Cell Biol 22:3527–3536

Olsen O-A (2001) Endosperm development: cellularization and cell fate specification. Annu Rev Plant Physiol Plant Mol Biol 52:233–267

Olsen O-A (2004a) Nuclear endosperm development in cereals and *Arabidopsis thaliana*. Plant Cell 16:S214–227

Olsen O-A (2004b) Dynamics of maize aleurone cell formation: the "surface" rule. Maydica 49:37–40

Opsahl-Sorteberg HG, Divon HH, Nielsen PS, Kalla R, Hammond-Kosack M, Shimamoto K, Kohli A (2004) Identification of a 49-bp fragment of the HvLTP2 promoter directing aleurone cell specific expression. Gene 341:49–58

Otegui M (2007) Endosperm cell walls: formation, composition, and functions (in this volume). Springer, Heidelberg

Randolph LF (1936) Developmental morphology of the caryopsis in maize. J Agric Res 53:881–916

Shen B, Li C, Min Z, Meeley RB, Tarczynski MC, Olsen O-A (2003) sal1 determines the number of aleurone cell layers in maize endosperm and encodes a class E vacuolar sorting protein. Proc Natl Acad Sci USA 100:6552–6557

Sheridan WF, Neuffer MG (1982) Maize developmental mutant embryos unable to form leaf primordia. J Heredity 73:319–329

Shiraha H, Glading A, Chou J, Jia Z, Wells A (2002) Activation of m-calpain (calpain II) by epidermal growth factor is limited by protein kinase A phosphorylation of m-calpain. Mol Cell Biol 22:2716–2727

Stewart A, Nield H, Lott JNA (1988) An investigation of the mineral content of barley grains and seedlings. Plant Physiol 86:93–97

Swanson SJ, Bethke PC, Jones RL (1998) Barley aleurone cells contain two types of vacuoles. Characterization of lytic organelles by use of fluorescent probes. Plant Cell 10:685–698

Vicente-Carbajosa J, Carbonero P (2005) Seed maturation: developing an intrusive phase to accomplish a quiescent state. Int J Dev Biol 49:645–651

Walbot V (1994) Overview of key steps in aleurone development. In: Freeling M, Walbot V (eds) The maize handbook. Springer, New York, pp 78–80

Wang C, Barry JK, Min Z, Tordsen G, Rao AG, Olsen O-A (2003) The calpain domain of the maize DEK1 protein contains the conserved catalytic triad and functions as a cysteine proteinase. J Biol Chem 278:34467–34474

Watanabe M, Tanaka H, Watanabe D, Machida C, Machida Y (2004) The ACR4 receptor-like kinase is required for surface formation of epidermis-related tissues in *Arabidopsis thaliana*. Plant J 39:298–308

Wisniewski JP, Rogowsky PM (2004) Vacuolar H^+-translocating inorganic pyrophosphatase (Vpp1) marks partial aleurone cell fate in cereal endosperm development. Plant Mol Biol 56:325–337

Young TE, Gallie DR (2000) Programmed cell death during endosperm development. Plant Mol Biol 44:283–301

Plant Cell Monogr (8)
O.-A. Olsen: Endosperm
DOI 10.1007/7089_2007_109/Published online: 24 March 2007

The Embryo Surrounding Region

M. Cosségal · V. Vernoud · N. Depège · P.M. Rogowsky (✉)

UMR 5667 CNRS-INRA-ENSL-UCBL, IFR128 BioSciences Lyon-Gerland, ENS-Lyon, 46 Allée d'Italie, 69364 Lyon Cedex 07, France
Peter.Rogowsky@ens-lyo.fr

Abstract There is converging evidence in maize, wheat, barley, *Arabidopsis* and other species that the endosperm in proximity of the embryo is cytologically different from the remaining endosperm. Gene expression restricted to this embryo surrounding region (ESR) reinforces the notion of a specialized endosperm domain at least in maize and *Arabidopsis*. The ESR is a dynamic structure that is set apart prior to cellularisation and starts to disappear with the onset of reserve accumulation in the developing seed. During later developmental stages it is frequently succeeded by a liquid filled space around the embryo. While the cytological characteristics of the regions surrounding the embryo are quite similar between the species analyzed, their functional equivalence has not yet been established. Possible functions of the ESR include nutrition or defense of the embryo as well as signaling between the embryo and the endosperm.

1
Introduction

The term "embryo surrounding region" (ESR) was first used by Opsahl-Ferstad et al. (1997) to describe a small region of the maize endosperm adjacent to the embryo. The ESR was defined as a group of cells that all share expression of the *Esr* gene family, a criterion clearly distinguishing these cells from those of the embryo or other domains of the endosperm, namely the starchy endosperm, the basal endosperm transfer layer (BETL) or the aleurone layer (Fig. 1; Royo et al., in this volume; Brown and Lemmon, in this volume; Becraft, in this volume). As pointed out by Opsahl-Ferstad et al., the same group of cells had previously been described as cytologically different from other endosperm cells but had not been given a particular name. ESR cells are smaller than neighbouring cells, have a very dense cytoplasm and contain a mass of highly ordered rough endoplasmic reticulum (Schel et al. 1984). The ESR is a dynamic structure that surrounds the entire embryo at very early stages, i.e. at 4 days after pollination (DAP). As the embryo elongates, the ESR forms a cup around the suspensor but no longer surrounds the embryo proper (7 DAP). Upon further development of the embryo the ESR is restricted to the lower part of the suspensor between 9 and 12 DAP. Concomitant with the programmed cell death of the suspensor the ESR starts to

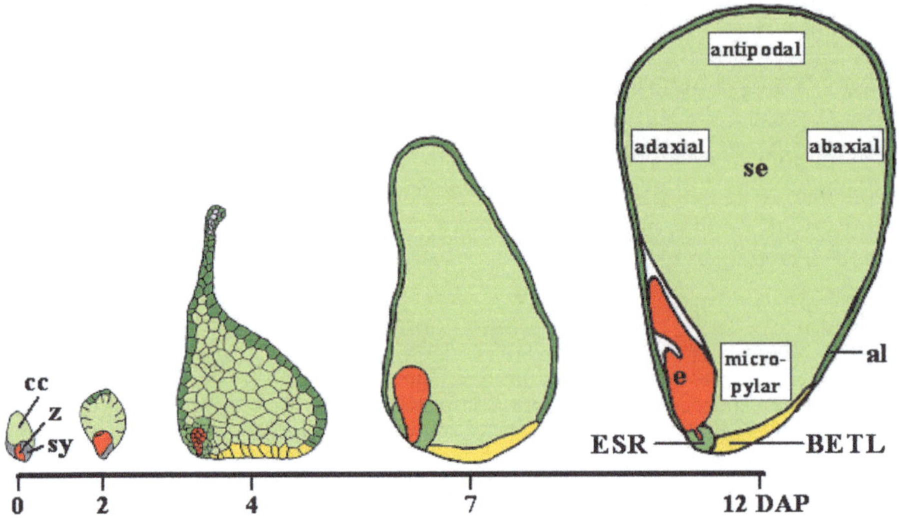

Fig. 1 The ESR during maize endosperm development. The ESR is coloured in *green* in schematic drawings of the maize endosperm from the time of fertilization (0 DAP) to stage 1 (12 DAP). al, aleurone layer (*dark green*); BETL, basal endosperm transfer layer (*yellowish green*); e, embryo (*red*); ESR, embryo surrounding region (*green*); cc, central cell; se, starchy endosperm (*light green*); sy, synergid; z, zygote

disappear from 12 DAP onwards at the profit of the neighbouring embryo and endosperm tissues (Fig. 1).

In other species several reports confirm differences between the bulk of the endosperm and the part next to the embryo both before and after cellularisation. In this review we will first introduce the present knowledge of the ESR in maize and then discuss the specialisations in other species in the light of the maize data.

2
ESR in Corn

At present no clear-cut role can be attributed to the ESR in the absence of functional studies involving for example an ablation of the ESR or alterations of its size or structure in mutants or transgenic plants. Nevertheless, there are strong leads from the interpretation of cytological data and the inferred or demonstrated function of genes specifically expressed in the ESR. They imply the ESR either in metabolism, defence or signalling. We will discuss the available data for each of these potential functions after a brief description of the cytological characteristics of ESR cells.

2.1 Cytological Description

The precise onset of ESR formation during maize endosperm development remains unknown. The fact that the cellularisation of the endosperm coenocyte is consistently initiated next to the embryo (Randolph 1936) may be an indication for a regionalisation prior to cellularisation. Unpublished observations of cell walls that appear to be forming around the embryo in the coenocytic endosperm led to the hypothesis that the ESR may form through a mechanism that permits a functional phragmoplast to form near the embryo (Olsen 2001). Immediately after cellularisation (4 DAP) the ESR can be clearly distinguished from other endosperm cells by a haematoxylin stain as nicely documented for the first time in 1949 in photomicrographs (Fig. 2A,B) that did not solicit any specific comment by its author (Kiesselbach 1949). The cells of the ESR remain very prominent up to 12 DAP. Soon after they are crushed and absorbed and have disappeared together with the suspensor by 15 DAP (Kiesselbach and Walker 1952).

The cytological characteristics of ESR cells were first described in the later works of Kiesselbach who mentions that these cells "are small, densely filled with cytoplasm and have large nuclei" and that they form "a mass of dense nourishing cells" (Kiesselbach and Walker 1952). Thirty years later a much more detailed study by Schel et al. (1984) based on both light and electron microscopy confirmed and extended these observations. ESR cells are thin-walled and contain numerous small vacuoles. They are interconnected by plasmodesmata and the outer nuclear membrane frequently shows sphere-like outgrowths. The endoplasmic reticulum is very abundant, is mainly of the rough type, frequently shows large intracisternal spaces and is organised in parallel arrays, in particular close to the embryo. The dictyosomes are characterised by the presence of numerous vesicles on their distal side. The plasma membrane shows infolds that are clearly different from the invaginations of BETL cells (see Royo in this volume); they concern only the membrane and not the cell wall, are spherical rather than lamellate in appearance, and contain osmiophilic material. Finally, a highly ordered arrangement of mitochondria close to the embryo has been noted. Taken together these characteristics suggest that ESR cells are metabolically very active, that they synthesise a high amount of membranes and proteins and that they provide nutrients or signals to the embryo via exocytosis (Schel et al. 1984).

Contrary to the numerous symplastic connections between ESR cells the connection between the ESR and the embryo is mainly apoplastic. While plasmodesmata exist at very early developmental stages between the syncytial endosperm and cells of the suspensor (but not of the embryo proper), the symplastic boundary rapidly extends from the embryo proper to the suspensor at the time of endosperm cellularisation. At 4 DAP there is no longer

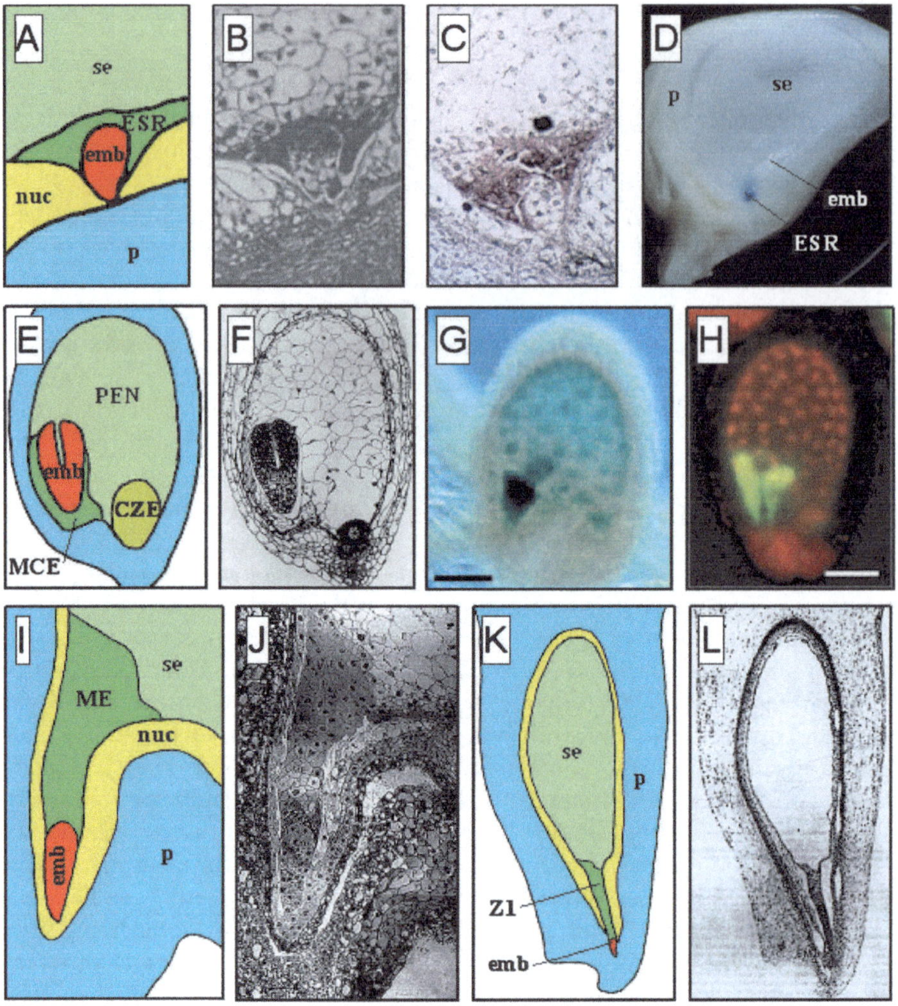

symplastic contact between the endosperm and the embryo and all exchanges have to take place via the apoplast (van Lammeren 1987).

2.2 ESR and Metabolism

Two lines of evidence support a role of the ESR in the nutrition of the embryo: the cytological data described above (Schel et al. 1984) and the specific expression of an invertase inhibitor (Bate et al. 2004). The ESR may be involved in transport of nutrients from the BETL to the embryo as indicated by an asymmetric organelle distribution. Regularly aligned mitochondria,

◄ **Fig. 2** Distinct endosperm regions in proximity of the embryo. **A** Schematic drawing of (**B**). **B** 5 DAP maize caryopsis stained with haematoxylin. The endosperm cells surrounding the embryo are smaller than in the remaining endosperm and absorb much stronger the haematoxylin dye (from Kiesselbach 1949). **C** 5 DAP maize caryopsis hybridised with *ZmAE3* probe. The purple signal is limited to the ESR (from Magnard et al. 2000). **D** $Prom_{ZmEsr2}$::Gus activity in a 12 DAP maize caryopsis. GUS activity is limited to the region around the basis of the embryo (from Bonello et al. 2000). **E** Schematic drawing of (**F**). **F** Longitudinal section of a torpedo stage *Arabidopsis* seed showing lysed cells or small cells close to the embryo, large cells in the central endosperm and absence of cellularisation in the chalazal endosperm (from Brown et al. 1999). **G** $Prom_{MINI3}$::Gus activity in 3 DAP *Arabidopsis* seed. GUS activity is only detected in the region surrounding the basis of the globular embryo (from Luo et al. 2005). **H** $Prom_{AtSuc5}$::mGFP5-ER activity observed by standard fluorescence microscopy in 5 DAF *Arabidopsis* seed. The green fluorescence is restricted to a region surrounding the basal part of the globular embryo (from Baud et al. 2005). **I** Schematic drawing of (**J**). **J** Embryo and ME in toluidine blue stained epoxy section of wheat caryopsis 15 days after anthesis (from Smart and O'Brien 1983). Cells in the ME are smaller and more stained than in the adjacent part of the endosperm. **K** Schematic drawing of (**L**). **L** Global view of barley caryopsis at 3 DAP stained with periodic acid-Schiff and aniline blue black (from Engell 1989). Only zone 1 of the endosperm adjacent to the embryo is cellularised. CZE, chalazal endosperm; emb, embryo; ESR, embryo surrounding region; MCE, micropylar endosperm; ME, modified endosperm; nuc, nucellus; se, p, pericarp; PEN, peripheral endosperm; starchy endosperm; Z1, zone 1

a high density of vesicles and membrane infolds are found in ESR cells mainly on the side next to the embryo. ESR cells may also actively synthesise nutrients as indicated by abundant rough endoplasmic reticulum (Schel et al. 1984).

A hint concerning the nature of the transported or newly synthesised substances comes from the study of genes expressed in the ESR. The gene *ZmINVINH1* coding for an invertase inhibitor is the only one of three *ZmINVINH* genes identified in the maize genome that is expressed specifically in the ESR (Bate et al. 2004). The function of ZmINVINH1 has been demonstrated in vitro; recombinant protein reduces invertase activity present in crude extracts of maize kernels and this inhibition is attenuated by pre-incubation with sucrose. Biochemical fractionation of maize kernels followed by Western blot as well as subcellular localisation of ZmINVINH1::GFP fusion proteins in onion cells indicate that ZmINVINH1 is exported to the apoplast. While ZmINVINH1 has no glycosylation signals, it is retained on concanavalin A columns, probably by binding to invertases that are known to be glycosylated proteins. The apoplastic localisation suggests that ZmINVINH1 regulates the activity of a cell wall rather than a vacuolar invertase. However, none of the four cell wall invertases INCW1 to INCW4 is known to be expressed in the ESR (Chourey et al. 2005; Kim et al. 2000), contrary to IVR2, one of the two vacuolar invertases, that is strongly expressed in different parts of the young caryopsis including the ESR (Bate et al. 2004). Since ZmINVINH1 inhibits the activity of both soluble (vacuolar) and insoluble (cell wall) invertases, it may not be directed against any particular enzyme but may represent an insurance against any invertase activity in the apoplast of the ESR, resulting in a protection of sucrose from cleavage (Bate et al. 2004). Sucrose is known to be both the major carbon source of the developing maize kernel and an important signal molecule in plant seed development (Wobus and Weber 1999). Additional experiments are needed to clarify which one if any of these roles sucrose is playing in the ESR and whether sucrose is one of the molecules transported by or produced in the ESR.

2.3 ESR and Defense

The protection of the developing seed is of strategic interest for the propagation of all plant species. In the maize kernel, the symplastic isolation of the endosperm and the embryo from each other and from the mother plant provides a first barrier of protection, especially against viruses. The most obvious strategy for pathogens to penetrate into the seed is to follow the nutrient flow, i.e. to pass from the maternal tissue into the endosperm via the interface pedicel/BETL and from the endosperm into the embryo via the interface ESR/suspensor. Consequently a role of the ESR in defence has been proposed at its re-discovery (Opsahl-Ferstad et al. 1997) and substantiated in a first instance by the ESR-specific expression of the gene *ZmAE3* (Fig. 2C) encoding a small, hydrophilic protein with similarity to basal layer antifungal proteins (Magnard et al. 2000; Sevilla-Lecoq et al. 2003). A broad-range activity against filamentous fungi including several plant pathogens has been demonstrated in vitro for BAP2, the founding member of the family (Serna et al. 2001).

More direct evidence comes from the functional characterisation of *ZmEsr6*, a gene strongly expressed in the ESR between 5 and 16 DAP and coding for a defensin-like protein (Balandin et al. 2005). The mature protein of 52 amino acids shares eight conserved Cys residues and most of its tertiary structure with canonical defensins, even though it has some atypical insertions and deletions of amino acids at its N-terminus. Its antimicrobial activity was demonstrated in vitro; the growth of three bacterial and three fungal pathogens was inhibited by recombinant ZmEsr6 protein produced in *E. coli*. *ZmEsr6* transcript was only detected in the kernel, where it was found mainly in the ESR but, towards the end of the temporal expression window, also in a small region at the abgerminal end of the BETL. Interestingly, the ZmEsr6 protein was immuno-localised not only in the regions of its synthesis but also in the pedicel region adjacent to these regions. A gradient of signal intensity indicates that the protein is produced in the endosperm at both extremities of the BETL and then diffuses into the apoplast of the maternal placentochalaza (Balandin et al. 2005). Therefore, ZmEsr6 may provide active protection not only of the embryo but also of the endosperm.

2.4 ESR and Signaling

Very young maize embryos require either the presence of surrounding tissues or the presence of nurse cells for their correct differentiation and morphogenesis in vitro (Leduc et al. 1996; Mol et al. 1993). Theoretically, the signal molecules responsible for proper embryo development have to be synthesised in or transit via the ESR that completely envelops the embryo at early stages. Interestingly, the proteins encoded by the highly homologous genes *ZmEsr1*, *ZmEsr2* and *ZmEsr3* show sequence similarity with CLV3 (Bonello et al. 2002), the ligand of the disulfide-linked CLV1/CLV2 receptor complex that restricts the stem cell population in the shoot apical meristem of *Arabidopsis* (Clark 2001). Although the sequence similarity is limited to a stretch of 14 amino acids, it is likely significant for three reasons: Firstly, the conserved region is probably a functional domain of CLV3 because the only two existing point mutations are due to substitutions in this region (Fletcher et al. 1999). Secondly, in the *Arabidopsis* genome this region is shared by the family of 25 CLE (CLV3/ESR-related) proteins that are all small hydrophilic proteins with a signal peptide (Cock and McCormick, 2001). Thirdly, the phenotype of plants ectopically expressing certain *Cle* genes can be mimicked by the in vitro application to wildtype plants of synthetic peptides of 14 amino acids corresponding to the CLE motif (Fiers et al. 2005). Consequently, the proteins Esr1 to Esr3 could be endosperm derived ligands that trigger a protein kinase-mediated signal transduction pathway in the embryo.

2.5 ESR and the Embryonic Cavern

We would like to contribute to the ongoing discussion on ESR function by proposing that the ESR might be involved in the formation of the embryonic cavern, a liquid filled space between the embryo and the endosperm. In fact, the embryonic cavern progressively forms around the parts of the embryo that emerge from the ESR to fully surround the embryo as soon as the ESR has disappeared. This progression is reflected during the micro-dissection of embryos. While proembryos (4 DAP) are extremely difficult to separate from the surrounding endosperm tissue, stage 1 embryos (12 DAP) can be isolated easily without a scalpel by simply pushing them out of the endosperm. Transition stage embryos (7 DAP) show an intermediate behaviour; the suspensor is deeply anchored in the endosperm and needs to be separated with a scalpel, while the embryo proper can easily be recovered. The function of the embryonic cavern is not really known, but it forms even in the absence of normal embryos and remains largely empty in maize *embryo specific* mutants (Clark and Sheridan 1991), while in rice *reduced embryo* mutants the available space is taken up by the endosperm (Hong et al. 1996). Interestingly, several of the genes specifically expressed in the ESR have no close counterparts in rice.

2.6 ESR-Specific Promoters

For several of the genes showing expression in the ESR, the promoters have been examined in the search for an ESR box responsible for ESR-specific transcription. Active promoters confirmed by the expression pattern of promoter-*Gus* fusions have been isolated for *ZmEsr1* to 3 (Fig. 2D) (Bonello et al. 2000), *ZmAE3* (Sevilla-Lecoq et al. 2003) and *ZmEBE1* and 2 (Magnard et al. 2003). Upstream sequences are also available for *ZmAE1* (Sevilla-Lecoq et al. 2003) and *ZmINVINH1* (Bate et al. 2004). The 18 bp AE box aATGaaTTAATGNTt-tAc is not only conserved in *ZmAE* and *ZmEsr* promoters but also in *Betl* promoters, making it unlikely that it is solely responsible for ESR-localised expression (Sevilla-Lecoq et al. 2003). The two short sequences AGCATA and T(A/T)AAAAT conserved between the *ZmEsr1* to 3 and the *ZmINVINH1* promoter are far better candidates (Bate et al. 2004). Further dissection of the promoters has been hampered by the very small size of the ESR that renders transient expression studies via biolistics virtually impossible. Confirmation of the above candidate *cis* elements will be dependent on stable transformation with suitable reporter gene fusions.

2.7 Conclusion

A small region of the maize endosperm in proximity of the embryo is clearly distinct from the other parts of the endosperm based upon cytological observations and marker gene expression. This region, baptised ESR, may play a role in the nutrition or defence of the embryo, in embryo/endosperm signalling or in the formation of the embryonic cavern. These proposed functions are not mutually exclusive.

3 The Embryo Surrounding Region in Other Species

Since the term ESR has been defined in maize and has been coined as recently as 1997, it is not surprising that there are almost no publications mentioning an ESR in other species, the only exception being *Arabidopsis*. Obviously this does not preclude the existence of an ESR in other species and below we examine whether there are cells close to the embryo that share cytological characteristics with the ESR, or genes with expression patterns restricted to a sub-territory of the endosperm in vicinity of the embryo.

3.1
ESR in *Arabidopsis*

Even though there are at least two recent publications that mention an "embryo surrounding region" in *Arabidopsis* (Boisnard-Lorig et al. 2001; Luo et al. 2005), it remains to be examined, whether this region is functionally equivalent to the ESR defined in maize. In the first publication the term "dense cytoplasmic embryo surrounding region" is used as a synonym for the micropylar endosperm (MCE). The MCE is one of the three zones of the *Arabidopsis* endosperm defined on the basis of morphological observations by Mansfield and Briarty (Mansfield and Briarty 1990a,b), the other two being the peripheral endosperm (PEN) and the chalazal endosperm (CZE; Fig. 2E,F). Observations in the light and electron microscope show that prior to cellularisation the nuclei in the MCE and PEN are smaller than in the CZE and that most of the cytoplasm of the syncytium is located in the MCE rather than the PEN or CZE. Similar to the cytoplasm of ESR cells in maize, the syncytial cytoplasm of the MCE shows abundant ER and almost no vacuoles (Brown et al. 2003; Mansfield and Briarty 1990a). It is also rich in plastids, mitochondria and ribosomes, while no particular comments were made on the number of these organelles in maize. In *Arabidopsis* cellularisation starts in the MCE close to the embryo, as it does in the ESR of maize (Mansfield and Briarty 1990b; Berger et al., in this volume). Little attention has been paid to the MCE after cellularisation, which may be partly due to the fact that the inner-most part closest to the embryo already starts to degenerate at the late heart stage of the embryo when the chalazal end of the PEN and the CZE are not yet cellularised (Mansfield and Briarty 1990b). Although intended to illustrate other aspects of endosperm development, published longitudinal sections of partially cellularised endosperm seem to indicate that the cells close to the embryo (MCE) are smaller and less vacuolated than cells in the PEN (Brown et al. 1999).

In the second publication that mentions an embryo surrounding region in *Arabidopsis*, the term is used with respect to the expression zone of a *Gus* reporter gene driven by the *Mini3* (*Miniseed3*) promoter. The *Mini3* gene encodes a transcription factor of the WRKY family that is necessary for normal seed size. In the developing seed the *Mini3* promoter shows weak activity in the embryo and in the nuclei of the PEN and strong activity in a zone surrounding the basis of the globular embryo (Fig. 2G). The latter expression zone seems to coincide with the MCE, even though the authors do not specifically address this point (Luo et al. 2005). Another example for preferential gene expression in the MCE is the sucrose transporter *AtSuc5* (Baud et al. 2005). A fusion of its promoter to the *Gfp* reporter gene shows that in the developing seed expression is restricted to the endosperm (Fig. 2H). While the expression zone is identical to that of *Mini3* up to the globular embryo stage, it expands at the torpedo embryo stage throughout the PEN (weak) into the CZE (strong) before disappearing at the upturned-U embryo stage (Baud et al. 2005). A similar dynamic expression pattern was observed in the enhancer trap line G222; again expression was first restricted to a zone close to the embryo at early developmental stages, before extending throughout the endosperm during later stages (Ingouff et al. 2005). In contrast, the expression remained exclusively micropylar in enhancer trap line N9185 (Ingouff et al. 2005). Several other reporter lines are characterised by strong expression in both the MCE and CZE; for the interpretation of these lines it has to be taken into account that the signal in the MCE or CZE may be enhanced by the fact that the GFP is frequently addressed to the ER and that these two zones are richer in ER than the PEN between them (Sorensen et al. 2001). The specific expression of several genes in a region surrounding the embryo indicates that this region, which likely equals the MCE, is not only cytologically but also genetically different from the other two zones of the *Arabidopsis* endosperm. Since none of the genes specifically expressed in the MCE of *Arabidopsis* has been characterised in maize and none of the ESR-expressed maize genes has been studied in *Arabidopsis*, it is presently impossible to draw any conclusion as to a functional equivalence of the two regions based upon the expression pattern of marker genes. Nevertheless, the specific expression of a sucrose transporter in the MCE of *Arabidopsis* and of an invertase inhibitor in the ESR of maize may not just be a coincidence and points at the need for a certain level of sucrose in the vicinity of the embryo.

The MCE, PEN and CZE can be distinguished from each other not only by their morphology and marker gene expression, they also represent distinct mitotic domains of the endosperm (Boisnard-Lorig et al. 2001; Berger et al., in this volume). While all endosperm nuclei divide synchronously during the first three rounds of division, the timing of subsequent divisions differs between the three zones. This has been demonstrated by the observation of mitotic figures as well as the expression of a *Gus* reporter gene under the control of a *Cyclin B1;1* promoter (Boisnard-Lorig et al. 2001; Brown et al. 2003). No comparable analysis has been done in maize, making it again impossible to draw any conclusion as to a functional equivalence of the *Arabidopsis* MCE with the maize ESR.

In conclusion, in *Arabidopsis* the part of the endosperm closest to the embryo is clearly different from the remaining endosperm both prior and after cellularisation. This part has been called micropylar endosperm (MCE) or embryo surrounding region (ESR). The equivalence with the ESR in maize remains to be demonstrated, despite some morphological similarities between the *Arabidopsis* MCE prior to cellularisation and the maize ESR after cellularisation. There are presently no cues as to the function of the MCE, other than the specific expression of a sucrose transporter.

3.2
ESR in Wheat

In wheat no "embryo surrounding region" has been mentioned in the literature to our knowledge. However, the endosperm in the immediate vicinity of the developing embryo has been described as "radically different" from the remaining endosperm and has been named "modified endosperm" (ME) by Smart and O'Brien (1983; Fig. 2I,J). The ME of wheat shares many characteristics with the ESR of maize. Firstly, it is the zone where cellularisation starts and where it is first complete. Secondly, the cells are densely cytoplasmic, have very few vacuoles and do not accumulate starch or protein bodies. Thirdly, the cytoplasm contains large quantities of rough endoplasmic reticulum that swells and contains amorphous contents, especially close to the embryo (Smart and O'Brien 1983). As pointed out above, several of these features also apply to the MCE of *Arabidopsis*. Another similarity with the MCE of *Arabidopsis* lies in the fact that the ME cells close to the developing embryo start to collapse and become digested rather early on in development (Smart and O'Brien 1983). The only major difference with maize and *Arabidopsis* is the position relative to the embryo. While the maize ESR and the *Arabidopsis* MCE surround primarily the suspensor and not the embryo proper, the bulk of the wheat ME is located around and on top of the embryo proper. It remains to be elucidated whether this position reflects functional differences or anatomical constraints of the wheat kernel.

To our knowledge no gene marking specifically or preferentially the ME of wheat has been identified. A recent systematic spatial analysis of gene expression during wheat caryopsis development by high-throughput in situ hybridisation did not really address the question, because the sections used did not include the embryo or ME. Therefore, it is not surprising, that only genes specifically expressed in the central endosperm, aleurone or modified aleurone were discovered (Drea et al. 2005). The modified aleurone or groove aleurone is distinct from the modified endosperm (ME) and designates the region next to the nucellar projection, which is involved in nutrient transfer from the mother plant into the seed.

A re-construction of the three-dimensional distribution of nuclei in the developing wheat endosperm identified three distinct populations in the syncitium: a group near the zygote in what seems to be the future ME, a group in the dorsal region and a group in the ventral region (Wegel et al. 2005). The zygote-associated nuclei formed a stem-like structure that connected this population with the dorsal but not the ventral nuclei. Nuclear divisions seem to be coordinated within but not between the three populations. The endosperm nuclei at the embryo interface were the earliest to stop dividing and to cellularise (Huber and Grabe, 1987a,b; Wegel et al. 2005).

In conclusion, in wheat a specialised zone of the endosperm has been described in the vicinity of the embryo. It exhibits striking similarities with the

ESR of maize and the MCE of *Arabidopsis*, but the functional equivalence of the three zones needs to be demonstrated.

3.3 ESR in Barley

As in wheat, we are not aware of a reference to an "embryo surrounding region" in the barley caryopsis. However, as in maize, *Arabidopsis* and wheat, a small part of the endosperm situated close to the embryo that has attracted particular attention has been named zone 1 (Fig. 2K,L). It can be distinguished from the remaining endosperm by several criteria (Brown et al. 1994; Engell 1989). Prior to cellularisation this region is richer in cytoplasm and has a higher density of nuclei than the rest of the syncitium. At cellularisation it is the earliest zone to cellularise, and after cellularisation the cells of this domain are smaller in size, have less vacuoles and denser cytoplasm than in the adjacent zone, while the nuclei are smaller and more roundish. The cells closest to the embryo show a tendency to lyse (Engell 1989). At the ultrastructural level the cells of this zone are rich in ribosomes, rough ER and dictyosomes with large secretion vesicles. There are no plasmodesmata connecting these endosperm cells to the embryo (Norstog 1972). All of these characteristics are strongly reminiscent of the ESR in maize, the MCE in *Arabidopsis* and the ME in wheat and suggestive of a particular function for this zone of the endosperm.

4 Conclusion

We present here evidence that the endosperm in proximity of the embryo is cytologically different from the remaining endosperm in maize, wheat, barley and *Arabidopsis*. This review is not exhaustive, and similar embryo surrounding regions likely exist in many other species. Clear indications exist for example in *Brassica napus* (van Lammeren et al. 1996) or *Coronopus didymus* (Nguyen et al. 2001). In all cases the ESR is the starting point for endosperm cellularisation. It is a dynamic structure that disappears early on during seed development and is frequently succeeded by a liquid filled space around the embryo. While the cytological characteristics of the ESR are frequently quite similar in the different species, their functional equivalence still needs to be established.

The specific or preferential expression of genes in the ESR of maize or *Arabidopsis* clearly indicates that the genetic program of the ESR is different from that of other endosperm regions. The nature of the ESR-expressed genes provides very limited information with regard to the function(s) of the ESR. The two *Arabidopsis* mutants with lesions in ESR-expressed genes show a transient reduction in fatty acid concentration and a small seed phenotype, respectively (Baud et al. 2005; Luo et al. 2005).

On the basis of the interpretation of the cytological features and the demonstrated or inferred function of ESR-expressed genes, several possible functions of the ESR have been proposed, including nutrient transfer from the endosperm to the embryo, defence of the embryo, signalling between the embryo and the endosperm or the formation of the embryonic cavern. The nutritional role may be complementary to that of the suspensor. In the species mentioned above the suspensor is rather small and the ESR may fulfil part of the nutrient transfer role attributed to the suspensor in species with much larger suspensors, such as *Phaseolus coccineus* (Yeung and Clutter 1979). The poor conservation of many ESR-expressed genes across species is an argument against a conserved role of the ESR in nutrition or signalling and certainly more compatible with the hypothesis of specific defence mechanisms adapted to fight against particular pathogens of each species. Finally, it needs to be stressed that the proposed functions are not mutually exclusive.

References

Balandin M, Royo J, Gomez E, Muniz LM, Molina A, Hueros G (2005) A protective role for the embryo surrounding region of the maize endosperm, as evidenced by the characterisation of ZmESR-6, a defensin gene specifically expressed in this region. Plant Mol Biol 58:269–282

Bate NJ, Niu X, Wang Y, Reimann KS, Helentjaris TG (2004) An invertase inhibitor from maize localizes to the embryo surrounding region during early kernel development. Plant Physiol 134:246–254

Baud S, Wuilleme S, Lemoine R, Kronenberger J, Caboche M, Lepiniec L, Rochat C (2005) The AtSUC5 sucrose transporter specifically expressed in the endosperm is involved in early seed development in *Arabidopsis*. Plant J 43:824–836

Boisnard-Lorig C, Colon-Carmona A, Bauch M, Hodge S, Doerner P, Bancharel E, Dumas C, Haseloff J, Berger F (2001) Dynamic analyses of the expression of the HISTONE::YFP fusion protein in arabidopsis show that syncytial endosperm is divided in mitotic domains. Plant Cell 13:495–509

Bonello JF, Opsahl-Ferstad HG, Perez P, Dumas C, Rogowsky PM (2000) *Esr* genes show different levels of expression in the same region of maize endosperm. Gene 246:219–227

Bonello J-F, Sevilla-Lecoq S, Berne A, Risueno M-C, Dumas C, Rogowsky PM (2002) Esr proteins are secreted by the cells of the embryo surrounding region. J Exp Bot 53:1559–1568

Brown RC, Lemmon BE, Nguyen H (2003) Events during the first four rounds of mitosis establish three developmental domains in the syncytial endosperm of *Arabidopsis thaliana*. Protoplasma 222:167–174

Brown RC, Lemmon BE, Nguyen H, Olsen OA (1999) Development of endosperm in *Arabidopsis thaliana*. Sex Plant Reprod 12:32–42

Brown RC, Lemmon BE, Olsen O-A (1994) Endosperm development in barley: Microtubule involvement in the morphogenetic pathway. Plant Cell 6:1241–1252

Chourey PS, Jain M, Li QB, Carlson SJ (2005) Genetic control of cell wall invertases in developing endosperm of maize. Planta 1–9

Clark JK, Sheridan WF (1991) Isolation and characterisation of 51 *embryo-specific* mutations of maize. Plant Cell 3:935–951

Clark SE (2001) Cell signalling at the shoot meristem. Nat Rev Mol Cell Biol 2:276–284

Cock JM, McCormick S (2001) A large family of genes that share homology with CLAVATA3. Plant Physiol 126:939–942

Drea S, Leader DJ, Arnold BC, Shaw P, Dolan L, Doonan JH (2005) Systematic spatial analysis of gene expression during wheat caryopsis development. Plant Cell 17:2172–2185

Engell K (1989) Embryology of barley: Time course and analysis of controlled fertilization and early embryo formation based on serial sections. Nord J Bot 9:265–280

Fiers M, Golemiec E, Xu J, van der Geest L, Heidstra R, Stiekema W, Liu CM (2005) The 14-amino acid CLV3, CLE19, and CLE40 peptides trigger consumption of the root meristem in *Arabidopsis* through a CLAVATA2-dependent pathway. Plant Cell 17:2542–2553

Fletcher JC, Brand U, Running MP, Simon R, Meyerowitz EM (1999) Signaling of cell fate decisions by CLAVATA3 in *Arabidopsis* shoot meristems. Science 283:1911–1914

Hong LS Kitano H, Satoh H, Nagato Y (1996) How is embryo size genetically regulated in rice? Dev Suppl 122:2051–2058

Huber AG, Grabe DF (1987a) Endosperm Morphogenesis in Wheat—Termination of Nuclear Division. Crop Sci 27:1252–1256

Huber AG, Grabe DF (1987b) Endosperm Morphogenesis in Wheat—Transfer of Nutrients from the Antipodals to the Lower Endosperm. Crop Sci 27:1248–1252

Ingouff M, Haseloff J, Berger F (2005) Polycomb group genes control developmental timing of endosperm. Plant J 42:663–674

Kiesselbach TA (1949) Reproduction and kernel development in the structure and reproduction of corn. University of Nebraska, Lincoln, Nebraska, pp 63–83

Kiesselbach TA, Walker ER (1952) Structure of Certain Specialized Tissues in the Kernel of Corn. Am J Bot 39:561–569

Kim JY, Mahe A, Guy S, Brangeon J, Roche O, Chourey PS, Prioul JL (2000) Characterization of two members of the maize gene family, Incw3 and Incw4, encoding cell-wall invertases. Gene 245:89–102

Leduc N, Matthys-Rochon E, Rougier M, Mogensen L, Holm P, Magnard JL, Dumas C (1996) Isolated maize zygotes mimic *in vivo* embryonic development and express microinjected genes when cultured *in vitro*. Dev Biol 177:190–203

Luo M, Dennis ES, Berger F, Peacock WJ, Chaudhury A (2005) MINISEED3 (MINI3), a WRKY family gene, and HAIKU2 (IKU2), a leucine-rich repeat (LRR) KINASE gene, are regulators of seed size in *Arabidopsis*. Proc Natl Acad Sci USA 102:17531–17536

Magnard JL, Le Deunff E, Domenech J, Rogowsky PM, Testillano PS, Rougier M, Risueno MC, Vergne P, Dumas C (2000) Genes normally expressed in the endosperm are expressed at early stages of microspore embryogenesis in maize. Plant Mol Biol 44:559–574

Magnard JL, Lehouque G, Massonneau A, Frangne N, Heckel T, Gutierrez-Marcos JF, Perez P, Dumas C, Rogowsky PM (2003) ZmEBE genes show a novel, continuous expression pattern in the central cell before fertilization and in specific domains of the resulting endosperm after fertilization. Plant Mol Biol 53:821–836

Mansfield SG, Briarty LG (1990a) Development of the free-nuclear endosperm in *Arabidopsis thaliana* (L). Arab Inf Serv 27:53–64

Mansfield SG, Briarty LG (1990b) Endosperm cellularization in *Arabidopsis thaliana* (L). Arab Inf Serv 27:53–64

Mol R, Matthys-Rochon E, Dumas C (1993) In vitro culture of fertilized embryo sacs of maize: Zygotes and two-celled proembryos can develop into plants. Planta 189:213–217

Nguyen H, Brown RC, Lemmon BE (2001) Patterns of cytoskeletal organization reflect distinct developmental domains in endosperm of Coronopus didymus (*Brassicaceae*). Int J Plant Sci 162:1–14

Norstog K (1972) Early development of the barley embryo: fine structure. Am J Bot 59:123–132

Olsen OA (2001) ENDOSPERM DEVELOPMENT: Cellularization and Cell Fate Specification. Annu Rev Plant Physiol Plant Mol Biol 52:233–267

Opsahl-Ferstad HG, Le Deunff E, Dumas C, Rogowsky PM (1997) ZmEsr, a novel endosperm-specific gene expressed in a restricted region around the maize embryo. Plant J 12:235–246

Randolph LF (1936) Developmental morphology of the caryopsis in maize. J Agric Res 53:882–916

Schel JHN, Kieft H, Lammeren AAM (1984) Interactions between embryo and endosperm during early developmental stages of maize caryopses (*Zea mays*). Can J Bot 62:2842–2853

Serna A, Maitz M, O'Connell T, Santandrea G, Thevissen K, Tienens K, Hueros G, Faleri C, Cai G, Lottspeich F, Thompson RD (2001) Maize endosperm secretes a novel antifungal protein into adjacent maternal tissue. Plant J 25:687–698

Sevilla-Lecoq S, Deguerry F, Matthys-Rochon E, Perez P, Dumas C, Rogowsky PM (2003) Analysis of ZmAE3 upstream sequences in maize endosperm and androgenic embryos. Sex Plant Reprod 16:1–8

Smart MG, O'Brien TP (1983) The development of the wheat embryo in relation to the neighbouring tissues. Protoplasma 114:1–13

Sorensen MB, Chaudhury AM, Robert H, Bancharel E, Berger F (2001) Polycomb group genes control pattern formation in plant seed. Curr Biol 11:277–281

van Lammeren AAM (1987) Embryogenesis in *Zea mays* L A structural approach to maize caryopsis development in vivo and in vitro. Agricultural University Wageningen, Wageningen

van Lammeren AAM, Kieft H, Ma F, van Veenendaal WLH (1996) Light microscopical study of endosperm formation in *Brassica napus* L. Acta Soc Bot Pol 65:267–272

Wegel E, Pilling E, Calder G, Drea S, Doonan J, Dolan L, Shaw P (2005) Three-dimensional modelling of wheat endosperm development. New Phytol 168:253–262

Wobus U, Weber H (1999) Sugars as signal molecules in plant seed development. Biol Chem 380:937–944

Yeung EC, Clutter ME (1979) Embryogeny of *Phaseolus coccineus*: the ultrastructure and development of the suspensor. Can J Bot 57:120–136

Plant Cell Monogr (8)
O.-A. Olsen: Endosperm
DOI 10.1007/7089_2007_110/Published online: 24 March 2007

Transfer Cells

J. Royo · E. Gómez · G. Hueros (✉)

Dpto. Biología Celular y Genética, Universidad de Alcalá,
28871 Alcalá de Henares, Madrid, Spain
gregorio.hueros@uah.es

Abstract Transfer cells are a specialized cell type that can be found in many exchange surfaces in plants, including branching points, hydathodes, or phloem loading and unloading areas. The developing seeds are strong sinks with no symplastic connection with the mother plant, as they are distinct individuals. Consequently, during the grain filling phase seed storage organs must develop specialized structures to facilitate nutrient uptake from the apoplastic space. The differentiation of an endosperm transfer cell layer is thus expected, and found, in seeds in which the endosperm acts as storage organ, namely monocot and particularly cereal seeds. We will review here the current knowledge on the cytological and molecular aspects of the differentiation and function of the cereal endosperm transfer cells. The increasing evidence suggesting an additional role in defense for the endosperm transfer cells will also be discussed.

1 Introduction

In a broad functional sense the term transfer cells (TC) includes any cell type involved in the transport of solutes or solvents from or to the vascular tissues. Very often, however, the term is restrictively applied to cells showing morphological adaptations for transport. The basic adaptation consists in the development of extensive cell wall ingrowths that increase the surface of plasma membrane available for transport (Pate and Gunning 1972; Thompson et al. 2001). In many cases cell wall ingrowths are developed only on one side of the cells, given a general appearance resembling that of the animal epithelial cell covering intestinal absorptive surfaces. As the studies in TC were extended to more tissues and physiological situations it became clear, however, that the presence of cell wall ingrowths is not the best criterion for identifying TC. The extent of cell wall ingrowth development in a particular tissue is dependent both on the species being examined and on the developmental stage of the cells. Furthermore, the TC phenotype is strongly influenced by the physiological stage (see below), as the morphology of these cells responds dynamically to the transport requirements of the surrounding tissues. Finally, the analysis of transgenic plants containing reporter constructs using TC-specific promoters has shown that cells lacking any TC morphology can correctly interpret TC regulatory signals.

TC have been found in all plant taxonomic groups as well as in algae and fungi (Pate and Gunning 1972; Gunning 1977). Within the plant, they have been found in almost every exchange surface examined (for a recent review, see Offler et al. 2002), including the loading and unloading areas of vascular tissues (where they are referred to as companion cells). TC at the vascular parenchyma in stem nodes are good examples of the dynamic nature of TC development; these areas transiently develop TC to facilitate nutrient exchange whilst the vascular connections are being established (apices of young seedlings, Pate et al. 1970; branching points, our group, unpublished observations). Possibly the best-characterized TC areas are developed at the symplastic discontinuities that appear between different individuals, including the association with rhizobium or mycorrhiza (Gunning et al. 1974; Allaway et al. 1985), feeding structures created by nematode infection, gametophyte/sporophyte interface in mosses (Ligrone and Gambardella 1988), and the maternal/filial border at the nutrient uptake areas of the seeds (Cochrane and Duffus 1980; Kiesselbach 1949, 1980; Offler and Patrick 1993; Talbot et al. 2001). The site of TC localization in the seed is determined by the compartment that will be used for nutrient storage, which will be the strongest sink of the seed. In the majority of dicotyledonous species, the cotyledons are the preferred organ for storage product accumulation. In these species the cotyledonary epidermis is very often converted into a TC layer during the seed filling phase. In this system TC are readily accessible, which facilitates the manipulation of their environment; this type of experiment has provided important biochemical (Offler et al. 1997; Farley et al. 2000; Weber et al. 1997, 2005) and genetic data (Borisjuk et al. 2002) on the transport function and developmental regulation of TC. In this chapter, however, we will concentrate on reviewing the current knowledge on endosperm TC development and function that has been obtained from the analyses of experimental systems in which the endosperm is the main storage tissue, namely the cereal caryopsis. A comparison between the *Arabidopsis* endosperm domains along the anterior–posterior axis (see the chapter by Berger et al., in this volume) with the corresponding domains in the maize endosperm points toward the *Arabidopsis* posterior endosperm as the TC equivalent, both tissues having equivalent positions with respect to the embryo and maternal vascular terminals. No TC morphology can be observed, however, in the *Arabidopsis* posterior endosperm and, when introduced in *Arabidopsis*, maize TC-specific promoters do not function in the developing endosperm, whilst they are functional in TC located in other domains of the adult plant.

2 Morphological Aspects of the Endosperm Transfer Cell Layer

The development of the cereal endosperm has been reviewed in other chapters in this volume; the differentiated endosperm contains four major cell types: starchy endosperm, aleurone (see the chapter by Becraft, in this volume), cells of the embryo surrounding region (ESR; see the chapter by Cosségal et al., in this volume), and TC. Aleurone and TC form the endosperm epidermis, which arises after the first, centripetal, periclinal division (see the chapter by Brown and Lemmon, in this volume). TC are thus a modified epithelium that differentiates at the areas adjacent to the maternal vascular terminals. The anatomical organization of the endosperm TC layer is determined by the organization of nutrient unloading areas in the seed. Morphologically distinct TC have been found in many grain caryopses (Kiesselbach and Walker 1952; Rost and Lersten 1970; Zee and O'Brien 1971; Wann-Neng 2004) but not in rice (Bechtel and Pomeraz 1977; Krishnan and Dayanandan, 2003).

2.1 Maize

In maize the vascular terminals form a cup-shaped cushion at the base of the endosperm. A least ten layers of crushed, dead maternal cells, the placento-chalazal zone (Kladnick et al. 2004), extend between the phloem terminals and the endosperm. The endosperm epithelial cells facing the placento-chalazal area differentiate into TC (Fig. 1a) to form the basal endosperm transfer cell layer (BETL, Thompson et al. 2001). At 16 days after pollination (16 DAP) the BETL is fully developed and occupies 65–70 rows of cells in width and three to six cells in depth (Gao et al. 1998). The shape and development of the BETL can be visualized in transgenic maize lines transformed with reporter constructs directed by TC-specific promoters (Fig. 1b). The BETL is the only exchange surface available between maternal and filial tissues, since the entire endosperm, except for the TC region, is surrounded by a cuticular layer (Davis et al. 1990).

TC are variable in their shapes and in the proliferation of cell wall ingrowths, this variability reflecting the developmental stage of each individual cell. The shape ranges from elongate to somewhat square when viewed in longitudinal section. The density of wall ingrowths varies from the most basal TC, having the greatest amount of cell wall proliferation, to the most internal cells, which show progressively less wall proliferation, until there is a complete transition from the TC to endosperm cells. At an intermediate developmental stage (10 DAP in the scheme of Fig. 1c) a gradient of TC development can be observed, with the most differentiated cells positioned at the basal part near, but not at, the abgerminal side of the endosperm. The differentiation stage of each individual cell reflects its position respective to the axis of the vascular bundle. Cells at the abgerminal pole of the BETL are morphologically indistinguishable from aleurone cells, except that they express TC-specific genes (see below).

The model TC is about three times as long as it is wide, and has a large accumulation of cell wall material in the basal part of the cell, which can give the

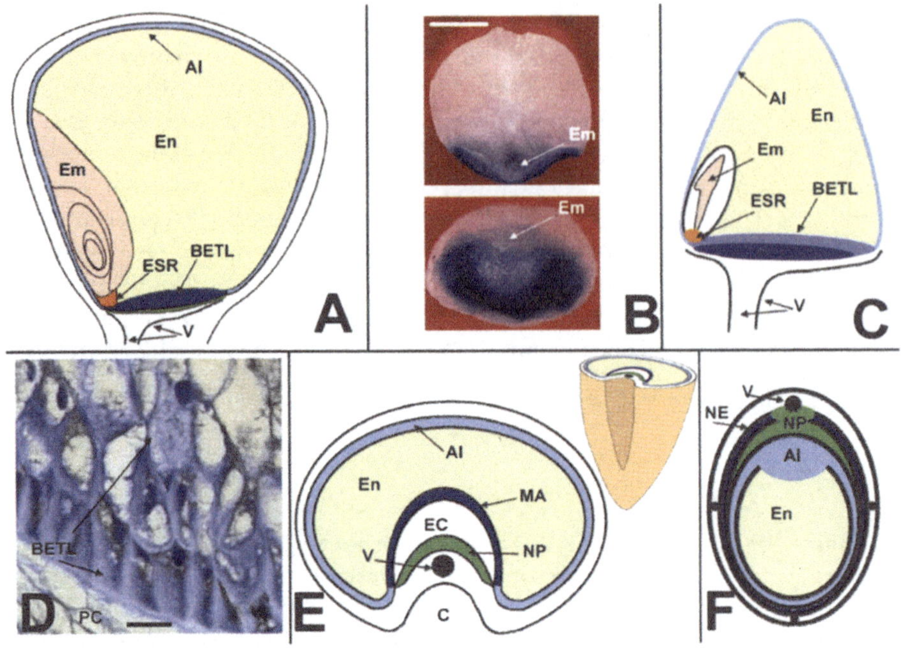

Fig. 1 **a** Schematic representation of a sagittal section of a maize kernel at 16 days after pollination (DAP). The aleurone (Al) layer surrounds the endosperm except for the cells at the basal part, which differentiate into a basal endosperm transfer layer (BETL). TC differentiate right in front of the vascular terminals (V); ESR, embryo surrounding region; En, endosperm; Em, embryo. Nucellar cells at the placento-chalaza are shown in *green*. See Fig. 2a for a real image of a maize kernel sagittal section. **b** Histochemical staining of a transgenic endosperm BETL1 promoter-GUS at 10 DAP. The endosperm was dissected out from the kernel and stained for GUS (*blue*), providing a 3D image of the BETL. The *upper panel* shows a lateral view from the abgerminal side; the *lower panel* shows a basal view. The BETL occupies all the base of the endosperm. Note the weaker color intensity in the center of the BETL; this effect is caused by the advanced maturation of the TC, which are densely filled by cell wall material and thus deprived of cytoplasmic contents. The same effect is also apparent in in situ hybridization (Fig. 2a) and immunolocalization (Figs. 1d and 2c) experiments. The *scale bar* represents 1 mm. **c** Schematic representation of the TC differentiation gradients observed in the BETL. The differentiation gradient is established from the slightly abgerminal center of the BETL (*darkest blue*) toward the edges of the basal and inner cell layers. **d** Azure B staining of the BETL. Note the thick cell walls of the most basal cells; the *brown-black color* in the cells shows the immunolocalization of a BETL-specific protein; PC, placento-chalaza. The *scale bar* represents 50 μm. **e** Transverse section of a wheat/barley kernel; the color codes are the same as those used in **a**. The *inset* shows how the kernel should be sectioned to obtain this point of view. In these systems the TC are often called modified aleurone (MA). The apoplastic barrier between maternal and filial tissues is most evident by the presence of the endosperm cavity (EC) in which the nutrients are released by maternal TC differentiated at the nucellar projection (NP). **c** is the crease of the grain. **f** Transverse section of a rice seed. In this system nutrients are accessible to all the endosperm surface, surrounded by a nucellar epidermis (NE); the aleurone shows some extra cell layers in front of the vascular bundle

appearance of a very thick wall. This is caused by the wall material extending from the primary wall as anastomosing lamellae; when the plane of section is just internal to the primary wall, it gives the ingrowths a ribbon-like appearance. These Y-shaped cell wall ingrowths resemble the wall thickenings of xylem elements and are characteristic of the TC associated with vascular tissues in monocots. By contrast, most TC types in dicots develop cell wall ingrowths that consist of papillate projections, which might form a complex reticule (Offler et al. 2002).

In a mature TC the wall extensions cover almost completely the inner cell space, the central portion is filled with cytoplasm, containing vesicles, and a polymorphic nucleus in the upper half of the cell. At this stage TC show a solid appearance under the light microscope; the ingrowths stain differentially with cell wall stains such as calcofluor white (light blue fluorescence) or azure B, which stains the secondary growth of the cell wall blue (Fig. 1d).

A detailed study of the BETL TC morphology under the electron microscope was conducted by Davis and coworkers (1990). In cross section, cell wall ingrowths appear as densely packed lobes that surround small interstices containing cytoplasm with mitochondria, and rough and smooth endoplasmic reticulum. The ingrowths tend to be attached to the primary cell wall in localized areas that alternate with portions of normal primary wall. These portions usually contain numerous plasmodesmata connecting the cytoplasm of the adjacent TC. Plasmodesmata are not present where the ingrowths occur or in the basal walls of TC in contact with the maternal crushed nucellus. The density of plasmodesmata is higher in the upper walls of the TC and in the TC that are deeper in the endosperm, suggesting that they play an important role in the radial distribution of metabolites.

Endosperm cells positioned immediately above the TC layer do not develop as the rest of the starchy endosperm. They are very often elongated in the same axis as the TC, whilst starchy endosperm cells are typically rounded, and do not accumulate starch until very late in development. These cells are sometimes referred to as transmitting tissue (Becraft 2001), reflecting the fact that they serve as a symplastic route for the assimilates toward the crown, where they are used to build storage products.

2.2 Wheat and Barley

In temperate cereal kernels, nutrients are unloaded from a single vascular bundle that runs dorsally parallel to the crease, along the length of the grain (Fig. 1e). Nutrients are unloaded from the vascular tissues into the nucellar projection, which also develops modified cells, in this case specialized in the secretion of nutrients into the endosperm cavity. The aleurone cells that surround the endosperm cavity have thicker cell walls than the peripheral aleurone cells and have been termed "modified aleurone cells" (Olsen 1992,

Drea et al. 2005). Although cells of the modified aleurone do not develop cell wall ingrowths to the same extent as those formed by the maize endosperm TC, they are undoubtedly functional TC homologous to the maize BETL, and have been shown to express maize BETL-specific orthologous genes and recognize the regulatory signals present in BETL-specific promoters (see below). Together, the nucellar and aleurone TC form a transport system connecting the phloem to the endosperm, thereby facilitating the transport of sucrose from the vascular system to the endosperm.

2.3 Rice

In rice there are no morphologically distinct endosperm TC (Fig. 1f), which might reflect the absence of a discrete entry region for nutrients into the endosperm (Bechtel and Pomeraz 1977; Oparka and Gates 1981; Krishnan and Dayanandan 2003). The main vascular bundle runs longitudinally as described for wheat and barley, and the nutrients are also released symplastically into the nucellar projection. From the nucellar projection, however, nutrients move symplastically into the nucellar epidermis, a remnant of the nucella composed of a single cell layer that completely surrounds the endosperm. Cells of the nucellar epidermis develop thicker cell walls, which prevent the tissue from being crushed during endosperm growth. Nutrients are then released into the apoplastic space that surrounds the endosperm. All endosperm epidermal cells (aleurone) are thus involved in nutrient uptake; consistent with this model, there are four to five additional layers of aleurone cells (see the chapter by Becraft, in this volume) in front of the nucellar projection.

3 Molecular Biology of the Endosperm Transfer Cells

3.1 Transfer Cell Genes and Promoters

Whilst the TC that cover the cotyledon surface of legumes are an ideal system for physiological experimentation, the endosperm TC cannot be accessed without damaging the kernels. This is one of the main reasons explaining the wide use of molecular biology techniques to study the development and function of TC. DNA probes and antibodies have been used to explore the expression of transport-related genes and to identify TC-specific transcripts, which have subsequently been used to infer functions for TC. The later advances in TC development and function are thus presented below in the light of the genes that have been found to be involved in these processes.

A remarkable feature of most TC-specific genes identified to date is the very low conservation exhibited in other plant species, even between close relatives. There are no clear orthologs in wheat for most of the maize TC-specific genes identified and vice versa. This phenomenon might be related to the existence of a large number of species-specific features that would be obscured by the obvious morphological similitude displayed by all TC. Alternatively, the proposed role (see below) in defense for most of the TC-specific genes identified so far would explain this lack of similarity, since defense-related peptides are characterized by their rapid evolutionary change thought to follow the challenges imposed by species-specific pathogens.

3.1.1
TC Genes and the Transport Function

The number of maize endosperm-specific genes has been estimated as 5500 (Lai et al. 2004), 78% of them showing clear orthologs in the rice genome. The only large-scale analysis of tissue specificity available was carried out in wheat (Drea et al. 2005) using in situ hybridization analyses 3, 6, and 9 days after anthesis (DAA) kernel sections. In this study 76 out of 665 genes examined (expression data can be viewed at: http://bioinf.scri.sari.ac.uk/cgi-bin/insitu/home) were found to be expressed in the modified aleurone, and among them 32 were found to be expressed exclusively in this tissue within the kernel. The function of the majority of these genes is currently unknown and, more surprisingly, the list does not contain genes involved in transport processes. A similar result has been found in the screening efforts conducted to identify TC-specific genes in barley and maize (see below). This indicates that the morphology and transport adaptation of TC is, very likely, not determined by the expression of tissue-specific genes but by the regulation of the timing of expression, location, and activity of the gene products. Significantly, sucrose transporters and H^+/ATPase pumps have been found to accumulate, but not to be specific, at the TC of wheat (Bagnall et al. 2000) and barley (Weschke et al. 2000). In the study performed in wheat, the use of antibodies directed against the sucrose transporter, the H^+/ATPase pump, and a sucrose binding protein allowed the authors to detect the selective accumulation of these proteins in the TC, whilst the transcripts were also detected in the aleurone layer, pointing to the existence of a posttranscriptional regulatory mechanism.

Invertases, the enzymes that cleave sucrose into glucose and fructose, are essential genes in determining the transport pathway operating in the storage organs; they have also been implicated in the sugar-regulated control of storage organ development (see below). In maize the gene *Miniature-1* encodes a cell wall invertase, CWI-2, highly expressed in the BETL between 12 and 18 DAP, when the massive entry of metabolites in the developing endosperms takes place (Chourey et al. 2006). *Miniature-1* mutants are characterized by the premature degeneration of the maternal tissues at the placento-chalaza, presumably due to changes in the osmotic pressure (Miller and Chourey 1992, Cheng et al. 1996). Nutrient uptake seems to be interrupted and the *mn1* kernels weigh only a fraction of the wt kernel. In wheat, (Drea et al. 2005) several genes encoding extracellular invertase inhibitors have been identified as TC-specific, suggesting a mechanism for the TC control of the assimilates available for transport. In maize, no such gene has been identified as TC-specific, although an ESR-specific CW invertase inhibitor has been reported (Bate et al. 2004). In barley, a detailed study (Weschke et al. 2003) showed co-ordinated expression of CW invertases and hexose transporters to determine the variant sugar ratios present in different parts of the developing seed. In this study genes expressed in endosperm TC were shown also to be expressed in the maternal tissues at different developmental stages.

Sucrose synthase (SS) is another enzyme thought to be essential in sucrose metabolism; the enzyme is responsible for the intracellular breakdown of sucrose. SS activity was detected in situ in the TC of maize but immunolocalization experiments failed to detect the protein (Wittich and Vreugdenhil 1998), even in minute amounts (Chen and Chourey 1989; Heinlein and Starlinger 1989), suggesting that the sucrose taken up by TC is either transported intact toward the starch synthesizing tissues or is metabolized by non-SS pathways.

The search for TC identity genes has produced more promising results in the case of the genes involved in cell wall synthesis; Drea and coworkers (2005) identified 14 TC-specific genes presumably implicated in the synthesis and modification of the cell wall. This points to a gene-based program determining the differentiation of the specialized cell wall of TC, in contrast to what was concluded from earlier biochemical studies reporting that no difference in the overall composition of cell walls could be found when comparing cells with and without TC morphology (De Witt et al. 1999).

3.1.2 TC Genes and Defense

We have followed several differential screening strategies for the identification of TC-specific genes in maize, the structure of the maize kernel (Fig. 1a) making it especially suited to this type of strategy. The first set of genes identified, termed *BETL1*, *2*, *3*, and *4* (Hueros et al. 1995, 1999b), proved to be extremely useful as molecular markers to monitor the development of TC. They are highly expressed and highly specific for TC (Fig. 2a), allowing TC to be marked before they start to develop their typical morphology, such as at the edges of the BETL (Fig. 2b). BETL markers were also instrumental in the demonstration of the existence of isolated patches of TC in tetraploid ($4n$) endosperms (Hueros et al. 1999b), thought to lack this type of cell based on cytological evidence (Charlton et al. 1995). The lack of a continuous BETL

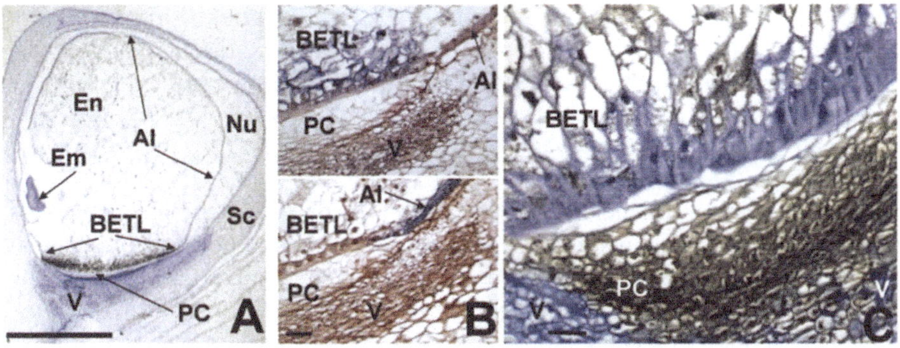

Fig. 2 **a** In situ hybridization experiment in a 10-DAP maize kernel. The probe, a 35S-labeled BETL-3 antisense RNA, labels (*black dots*) specifically the TC. Note that the signal intensity is higher at the edges of the basal endosperm transfer layer (BETL). At this stage, the endosperm has not yet occupied the nucellar space; there are some remains of nucella (Nu) at the abgerminal side and the BETL does not occupy all the basal region in front of the vascular tissue (V). En, endosperm; Em, embryo; Al, aleurone; PC, placento-chalaza; Sc, seed coat. The *scale bar* represents 1 mm. **b** In situ hybridization with DIG-labeled riboprobes specific for TC (*upper panel*) and aleurone (*lower panel*) in 16-DAP maize kernel sections. Although TC morphology is not well developed at the abgerminal edge of the BETL, the border between TC and aleurone is neatly defined by these markers (*purple*). The *scale bar* represents 50 μm. **c** Immunolocalization of BETL-2 in a 16-DAP maize kernel section. Note the accumulation of the BETL-2 protein in the placento-chalaza, even at levels higher than in the TC. The *scale bar* represents 50 μm

in mutants affected in endosperm development has also been demonstrated using BETL markers (Costa et al. 2003).

The proteins encoded by the genes *BETL1–4* share some structural features: they are hydrophilic cysteine-rich cell wall polypeptides highly expressed between 8 and 16 DAP, a developmental window in which TC are differentiating. The same features are shared by *MEG-1* (Gutierrez-Marcos et al. 2004), another maize TC-specific gene which belongs to a small gene family, some members of which are not endosperm-specific. Interestingly, *MEG-1* shows imprinting, and only the maternally inherited allele is expressed during the first developmental stages (see the chapter by Penterman et al., in this volume). The same protein features and parental imprinting were found for *ZmEBE2* (Magnard et al. 2003), although the expression pattern of this gene differs from that of the other BETL genes. *ZmEBE2* is already expressed at the central cell before fertilization, it is expressed both in the TC and the ESR up to 9 DAP, and finally it is exclusively expressed in the BETL between 9 and 15 DAP. *ZmEBE* orthologs have been identified in wheat, where they are specifically expressed at the modified aleurone (Drea et al. 2005). Similar to *ZmEBE2*, *ZmAE1* (Magnard et al. 2000) is expressed in both the ESR (see the chapter by Cosségal et al., in this volume) and the BETL between 3 and 12 DAP, being TC-specific between 12 and 18 DAP. Interestingly, *ZmAE1* expression was detected in the embryoid structures produced in microspore cultures. In barley, the first TC-specific gene identified, *END-1* (Doan et al. 1996), also encodes a small hydrophilic protein containing a signal peptide for secretion. *END-1* orthologs have been identified in wheat, *Arabidopsis*, and maize; in maize an *END-1* ortholog shows a typical BETL gene expression pattern (Fig. 2b, upper panel).

Over a dozen endosperm TC-specific genes identified in various cereals are highly expressed during the endosperm developmental and filling processes, and encode extracellular, small, hydrophilic proteins. What could be the function of these genes in the endosperm TC? Unfortunately, there is currently no genetic evidence supporting any of the functional roles that have been proposed. Indirect evidence, however, supports a role for these peptides in defense against pathogens. All of them show weak structural similarities to antimicrobial peptides; for instance, the position of cysteine residues in *BETL1* and *BETL3* is equivalent to that reported for canonical defensins, and additional genes with higher homologies to this group of antimicrobial peptides have been found to be TC-specific (our group, unpublished results). In the case of *BETL2*, it has been demonstrated that the mature protein possesses in vitro antifungal activity, and has it been renamed BAP-2 (basal layer antifungal protein 2; Serna et al. 2001). BAP-2 defines a new type of antimicrobial peptide that is activated through a posttranslational double processing. The removal of the signal peptide produces a noncytotoxic preprotein that must be further processed through the removal of the N-terminus by a serine protease, which renders the antifungal C-terminal mature protein. Interestingly, the mature BAP-2 preferentially accumulates in the placento-chalazal region, where it might contribute to form an antipathogen barrier (Fig. 2c); a similar behavior has been found for *BETL1* and the maize ortholog of END-1. Other endosperm domains might also contribute to the formation of this chalazal barrier designed to impede pathogen ingress in the developing seed, as has been reported for ESR-6, an ESR-specific defensin protein (Balandin et al. 2005). Other ESR-specific proteins also share structural features with BETL proteins (Opsahl-Ferstad et al. 1997; Bonello et al. 2002). The expression of defense systems in the TC is full of biological meaning: these cells must support a high exchange rate with the maternal tissues, and the development of mechanical barriers to impede the entrance of mother-borne pathogens into the seed is thus not an option. Demonstration of a role for the BETL proteins and related genes in defense, however, awaits additional, genetic, evidence. It also remains to be explained why TC-specific defense mechanisms appear to have been employed instead of those functioning in the adult plant tissues. The recent characterization of the barley gene *jekyll* (Radchuk et al. 2006), encoding a protein with structural features similar to those of the TC-specific genes discussed here, suggests that these peptides might have a previously unsuspected role in grain filling. *Jekyll* is essential for the differentiation of the nucellar TC of barley; downregulation of *jekyll* in transgenic plants impaired the autolysis of the nucellar cells and dramatically reduced the endosperm filling process.

An additional class of TC-expressed genes comprises those expressed in the TC during endosperm development and in the aleurone layer in germination. This class includes a receptor-like kinase also putatively involved in defense (Royo et al. 2006) and a wheat cathepsin B, a thiol protease presumably involved in storage protein mobilization (Cejudo et al. 1992; Dominguez and Cejudo 1998; Drea et al. 2005).

3.1.3 TC Genes and Regulatory Pathways

Since the TC differentiation involves the acquisition of a unique phenotype, it seems reasonable that genes regulating those processes are TC-specific. We have hence explored the TC-specific transcriptome in search of candidate regulatory genes, and have characterized two of them. *ZmTCRR-1* encodes a type A, TC-specific response regulator. These molecules are part of the signal transduction pathways known as two-component systems, present in microorganisms and plants and implicated in the perception of extracellular signals. The identification of *ZmTCRR* (Muñiz et al. 2006) points to the existence of a TC-specific signal transduction pathway. The nature of the signals that is postulated to be perceived by this pathway is currently unknown, but the ZmTCRR-1 protein localization suggests an implication of the coordination between BETL and transmitting tissue development. Contrary to previously known response regulator proteins, the ZmTCRR-1 protein accumulates in cells that do not transcribe the gene, namely the inner cell layers positioned above the BETL (Muñiz et al. 2006)

The second TC-specific regulatory gene identified, *ZmMRP-1* (Gómez et al. 2002), encodes a transcription factor expressed at the basal part of the endosperm already at the syncytial stage. After cellularization, the gene is expressed in the BETL during the lifespan of the tissue, although the highest level of transcript accumulation is found between 8 and 12 DAP, when the differentiation of the TC is most active. ZmMRP-1 belongs to the SHAQYF family of proteins, transcription factors containing a single MYB-related DNA binding domain. As discussed in Sect. 2.2, the interaction of ZmMRP-1 with the promoter sequences of a large number of TC-specific genes has been shown to be a major determinant of their TC specificity.

3.1.4 TC-Specific Promoters

TC specificity seems to reside in the promoter sequence located upstream of the coding region of the gene in all cases examined so far. Unfortunately, endosperm TC cannot be transiently transformed by particle bombardment, so that promoter validation always has to be achieved through the production of transgenic plants containing reporter (GUS) constructs driven by the corresponding promoter sequence. In this way, TC specificity has been demonstrated for the promoters of *BETL1* (Hueros et al. 1999a), *Meg-1* (Gutierrez-Marcos et al. 2004), and *ZmEBE2* (Magnard et al. 2003). Additional promoters validated in our group (unpublished results) include those from a maize homolog of *END-1* and the *ZmMRP-1* promoter. When tested in barley or wheat, the maize TC-specific promoters are functional in the modified aleurone.

In the case of *ZmMRP-1*, we have examined the promoter–GUS reporter construct in barley, *Arabidopsis*, and tobacco transgenic plants (unpublished results). Interestingly, the promoter is transiently active in the TC developing in different areas of the adult plant, and their expression coincides with the transient differentiation of these specialized cells, for instance, in the young branching points or the inter-cotyledonary internodes (Offler et al. 2002).

3.2 Transcriptional Regulation in Endosperm Transfer Cells

The identification of a TC-specific transcriptional activator, ZmMRP-1, prompted us to explore the possibility that this factor was responsible for the transactivation of the TC-specific promoters previously identified (see Sect. 3.1.4). Indeed, it was found (Gómez et al. 2002) that ZmMRP-1 strongly transactivates the promoters of *BETL1* and *BETL2*, although it had no effect on its own promoter or on those from *BETL3* and *BETL4*. Later on it was also shown that ZmMRP-1 transactivates the promoters of *Meg-1* (Gutierrez-Marcos et al. 2003) and *ZmTCRR-1* (Muñiz et al. 2006). These results point toward *ZmMRP-1* as a major determinant of TC specificity, and possibly identity. The interaction between the transcription factor and its target promoters is remarkably robust. TC-specific promoters are virtually silent in a variety of transient expression systems, including maize and *Arabidopsis* leaf, tobacco protoplast, onion epidermis, and yeast, but are transactivated by nearly two orders of magnitude by coexpression with ZmMRP-1. We have used this system to dissect the interaction ZmMRP-1–*BETL1* promoter, our results leading to the identification of a "transfer cell box", a sequence related to 5′-TATCTCTATCTC-3′ that is active in both orientations and is positioned 50–70 bp upstream of the TATA box (Barrero et al. 2006).

3.3 Regulation of Transfer Cell Differentiation

Based on cytological observations, a model describing the development of a functional TC and the contribution of various cellular components to the process has been proposed (Offler et al. 2002). We would like here to focus on the discussion of the mechanisms that trigger the differentiation of TC. Specification of cell fates in the cereal endosperm appears to occur via positional signaling; cells in peripheral positions, except over the main vascular tissues, assume aleurone cell fate. Cells over the main vascular tissue become TC, whilst all interior cells become starchy endosperm cells (Olsen 2001, 2004). The expression pattern observed for several barley (*END-1*, Doan et al. 1996) or maize (*ZmMRP-1*, Gómez et al. 2002; *MEG-1*, Gutierrez-Marcos et al. 2003) genes, already detected at the syncytial stage, indicates that TC cell differentiation occurs earlier than aleurone cell differentiation, which is initiated after the first periclinal division (Olsen et al. 1999). The exact correspondence between vascular terminals and BETL development (see, for example, Fig. 2a and c) strongly suggests that phloem-derived signals are initially responsible for the acquisition of TC fate in the endosperm epidermal cells exposed to them. TC would subsequently differentiate in response to a gradient of signals derived from the maternal pedicel region (Becker et al. 1999). This model would account for the observed gradient in TC morphology (Fig. 1c), with the most developed cells occupying the position immediately above the center of the phloem unloading area, whilst less differentiated cells appear at the edge of the basal endosperm or in inner cell layers. An alternative model to explain the specification of BETL cells was proposed whilst studying the *globby-1* mutant (Costa et al. 2003); following this hypothesis TC fate acquisition would be an irreversible event occurring during the syncytial developmental stage and transmitted in a lineage-dependent manner. Although this is a very attractive proposal, we have recently obtained experimental evidence demonstrating that endosperm TC fate specification is neither irreversible nor clonally transmitted. It is rather the result of the expression of TC-specific genes in endosperm epidermal cells, which need the continuous expression of these factors for the maintenance of the TC morphology (Gómez et al., unpublished results).

The flexibility of the acquisition of the TC morphology seems to be a general property of this type of cell, which responds rapidly to transport requirements. For instance, TC are developed in roots as a response to Fe and P deficiency (Schikora and Schmidt 2001, 2002), in the giant cells induced by nematodes (Jones and Gunning, 1976), and in the stem branching points (Gunning et al. 1970) while vascular connections are being established. In the nodal region of young seedlings, a correlation has been observed between assimilate supply and TC differentiation (Pate et al. 1970). In vitro experiments using the *Vicia faba* cotyledonary epidermis system have shown a direct influence of sugars in the culture medium on TC differentiation, which seems to be induced in low-sucrose media or media containing glucose and fructose (Offler et al. 1997; Weber et al. 1997). Significantly, we have observed a similar flexible behavior in the nodal regions of stems and fruit bases for maize TC-specific promoters, when analyzed in *Arabidopsis* and tobacco transgenic plants (our group, unpublished results).

4 Future Prospects

Within the last few years, the application of molecular biology techniques has provided significant contributions toward our understanding of the mechanisms governing endosperm TC differentiation. However, further research is needed to: (1) identify the maternal tissue-derived signal(s) inducing TC differentiation; (2) identify downstream genes responsible for the development of the TC morphology, particularly genes involved in the development of the specialized TC walls; and (3) clarify the role of TC in impeding the entry of pathogenic agents in the seeds and understand the maternal signals that modulate the activity of the TC, very likely allowing the integration of information on the metabolic stage of the mother plant into the process of grain filling.

Acknowledgements Research on transfer cells in the authors' laboratory is supported by National grants BIO2000-0848 and BIO2003-03721. J.R. is supported by a "Ramón y Cajal" contract of the Spanish MEC.

References

Allaway WG, Carpenter JL, Ashford AE (1985) Amplification of inter-symbiont surface by root epidermal transfer cells in the *Pisonia* mycorrhiza. Protoplasma 128:227–31

Bagnall N, Wang X-D, Scofield G, Furbank R, Offler C, Patrick J (2000) Sucrose transport-related genes are expressed in both maternal and filial tissues of developing wheat grains. Aust J Plant Physiol 27:1009–1020

Balandín M, Royo J, Gómez E, Muñiz LM, Molina A, Hueros G (2005) A protective role for the embryo surrounding region of the maize endosperm, as evidenced by the characterisation of *ZmESR-6*, a defensin gene specifically expressed in this region. Plant Mol Biol 58:269–282

Barrero C, Muñiz LM, Gómez E, Hueros G, Royo J (2006) Molecular dissection of the interaction between the transcriptional activator ZmMRP-1 and the promoter of *BETL-1*. Plant Mol Biol 62:655–658

Bate N, Niu X, Wang Y, Reimann K, Helentjaris T (2004) An invertase inhibitor from maize localizes to the embryo surrounding region during early kernel development. Plant Physiol 134:246–254

Bechtel DB, Pomeranz Y (1977) Ultrastructure of the mature ungerminated rice (*Oryza sativa*) caryopsis. The caryospis coat and the aleurone cells. Am J Bot 64:966–973

Becker H, Hueros G, Maitz M, Varotto S, Serna A, Thompson RD (1999) Domains of gene expression in developing endosperm. In: Cresti M, Cai G, Moscatelli A (eds) Fertilization in higher plants. Springer, Berlin, pp 361–375

Becraft PW (2001) Cell fate specification in the cereal endosperm. Cell Dev Biol 12:387–394

Becraft PW (2007) Aleurone cell development (in this volume). Springer, Heidelberg

Berger F (2007) Development biology of the *Arabidopsis thaliana* endosperm (in this volume). Springer, Heidelberg

Bonello JF, Sevilla-Lecoq S, Berne A, Risueño MC, Dumas C, Rogowsky PM (2002) ESR proteins are secreted by the cells of the embryo surrounding region. J Exp Bot 53:1559–1568

Borisjuk L, Wang TL, Rolletschek H, Wobus U, Weber H (2002) A pea seed mutant affected in the differentiation of the embryonic epidermis is impaired in embryo growth and seed maturation. Development 129:1595–1607

Brown RC, Lemmon BE (2007) The developmental biology of the cereal endosperm (in this volume). Springer, Heidelberg

Cejudo FJ, Murphy G, Chinoy C, Baulcombe DC (1992) A gibberellin-regulated gene from wheat with sequence homology to cathepsin B of mammalian cells. Plant J 2:937–948

Charlton WL, Keen CL, Merriman C, Lynch P, Greenland AJ, Dickinson HJ (1995) Endosperm development in *Zea mays*; implication of gametic imprinting and paternal excess in regulation of transfer layer development. Development 121:3089–3097

Chen Y-C, Chourey P (1989) Spatial and temporal expression of the two sucrose synthase genes in maize: immunohistological evidence. Theor Appl Genet 78:553–559

Cheng W-H, Taliercio E, Chourey P (1996) The *Miniature-1* seed locus of maize encodes a cell wall invertase required for normal development of endosperm and maternal cells in the pedicel. Plant Cell 8:971–983

Chourey P, Jain M, Li Q-B, Carlson S (2006) Genetic control of cell wall invertases in developing endosperm of maize. Planta 223:59–167

Cochrane MP, Duffus CM (1980) The nucellar projection and modified aleurone in the crease region of developing caryopses of barley (*Hordeum vulgare* L var *disitichum*). Protoplasma 103:361–375

Cosségal M, Vernoud V, Depège N, Rogowsky PM (2007) Embryo surrounding region (in this volume). Springer, Heidelberg

Costa L, Gutierrez-Marcos JF, Brutnell T, Greenland A, Dickinson H (2003) The *globby1-1* (*glo1-1*) mutation disrupts nuclear and cell division in the developing maize seed causing alterations in endosperm cell fate and tissue differentiation. Development 130:5009–5017

Davis RW, Smith JD, Cobb BG (1990) A light and electron microscope investigation of the transfer cell region of maize caryopses. Can J Bot 68:471–479

DeWitt G, Richards J, Mohnen D, Jones AM (1999) Comparative compositional analysis of walls with two different morphologies: archetypical versus transfer cell like. Protoplasma 209:238–245

Doan D, Linnestad C, Olsen O-A (1996) Isolation of molecular markers from the barley endosperm coenocyte and the surrounding nucellus cell layers. Plant Mol Biol 31:877–886

Dominguez F, Cejudo FJ (1998) Germination-related genes encoding proteolytic enzymes are expressed in the nucellus of developing wheat grains. Plant J 15:569–574

Drea S, Leader D, Arnold B, Shaw P, Dolan L, Doonan J (2005) Systematic spatial analysis of gene expression during wheat caryopsis development. Plant Cell 17:2172–2185

Gao R, Dong S, Fan J, Hu C (1998) Relationship between development of endosperm transfer cell and grain mass in maize. Biol Plant 41:539–546

Gómez E, Royo J, Guo Y, Thompson RD, Hueros G (2002) Establishment of cereal endosperm expression domains: identification and properties of a maize transfer cell-specific transcription factor, *ZmMRP-1*. Plant Cell 14:599–610

Gunning BES (1977) Transfer cells and their roles in transport of solutes in plants. Sci Prog Oxf 64:539–568

Gunning BES, Pate JS, Green LW (1970) Transfer cells in the vascular system of stems: taxonomy association with nodes, and structures. Protoplasma 71:147–171

Gunning BES, Pate JS, Minchin FR, Marks I (1974) Quantitative aspects of transfer cell structure in relation to vein loading in leaves and solute transport in legume nodules. Symp Soc Exp Biol 28:87–126

Gutierrez-Marcos JF, Costa L, Biderre-Petit C, Khbaya B, O'Sullivan D, Wormald M, Perez P, Dickinson H (2004) *maternally expressed gene1* is a novel maize endosperm transfer cell-specific gene with a maternal parent-of-origin pattern of expression. Plant Cell 16:1288–1301

Heinlein M, Starlinger P (1989) Tissue- and cell-specific expression of the two sucrose synthase isozymes in developing maize kernels. Mol Gen Genet 215:441–446

Hueros G, Varotto S, Salamini F, Thompson RD (1995) Molecular characterization of *BET1*, a gene expressed in the endosperm transfer cells of maize. Plant Cell 7:747–757

Hueros G, Gomez E, Cheikh N, Edwards J, Weldon M, Salamini F, Thompson RD (1999a) Identification of a promoter sequence from the *BETL1* gene cluster able to confer transfer-cell-specific expression in transgenic maize. Plant Physiol 121:1143–1152

Hueros G, Royo J, Maitz M, Salamini F, Thompson RD (1999b) Evidence for factors regulating transfer cell-specific expression in maize endosperm. Plant Mol Biol 41:403–414

Jones MGK, Gunning BES (1976) Transfer cells and nematode induced giant cells in *Helianthemum*. Protoplasma 87:273–279

Kiesselbach TA (1949) The structure and reproduction of corn. Res Bull Univ Nebr Coll Agric Exp Stat 161:2–98

Kiesselbach TA (1980) Reproduction and kernel development. In: The structure and reproduction of corn. University of Nebraska Press, Lincoln, pp 63–68

Kiesselbach TA, Walker ER (1952) Structure of certain specialized tissues in the kernel of corn. Am J Bot 39:561–569

Kladnick A, Chamusco K, Dermastia M, Chourey P (2004) Evidence of programmed cell death in post-phloem transport cells of the maternal pedicel tissue in developing caryopsis of maize. Plant Physiol 136:3572–3581

Krishnan S, Dayanandan P (2003) Structural and histochemical studies on grain filling in the caryopsis of rice (*Oryza sativa* L). J Biosci 28:455–469

Lai J, Dey N, Kim C-S, Bharti A, Rudd S, Mayer K, Larkins B, Becraft P, Messing J (2004) Characterization of the maize endosperm transcriptome and its comparison to the rice genome. Genome Res 14:1932–1937

Ligrone R, Gambardella R (1988) The sporophyte-gametophyte junction in bryophytes. Adv Bryol 3:225–274

Magnard J-L, Deunff E, Domenech J, Rogowsky P, Testillano P, Rougier M, Risueño MC, Vergne P, Dumas C (2000) Genes normally expressed in the endosperm are expressed at early stages of microspore embryogenesis in maize. Plant Mol Biol 44:559–574

Magnard J-L, Lehouque G, Massonneau A, Frangne N, Heckel T, Gutierrez-Marcos JF, Perez P, Dumas C, Rogowsky P (2003) *ZmEBE* genes show a novel, continuous expression pattern in the central cell before fertilization and in specific domains of the resulting endosperm after fertilization. Plant Mol Biol 53:821–836

Miller M, Chourey P (1992) The maize invertase-deficient *miniature-1* seed mutation is associated with aberrant pedicel and endosperm development. Plant Cell 4:297–305

Muñiz L, Royo J, Gómez E, Barrero C, Bergareche D, Hueros G (2006) The maize transfer cell-specific type-A response regulator *ZmTCRR-1* appears to be involved in intercellular signalling. Plant J 48:17–27

Offler CE, Patric JW (1993) Pathway of photosynthate transfer in the developing seed of *Vicia faba* L: a structural assessment of the role of transfer cells in unloading from the seed coat. J Exp Bot 44:711–724

Offler CE, Liet E, Suton EG (1997) Transfer cell induction in cotyledon of *Vicia faba*. Protoplasma 200:51–64

Offler CE, McCurdy DW, Patric JW, Talbot MJ (2002) Transfer cells: cells specialized for a special purpose. Annu Rev Plant Biol 54:431–454

Olsen O-A (1992) Histo-differentiation and molecular biology of developing cereal endosperm. Seed Sci Res 2:117–131
Olsen O-A (2001) Endosperm development: cellularization and cell fate specification. Annu Rev Plant Physiol Plant Mol Biol 52:233–267
Olsen O-A (2004) Nuclear endosperm development in cereals and *Arabidopsis thaliana*. Plant Cell 16:214–227
Olsen O-A, Linnestad C, Nichols S (1999) Developmental biology of the cereal endosperm. Trends Plant Sci 4:253–257
Oparka KJ, Gates P (1981) Transport of assimilates in the developing cayopsis of rice (*Oriza sativa*): ultrastructure of the pericarp vascular bundle and its connections with the aleurone layer. Planta 151:561–572
Opsahl-Ferstad HG, Le Deunff E, Dumas C, Rogowsky PM (1997) ZmEsr, a novel endosperm-specific gene expressed in a restricted region around the maize embryo. Plant J 12:235–246
Pate JS, Gunning BES, Miliken FF (1970) Function of transfer cells in the nodal regions of stems, particularly in relation to the nutrition of young seedlings. Protoplasma 71:313–334
Pate JS, Gunning BES (1972) Transfer cells. Annu Rev Plant Physiol 23:173–196
Penterman (2007) Imprinting in maize (in this volume). Springer, Heidelberg
Radchuk V, Borisjuk L, Radchuk R, Steinbiss H-H, Rolletschek H, Broeders S, Wobus U (2006) *Jekyll* encodes a novel protein involved in the sexual reproduction of barley. Plant Cell 18:1652–1666
Rost TL, Lersten NR (1970) Transfer aleurone cells in *Setaria lutenscens* (*Gramineae*). Protoplasma 71:403–408
Royo J, Gómez E, Balandín M, Muñiz LM, Hueros G (2006) ZmLrk-1, a receptor-like kinase induced by fungal infection in germinating seeds. Planta 223:1303–14
Schikora A, Schmidt W (2001) Acclimative changes in root epidermal cell fate in response to Fe and P deficiency: a specific role for auxin? Protoplasma 218:67–75
Schikora A, Schmidt W (2002) Formation of transfer cells and H^+-ATPase expression in tomato roots under P and Fe deficiency. Planta 215:304–311
Serna A, Maitz M, O'Connell T, Santandrea G, Thevissen K, Tienens K, Hueros G, Faleri C, Cai G, Lottspeich F, Thompson RD (2001) Maize endosperm secretes a novel antifungal protein into adjacent maternal tissue. Plant J 25:687–698
Talbot MJ, Franceschi VR, McCurdy DW, Offler CE (2001) Wall ingrowth architecture in epidermal cells of *Vicia faba* cotyledons. Protoplasma 215:191–203
Thompson RD, Hueros G, Becker H, Maitz M (2001) Development and functions of seed transfer cells. Plant Sci 160:775–783
Weber H, Borisjuk L, Heim U, Sauer N, Wobus U (1997) A role for sugar transporters during seed development: molecular characterization of a hexose and a sucrose carrier in Fava bean seeds. Plant Cell 9:895–908
Weschke W, Panitz R, Sauer N, Wang Q, Neubhn B, Weber H, Wobus U (2000) Sucrose transport into barley seeds: molecular characterisation of two transporters and implications for seed development and starch accumulation. Plant J 21:459–469
Weschke W, Panitz R, Gubatz S, Wang Q, Radchuk R, Weber H, Wobus U (2003) The role of invertases and hexose transporters in controlling sugar ratios in maternal and filial tissues of barley caryopses during early development. Plant J 33:395–411
Wittich PE, Vreugdenhil D (1998) Localization of sucrose synthase activity in developing maize kernels by *in situ* enzyme histochemistry. J Exp Bot 49:1163–1171
Zee SY, O'Brien TP (1971) Aleurone transfer cells and other structural features of the spikelet of millet. Aust J Biol Sci 24:391–395

Plant Cell Monogr (8)
O.-A. Olsen: Endosperm
DOI 10.1007/7089_2007_111/Published online: 24 March 2007

Arabidopsis as a Model for Understanding the Basics of Endosperm Development

F. Berger (✉) · J. N. Fitz Gerald · M. Ingouff

Chromatin and Reproduction Group, Temasek Life Sciences Laboratory, 1 Research Link, National University of Singapore, 117604 Singapore, Singapore
fred@tll.org.sg

Abstract *Arabidopsis thaliana* has emerged as a model species for understanding the basic mechanisms of endosperm development that are conserved in many Angiosperms. In this work, we focus on the genetic, molecular and cellular studies in *Arabidopsis* that have contributed to our current knowledge. Although initially syncytial, the endosperm differentiates several domains, the origin and function of which will be reviewed. This syncytial phase is followed by a cellular phase. We discuss the mechanisms controlling transitions between the two major phases. This work also emphasizes the major role played by endosperm in the control of seed size. Such regulation involves complex co-ordination with other seed components and epigenetic controls.

1 Introduction

The basic organization and functions of endosperm are conserved in most Angiosperm species (Maheshwari 1950; Brown and Lemmon, in this volume). A remarkable feature of endosperm development is the transition from a syncytial to a cellular mode of development (nuclear type), although in some species the syncytial phase is bypassed (cellular type) or the order between the two phases is reversed (helobial type) (Lopes and Larkins 1993; Maheshwari 1950; Olsen 2001). The overall pattern depends on a main antero-posterior axis and radial symmetry (Berger 1999, 2003; Olsen 2004). The endosperm is located at the interface between the embryo and the seed integument and as such is able to combine maternal and zygotic influences. This complicates the genetic controls of endosperm development that are also generally associated with epigenetic regulation. This work details the cellular and genetic controls of endosperm development. We also examine the role played by the endosperm as an integrator of seed size control. Other topics related to epigenetic control of endosperm development have been detailed in recent reviews (Berger 2004; Gehring et al. 2004; Köhler and Grossniklaus 2005; Penterman et al., in this volume).

2 Double-Fertilization: Harnessing the Female Gametophyte

In flowering plants, the gametophytic phase of the life cycle has been reduced to the production of gametes. Female gametogenesis takes place after meiosis and leads to production of the egg cell. Differentiation of the egg cell comprises its mitotic arrest in parallel with that of other cells of the gametophyte (the antipodals, the central cell and the synergids). Fertilization of the egg cell triggers the onset of embryogenesis while mitotic activity in the central cell is activated by the second sperm cell, leading to endosperm development. Although re-activation of the central cell has been considered equivalent to a fertilization event, the very fact that the central cell does not give rise to a further generation suggests that the endosperm may not be considered as a true fertilization product. In fact, the basic mechanisms for activation of the short-lived endosperm appear to be under controls that are independent from the regulation of its embryonic counterpart (Berger et al. 2006).

The first division in the endosperm takes place much earlier than the first embryonic division in *Arabidopsis* and Maize. Both male gametes are, however, delivered at the G2/M transition of the cell cycle (Rotman et al. 2005). This indicates that the cell cycle arrests in the egg cell and the central cell may originate from distinct mechanisms. This is supported by the demonstration of the expression of the *RETINOBLASTOMA RELATED* (*RBR*) gene in the central cell but not in the egg cell (Ingouff et al. 2006). As RBR controls the G1/S transition, the central cell is probably arrested at the G1/S checkpoint, while the egg cell is still in G1 phase.

Another line of evidence in support of the conclusion that the endosperm does not have a zygotic identity is the differential controls allowing release of cell cycle arrest in the egg cell and in the central cell. In the central cell, loss-of-function in genes encoding members of the conserved Polycomb-Group complex FERTILIZATION INDEPENDENT SEED (FIS) bypass the cell cycle arrest (Chaudhury et al. 1997). This leads to autonomous endosperm development resulting in seed-like structures that do not contain an embryo. In contrast, ill-timed onset of division in the egg cell leading to non-viable parthenogenetic embryos has been observed only in mutants for the gene encoding the WD40 protein MULTICOPY SUPPRESSOR OF IRA 1 (MSI1) (Guitton and Berger 2005). Since MSI1 is associated with the FIS complex *msi1* ovules also display autonomous development (Guitton et al. 2004; Köhler et al. 2003a). However, the unique occurrence of parthenogenesis in *msi1* suggests a larger role for MSI1, which is also associated to other chromatin remodeling complexes (Hennig et al. 2005). The distinct pathways for parthenogenesis and autonomous endosperm development likely outline the difference between the origin of the cell cycle arrests of the egg cell and the central cell (Guitton and Berger 2005). MSI1 interacts with the plant homologues to RETINOBLASTOMA (Ach et al. 1997). In *Arabidopsis*, MSI1 together with the retinoblastoma homologue *RBR1* likely regulate progression of the cell cycle. However, *rbr1* null mutations do not produce autonomous endosperm but rather cause anomalous proliferation of the central cell lineage. Moreover, *rbr1* mutant embryo sacs do not appear to produce parthenogenetic embryo (Ingouff et al. 2006). In conclusion, unleashing cell division in the central cell requires Polycomb-group dependent controls whereas the egg cell arrest relies on a *FIS*-independent repression associated to MSI1. The difference between the egg cell and the central cell may point to the gametophytic nature of the central cell outlined in some evolutionary scenarios (Friedman 1998). Evolution of the seed plants is characterized by reduction of the length and importance of the gametophytic phase. We propose that this may be the result of specification of the egg cell causing precocious arrest of female gametophyte development. This would have caused cell cycle arrest of the gametophyte after the first few divisions producing a single syncytial central cell. Requirement for embryo-nurturing tissues derived from the gametophyte would have positively selected a "fertilization" of the central cell to re-initiate mitotic activity.

3
Main Features of Endosperm Development

3.1
Syncytial Phase

Once fertilized, the central cell immediately undergoes mitoses that are not followed by cell division. This leads to production of a multinucleated cell, also called a syncytium (Brown and Lemmon, in this volume). Syncytial development is typical of large rapidly growing cells and this mode of development is observed in other tissues associated with nutrient transport function such as the vascular system of plants, feeding cells induced by parasitic nematodes (Favery et al. 2004) and some cells in the mammalian placenta (Cross 2005). A series of eight syncytial divisions leads to approximately 200 nuclei mostly located at the periphery of the endosperm (Boisnard-Lorig et al. 2001; Brown et al. 1999; Mansfield and Briarty 1990b). Quasi-synchronicity of the syncytial divisions allows the classification of the syncytial phase into successive stages I to VIII (Ingouff et al. 2005b) (Table 1). The center of the syncytium is occupied by a single large vacuole (Mansfield 1990b). Each nucleus is surrounded by a mass of cytoplasm, the boundary of which is delimited by a dense cortical array of microtubules (Brown et al. 1999). Such a unit has been defined as a nucleo-cytoplasmic domain (NCD). The syncytial phase of the endosperm occurs during the major phase of growth of the seed from 1.5 to 4 Days After Pollination (DAP), while the major phase of proliferation of the embryo initiates only after 5 DAP (late heart stage).

Table 1 Endosperm developmental stages during seed development (for an overview of endoreduplication in the cereal endosperm, see Nguyen et al., in this volume). Collectively, endosperm with 26, 28 and 30 nuclei representing 22, 37 and 41%, respectively, of a given seed population are referred to as being at developmental stage VI (Ingouff et al. 2005b). The presence of larger nuclei thus initially defines the posterior endosperm pole. The next round of syncytial division is marked by the definition of a second mitotic domain at the anterior pole composed of five to eight nuclei surrounding the embryo (Boisnard-Lorig et al. 2001). Direct dynamic observation of mitosis as well as regional expression of the mitotic cyclins B1:1 and B1:2 show that these nuclei divide earlier than the nuclei in the peripheral mitotic domain. Thereafter, the cell cycle maintains an independent pace in each of the three domains. Divisions in the peripheral domain define further stages, VII (approximately 50 nuclei), VIII (100 nuclei) and IX (200 nuclei). Mutations in *FIS* genes disrupt mitotic domain organization (see Sect. 3.1) (Ingouff et al. 2005b). *FIS* genes presumably control the transition between developmental phases during endosperm development and fis loss-of-function prevents the exit of the juvenile stage prior to definition of mitotic domains. Besides this global molecular control by the FIS complex, the molecular or cellular mechanisms responsible for the limitation of the mitotic domains in the syncytial endosperm are not understood

Endosperm stage	Number of endosperm nuclei	Cellular events in endosperm	Embryo stage (Jürgens 1994)	Seed integuments	Time after pollination
I	1	Karyogamy Release of G2/M arrest	Zygote	Active cell divisions and Pro-	0
II	2	First syncytial mitotic division	Zygote	antho-cyanidin synthesis	6–8 h
III	4	–	Zygote		∼ 10 h
IV	8	–	Elongated zygote		∼ 12 h
V	14 in 25% seed 16 in 75% seed	Posterior miotic domain with large nuclei	1 cell embryo		∼ 24 h
VI	26 in 25% seed 28 in 25% seed 30 in 50% seed	–	1–2 cell embryo	Cell elongation	∼ 36 h
VII	∼ 50 nuclei	Anterior miotic domain	∼ quadrant		∼ 48 h
VIII	∼ 100 nuclei	Initiation of NCD migration	Octant-Globular		2.5–4 days
IX	∼ 100 cells	Cellularization	Globular-Triangular	Cell differentiation	∼ 4 days
X	∼ 300 cells	Conventional cytokinesis	Heart-Torpedo	of each cell layer	∼ 5 days
Mature	–	One cell layer remains	Torpedo-Mature		6–15 days

The pace of the first three syncytial divisions is approximately 2 to 3 hours per cell cycle and gradually increases to longer periods of 12 h during following cycles. The first divisions lead to eight endosperm nuclei evenly spaced along a curved tube bounded by the inner seed integument (Boisnard-Lorig et al. 2001). In a quarter of the seeds, during the fourth syncytial division, the posterior-most nucleus does not enter a further round of mitosis and instead enlarges as it presumably undergoes endoreduplication. In the remaining seeds, the fourth division produces 16 nuclei and at the fifth cycle of division, the two or four posterior-most nuclei supposedly enter endoreduplication while the other nuclei undergo a fifth synchronous mitosis leading to 30 or 28 endosperm nuclei (Ingouff et al. 2005b) (Table 1). To date no direct evidence is available for an underlying mechanism to directly support the involvement of endoreduplication.

3.2
Cellularization

The division cycle from stage VIII to IX marks the end of the syncytial phase as it is followed by cellularization in the anterior and peripheral domains, leading to isolation of most nuclei into individual cells. In *Arabidopsis* time-lapse recordings have shown that endosperm cellularization occurs immediately after the 8th mitotic division (Sørensen et al. 2002). Similarly, early embryogenesis in *Drosophila* is marked by the transition from a syncytial to a cellular state by cytokinesis occurring after a specific series of syncytial mitotic divisions (Edgar et al. 1986; Edgar and O'Farrell 1989). In *Drosophila*, cellularization of the syncytium is triggered once a specific ratio between the relative amount of nuclei and cytoplasm has been reached. In *Arabidopsis* the influence of a nucleo-cytoplasmic ratio on the onset of cellularization was suggested through the use of recessive mutants that are specifically deficient in either growth of endosperm or the surrounding maternal integuments. Reduction of endosperm growth with no alteration of the cell cycle pace causes early cellularization as the critical nucleo-cytoplasmic ratio is reached prematurely. Restriction of endosperm growth can result from loss-of-function of the *HAIKU (IKU)* class genes in *iku* mutants including *iku1*, *iku2*, (Garcia et al. 2003) and *mini3* (Luo et al. 2005). *HAIKU2* and *MINI3* encode a leucine rich repeat kinase and the transcription factor WRKY10, respectively, and are expressed in endosperm after fertilization (Luo et al. 2005). Expression studies suggest that *IKU1* activates the expression of *MINI3*, itself positively regulating *IKU2* transcription (Luo et al. 2005). The targets of the *IKU* pathway responsible for endosperm growth are not known but their regulation might depend on epigenetic controls since *iku* is phenocopied by increasing the dosage of maternal genome in a wild-type background (Garcia et al. 2003). Crosses between ovules from tetraploid plants and pollen from diploid plants produce seeds with reduced endosperm growth, precocious endosperm cellularization and reduced seed size (Scott et al. 1998). These results suggest the existence of maternally expressed imprinted genes encoding inhibitors of endosperm growth. Similar phenotypes were observed in seeds resulting from fertilization of wild-type ovules with pollen lacking the maintenance DNA methyltransferase MET1 (Adams et al. 2000). These results have suggested that MET1 dependent DNA methylation silences the paternal allele of imprinted endosperm growth inhibitors. Such genes could include the imprinted *FIS* genes (Spielman et al. 2001; Penterman et al., in this volume) or their targets as *met1* pollen also partially suppresses endosperm defects when used to fertilize *fis* mothers.

The process of cellularization itself is a variation of conventional cytokinesis with notable differences in cytoskeletal arrangement (Dickinson 2003; Olsen 2004; Otegui and Staehelin 2000a). As such, endosperm cellularization depends mostly on the functions essential for cytokinesis in vegetative cells (microtubules, actin and associated proteins, and vesicle trafficking) (Sørensen et al. 2002). For example, compromised microtubule assembly in *pilz* mutants (Mayer et al. 1999; Steinborn et al. 2002) leads to similar defects of cytokinesis in the endosperm and in the embryo. Similarly, loss of function in vesicle trafficking in mutants for the syntaxin KNOLLE (Lukowitz

Fig. 1 Cellularization and construction of the posterior pole in *Arabidopsis* endosperm. ▶ **a** The stages of endosperm cellularization. Although no pre-prophase band is observed during the cellularization of the syncytial endosperm, radial microtubule systems organize the NCDs **A** and provide even spacing between nuclei (Brown and Lemmon, 2001; Nguyen et al. 2002). When cellularization initiates **B**, Golgi-derived vesicles migrate along these microtubules to the site of de novo cell plate construction. Unlike standard cytokinesis, "mini-phragmoplasts" are derived from hour-glass shaped intermediates of fused vesicles. These structures are continuously associated with all stages of cellularization. Mini-phragmoplasts develop their own tubular networks and subsequently fuse to produce the wide tubular network **C** and subsequent expansion to the thin tubular network **D** (Otegui and Staelin 2000b). The last stage before proper cellularization is the further maturation of the thin tubular network to a planar fenestrated sheet **E**. Callose deposition is detected as early as **F** and persists after maturation which further distinguishes syncytial cellularization from conventional cytokinesis. **b** *Arabidopsis* seed at the late torpedo embryo stage. The endosperm is completely cellularized except at the posterior pole where the cyst develops. **c** Enlarged representation of the cyst at the posterior pole. Arrows indicate NCD and nodule movements directed towards the posterior pole. **d** 3D reconstruction of the architecture of the posterior of stage VII endosperm (quadrant stage embryo). Nuclei (*yellow*) are embedded in dense cytoplasm, forming the NCD. At this stage a few couples of NCDs have fused and form binucleate nodules (n) above the cyst (c). *Scale bar* represents 20 μm. **e** At the torpedo embryo stage, all nodules have been absorbed in the cyst (c), which contains a large number of nuclei (*red*). Above the cyst remains cellularized peripheral endosperm that will be absorbed upon embryo growth. *Scale bar* represents 100 μm. We thank Claire Lionnet (ENS-Lyon, France) for the 3D reconstructions obtained from confocal series of fixed Feulgen stained material (Guitton et al. 2004)

et al. 1996) causes defective cytokinesis in endosperm and embryo. Formins, as actin polymerization regulating factors are crucial for cytokinesis in yeast and in animals (Glotzer 2003) and have been recently reported to play a role in plant cytokinesis (Ingouff et al. 2005a). Analysis of formin function in *Arabidopsis* is complicated by gene duplication to a large gene family of 22 genes. A member of this family, AtFH5, is expressed in roots and localizes to the cell plate. However, null alleles *atfh5-1* and *atfh5-2* do not show defects in conventional cytokinesis as a probable redundancy with other formins. During seed development expression of AtFH5 is limited to the endosperm (Ingouff

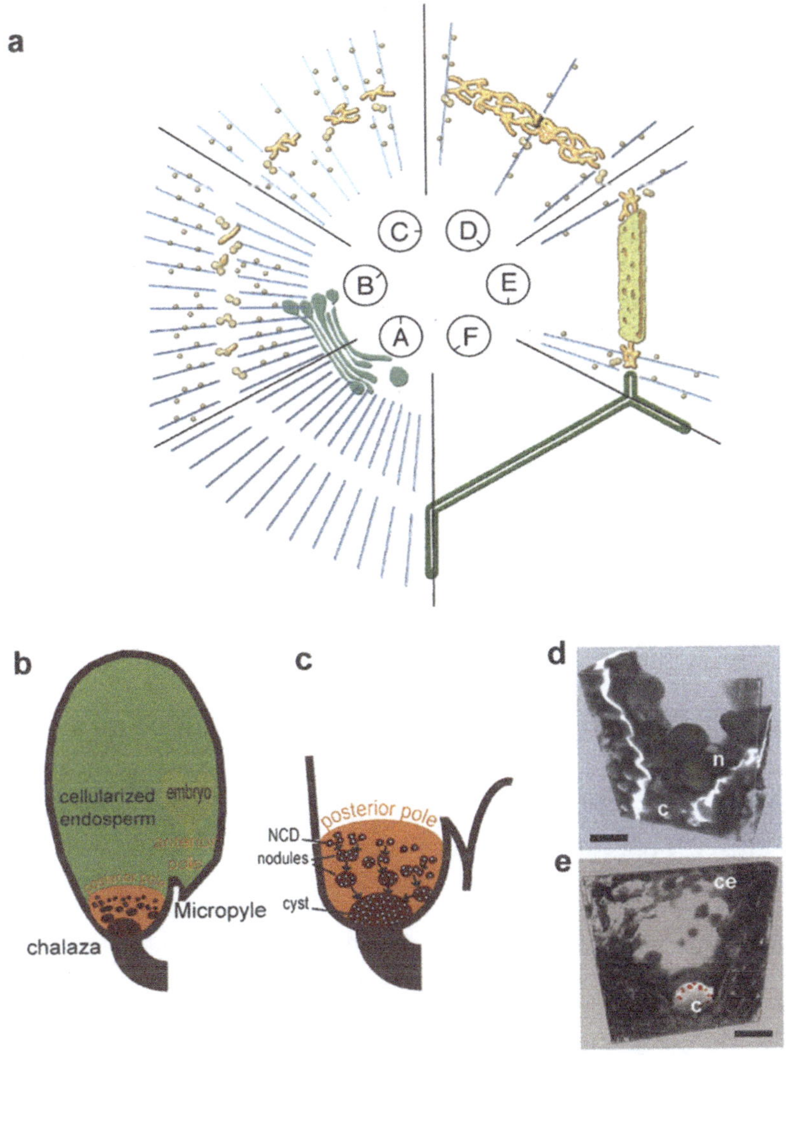

et al. 2005a). Loss-of-function of AtFH5 causes a delay in endosperm cellularization.

To date only one other mutation, *spätzle*, has been reported to be specifically defective in endosperm cellularization without effects on conventional cytokinesis (Sørensen et al. 2002). Hence, it is very likely that endosperm cellularization mostly depends on the same molecular components as cytokinesis. However, the defining feature of cellularization, de novo formation of cell walls around syncytial NCDs, most likely relies on distinct architectural organization of these conserved components. Each NCD is delimited by a dense array of microtubule defining the future division sites. Unlike cytokinesis, no preprophase band anticipates the spatial position of the new cell wall during cellularization (Brown et al. 1999; Otegui and Staehelin 2000b; Brown and Lemmon, in this volume). After the critical last syncytial mitosis a series of cell plates is formed between the sister nuclei as a result of a rather complex mechanism based on endoplasmic reticulum derived vesicles assembly into tubules and laminar sheets (Fig. 1a) (Otegui and Staehelin 2000b; Otegui et al. 2001; Otegui, in this volume). Other de novo assembly of cell plates takes place between non-sister nuclei defining hexagonal- or pentagonal-shaped boundaries around each NCD. Endosperm cell plates do not divide a cell into two daughter cells as in conventional cytokinesis, but arise simultaneously around all NCDs. This implies that some unknown co-ordination mechanisms ensure that each cell plate connects only once to its multiple neighbors forming three junctions, which is uncommon to conventional cytokinesis. Another idiosyncratic feature of cellularization lies in the apparent centripetal growth of cell plates. At cellularization the cell plates form from the center to the periphery in all directions as during conventional cytokinesis. Only one end of the growing cell plate abuts on the outer endosperm cell wall of maternal origin while the other side grows towards the center of the endosperm (Nguyen et al. 2002). This cell plate growth is sustained by further furrowing of cortical microtubule arrays towards the center of the endosperm similar to cellularization in cereals endosperm (Olsen 2004).

3.3
Cellular Phase and Maturation

After cellularization the cellular developmental phase comprises a small series of synchronous cell divisions leading to formation of four to six layers of endosperm cells depending on the position along the antero-posterior axis. Divisions persist in the central part of the peripheral endosperm and progress towards the posterior pole while the central vacuole recesses progressively until the entire peripheral endosperm becomes cellular.

In *Arabidopsis* and many other dicotyledonous species that produce seeds with a morphology comparable to beans, the endosperm does not persist with the exception of the cell layer immediately in contact with the integuments, the function of which remains to be defined. Reserves are stored in the embryo, which proliferates rapidly at the expense of the cellular endosperm after the heart stage. Starch synthesis takes place in the cellular endosperm and amyloplasts differentiate, although this phase has received very limited attention in *Arabidopsis* (Mansfield and Briarty 1990a).

Endosperm development ends with desiccation together with other seed components. Most of the endosperm cells loaded with reserves probably die, with the exception of the outer cell layer, called the aleurone layer in cereals (Becraft, in this volume). The differentiation of an aleurone layer in *Arabidopsis* endosperm remains to be established. Mechanisms controlling the arrest of endosperm proliferation are not understood and it may be affected in viviparous mutants characterized by a lack of arrest of embryo growth and immediate seed germination in the silique (Raz et al. 2001). A similar role of endosperm in the control of dormancy is mediated by ABA (Penfield et al. 2006). The genes Abscissic Acid Insensitive 3 and 5 are expressed in endosperm during seed maturation acting through at least one effector of ABA signaling, the NADPH oxidase RBOHD. Expression of ABI5 at the micropylar region where the emerging root tears open the seed coat may play a specific role, which remains to be analyzed further.

4
Patterning of Endosperm

The early appearance of mitotic domains suggests that the endosperm is not a uniform tissue. The cellularized endosperm contains several cell types based on expression studies and differences in structure. These structures are arranged along the axis running from the anterior pole where the pollen tube delivers the sperm cells (micropyle) to the posterior pole where maternal nutrients transit (chalaza) (Fig. 1b). Other cell types are arranged according to radial symmetry. These two axes of patterning are conserved amongst Angiosperm species.

4.1
Polar Organization

The antero-posterior axis pre-exists in the embryo sac and it is unknown if it is maternally transmitted to the endosperm. The influence of this polar axis is observed in the organization of the previously described syncytial divisions and the migration of NCDs (Fig. 1c). The first two divisions take place along the antero-posterior axis of the endosperm. The third division is perpendicular to the antero-posterior axis. Subsequently, the mitotic domains are established from anterior to posterior along this axis in the endosperm (Fig. 1b). This general organization is also marked by distinct cytological organization of the cytoskeleton in the three domains (Brown et al. 1999; Nguyen et al. 2002). The posterior domain, initially defined by two to four large nuclei in a pool of cytoplasm forming the posterior cyst, is later marked by a posterior-directed migration of NCDs (Guitton et al. 2004). After developmental stage VI, each syncytial division is followed by gradual migration of NCDs from the posterior part of the peripheral endosperm toward the posterior pole (Fig. 1c–e). Theses NCDs fuse to generate multinucleate nodules, which appear as large NCDs (Fig. 1d). Over time, individual NCDs and nodules migrate and fuse with the chalazal cyst. As cellularization takes place, all remaining NCDs fuse into the cyst, which gradually recedes while the entire peripheral endosperm becomes cellular (Fig. 1e).

The identity of the three domains along the antero-posterior axis is established by specific expression of genes and other markers (Table 2). In cereals several other genes have been characterized by their localized expression at the anterior embryo-surrounding pole (Cosségal et al., in this volume) and at the posterior (basal) pole (Royo et al., in this volume). The antero-posterior axis is thus potentially conserved, but homologous functions in the various domains remain to be identified between cereals and *Arabidopsis*.

The definition of the antero-posterior polarity is compromised in mutants for the Polycomb group (Pc-G) pathway *FIS* (Sørensen et al. 2001) including *MEA*, *FIS2*, *FIE*, *MSI1* and *DEMETER*, an activator of *MEA* and *FIS2* (Choi et al. 2002; Guitton et al. 2004; Köhler et al. 2003a; Jullien et al. 2006b). The expression of *MEA* and *FIS2* is limited to endosperm, providing a specific tissue identity to the *FIS* Polycomb group complex (Luo et al. 2000), while *FIE* and *MSI1* are expressed in other tissues (Hennig et al. 2003; Yadegari et al. 2000) and participate in other Pc-G protein complexes (Chanvivattana et al. 2004; Katz et al. 2004).

FIS loss-of-function causes a complex pleiotropic phenotype including lack of cellularization and an apparent loss of polar organization as the mitotic domains and posterior directed NCD migration are altered (Ingouff et al. 2005b). However, the posterior marker N9307 remains expressed at the posterior pole in *fis* endosperm and the general anterior-posterior axis although perturbed remains identifiable (Ingouff et al. 2005b). This implies that mutation in Pc-G members does not alter seed polarity itself but the ability to recognize or employ polarity signals during development. Thus, the effect of the *FIS* pathway on endosperm antero-posterior patterning is likely indirect and might result from a general heterochrony. In support of this hypothesis, *fis* endosperm retains juvenile traits such as the syncytial mode of division, the absence of NCD migration and a limited differentiation of the posterior pole. The origin of heterochronic endosperm development in the absence of Pc-G function likely originates from compromised maintenance of transcriptional states regulated by histone3K27 methylation by the FIS Pc-G. A very few targets of the FIS pathway have been identified. These include the type I MADS box domains transcription factor PHERES1 and likely a homologue of the cell cycle regulator SKIP1 (Köhler et al. 2003b). Loss-of-function of *PHERES1* has no obvious consequence on endosperm development. However, overexpression of *PHERES1* in *fis* mutant endosperm causes partial rescue of the phenotype and leads to enlarged endosperm with additional cell layers. Thus, the function of *PHERES1* accounts for a limited aspect of the phenotype observed in *fis* endosperm. Other key target genes will need to be identified before the mechanisms of FIS regulation are understood.

4.2
Maternal Control of Endosperm Polarity

The maternal effect of *mea* and *fis2* mutations most probably originates from imprinting affecting *MEA* and *FIS2* (Chaudhury et al. 1997; Grossniklaus et al. 1998; Jullien et al. 2006a, 2006b; Kiyosue et al. 1999; Penterman et al., in this volume). Loss-of-function of *MSI1* and *FIE* also yield maternal effects, but these are possibly linked to causes distinct from imprinting mechanisms. *FIE* is expressed in a complex temporal pattern in endosperm. According to a transcriptional reporter, *FIE* is maternally expressed in endosperm up to stage VII. *FIE* expression arrests until stage IX, after which *FIE* is expressed from both parental alleles in endosperm and in the embryo (Yadegari et al. 2000). Whether *FIE* is imprinted and maternally expressed from stage I to VII remain to be established. However, this is possible since a FIE::GFP transcriptional reporter is only maternally expressed during early endosperm development. *MSI1* expression is very unlikely to be imprinted since *msi1/msi1* endosperm displays a phenotype distinct from the phenotype resulting from *msi1* maternal inheritance (Guitton et al. 2004). Hence, *MSI1* paternal allele is likely expressed in endosperm, although it is currently very difficult to test this hypothesis since isolation of pure endosperm tissue during syncytial stages remains difficult. As an alternative to imprinting, maternal effects linked to loss-of-function of *MSI1* and *FIE* could originate from defects in the central cell.

4.3
Radial Organization

After cellularization, the outer endosperm cell layer undergoes a specific series of events leading to its differentiation in a specialized structure, the aleurone layer. This differentiation has been documented in detail in cereals but very limited data have been obtained for *Arabidopsis*. The *Defective kernel 1* (*Dek1*) gene encoding a cysteine proteinase (Lid et al. 2005) and the leucine rich repeat kinase encoding *Crinkly4* (*Cr4*) are important for aleurone differentiation in Maize (Becraft, in this volume). However, the function of the *Arabidopsis ACR4* homologue does not appear conserved in *Arabidopsis* endosperm. The *DEK1* homologue in *Arabidopsis* might have a general role in cell differentiation and the lack of aleurone cells in *dek1/dek1* endosperm does not necessary reflect a specific function in aleurone differentiation. In conclusion, the equivalence between the aleurone layer in cereal endosperm and the outer endosperm cell layer remains undetermined.

5
Interactions Between Endosperm and Other Components of the Developing Seed

5.1
Co-Ordination of Growth between Seed Components

In the seed, the endosperm is at the interface between the embryo and the seed integuments of maternal origin. In *Arabidopsis*, the major period when seed growth takes place coincides with the syncytial phase of endosperm development (see Table 1; Boisnard-Lorig et al. 2001; Garcia et al. 2003). During this phase, the globular stage embryo consists of a small group of cells and does not appear to contribute significantly to seed mass and volume (Jürgens 1994). Hence co-ordination of seed growth during the syncytial phase must occur primarily between seed integument and the endosperm (Chaudhury and Berger 2001; Haughn and Chaudhury 2005). After fertilization cell division is still ongoing in the seed integuments and together with cell elongation control

Table 2 Regional markers of endosperm (the star indicates that the expression of the marker starts after cellularization)

Endosperm	Marker name	Type	Refs.
Cap	–	–	–
	DFJ48	Enhancer trap	(Dubreucq et al. 2000)
	n.a.	Enhancer trap	(Liu et al. 2005)
	n.a.	Enhancer trap	(Liu et al. 2005)
	n.a.	Enhancer trap	(Liu et al. 2005)
Anterior	–	–	–
	AtSUC5	Promoter	(Baud et al. 2005)
	N9185*	Enhancer trap	(Ingouff et al. 2005b)
	G222*	Enhancer trap	(Ingouff et al. 2005b)
Posterior	–	–	–
	MEDEA-GUS	Promoter	(Luo et al. 2000)
	FIS2-GUS	Promoter	(Luo et al. 2000)
	AtIPT4/8	Promoter	(Miyawaki et al. 2004)
	AtFH5	Promoter	(Ingouff et al. 2005b)
	N9307	Enhancer trap	(Ingouff et al. 2005b)

the maximum growth achievable (Garcia et al. 2005). Co-ordination of integument cell elongation and endosperm growth has been genetically dissected. When the endosperm size is reduced by loss-of-function of the endosperm specific *HAIKU* genes, growth of seed integuments is prevented because cell elongation does not take place. Hence, a chemical or mechanical signal from the endosperm provokes cell elongation in the integuments. The capacity for cell elongation in the integuments is under a separate maternal control, the *TRANSPARENT TESTA GLABRA* (*TTG*) pathway involving anthocyanin and flavonoid metabolism (Debeaujon et al. 2003; Lepiniec et al. 2006), which might affect cell wall properties. If the *TTG* pathway is not active, cell elongation in the integuments is reduced causing reduced endosperm growth and seed size (Garcia et al. 2005). In contrast, if the cell number in the integuments is decreased experimentally by ectopic expression of the cell cycle inhibitor KIP1, endosperm and seed size remain unaffected. Instead, cell elongation in the integument adjusts to accommodate the degree of endosperm growth. In this case, the seed reaches a size expected according to the genetic composition of endosperm, thus demonstrating that endosperm dictates a final seed size, to which both integuments and embryo adjust (Garcia et al. 2005).

Embryo development is influenced by endosperm growth, at least from the trophic point of view. In *Arabidopsis* reduction of endosperm growth causes a reduction of cell proliferation in the embryo coordinated with the size imposed by the seed integuments. This is obvious when seed size is reduced in *iku* mutants and in crosses affecting the epigenetic status of endosperm (Garcia et al. 2005). Similarly, increasing the number of integument cells causes a symmetrical increase in endosperm and embryo growth (Canales et al. 2002; Schruff et al. 2006). Transcriptional control by APETALA2 likely controls some of the interactions between the endosperm and the seed integuments (Jofuku et al. 2005; Ohto et al. 2005). Seeds produced by *ap2* mutants are larger than wild type and bear larger embryos. Since *AP2* is expressed in all seed components, the mode of action of *AP2* remains unclear.

5.2
Hormonal Physiology of *Arabidopsis* Endosperm

Endosperm is known as a source of phytohormones (Lopes and Larkins 1993). Cytokinins were identified in maize endosperm (Yang et al. 2002). Coconut liquid syncytial endosperm is used to provide phytohormones to in vitro cell cultures (Ge et al. 2005). However, little is known about the role of phytohormones in *Arabidopsis* endosperm development. Specific expression of cytokinin-biosynthetic genes has been detected in the endosperm posterior pole suggesting a localized production of cytokinin in the cyst (Miyawaki et al. 2004). The triple mutant for the cytokinin receptors AHK2, AHK3 and CRE/AHK4 generate seeds that are 30% larger than the wild type (Riefler et al. 2006). The cellular mechanism responsible for this increase is unknown.

Functional analysis of abscisic acid (ABA)-biosynthetic genes showed that in parallel to the embryo the endosperm contributes to the production of ABA that is involved in the induction of seed dormancy (Lefebvre et al. 2006; Penfield et al. 2006). This function is likely linked to the ABA dependent pathway affecting dormancy and causing vivipary in Maize (Hattori et al. 1992).

It is likely that other classical phytohormones play a role during endosperm development in *Arabidopsis* since auxin plays a role in the control of cell division in maize endosperm (Lur and Setter 1993).

5.3
Seed Morphogenesis

Gross seed morphology is prefigured in part by ovule morphogenesis. An initial antero-posterior axis may be established in part by the asymmetrical expression of the homeobox gene *WUSCHEL* in the distal nucellus of the ovule. WUSCHEL signals the initiation of integument development from the chalaza (Gross-Hardt et al. 2002). In *Arabidopsis* as in many other species subsequent ovule development is asymmetrical with a more predominant growth and size in the abaxial integument compared to the adaxial integument, which develops close to the chalaza. The adaxial/abaxial asymmetry results from abaxial localization of the expression of the YABBI transcription factor INNER NO OUTER (INO), which is directed by the zinc-finger transcription factors SUPERMAN (Meister et al. 2002), ABERRANT TESTA GLAGRA (KANADI4) (McAbee et al. 2006) and NOZZLE/SPOROCYTELESS (Sieber et al. 2004). This early unequal development of the seed integuments is paralleled by accentuated cell elongation in the abaxial integument during the endosperm syncytial phase. The adaxial integument hardly expands by cell elongation during seed development. The asymmetrical elongation between adaxial and abaxial integuments is also paralleled with unequal cell division activity in the adjacent seed integuments (F. Berger, unpublished observations). *Superman* and *ats* mutant ovules are symmetrical and produce nearly spherical shaped endosperm and seeds. This effect rather suggests that in the wild-type seed, mechanical constraints from the endosperm accompany cell elongation in the integument and enhance the initial asymmetry between adaxial and abaxial integuments sides.

6
Conclusions

Experimental work on *Arabidopsis* endosperm during the past ten years has uncovered fascinating aspects of endosperm development and shown that it represents a valuable model for basic research. As a result of its small size, access to the transcriptome of *Arabidopsis* endosperm has been hampered. Enhancer trap strategies (Drea et al. 2005; Ingouff et al. 2005b; Stangeland et al. 2003; Vielle-Calzada et al. 2000), and isolation of mutants with endosperm-specific defects having been the main means for identification of genes expressed in endosperm. However, it is possible to obtain endosperm enriched fraction at the mature stage (Penfield et al. 2006). New technological developments such as laser-assisted micro-capture (Day et al. 2005) and mRNA tagging approaches (Roy et al. 2002) should grant access to pure fractions of endosperm tissue.

The origin of patterning and control of differentiation of the various domains along the antero-posterior axis are still not understood. It is possible that the cues directing female gametophyte polarity contribute to the definition of the antero-posterior endosperm axis and the embryo apical-basal polarity (Jürgens 2001). In such a case forward genetic analysis will lead to gametophytic lethality and other strategies should be used to uncover potential maternal effects on endosperm polarity.

Besides this fundamental aspect of research, the demonstration of the central role of endosperm in determination of final seed size and its interaction with other seed components has shed light on new pathways to modulate seed development of interest to breeders. The gradual shift between model species used to study endosperm has also played an essential role in the development of this research field and it is forecasted that studies on endosperm will develop largely using rice in future years.

References

Ach RA, Taranto P, Gruissem W (1997) A conserved family of WD-40 proteins binds to the retinoblastoma protein in both plants and animals. Plant Cell 9:1595–1606

Adams S, Vinkenoog R, Spielman M, Dickinson HG, Scott RJ (2000) Parent-of-origin effects on seed development in *Arabidopsis thaliana* require DNA methylation. Development 127:2493–2502

Baud S, Wuilleme S, Lemoine R, Kronenberger J, Caboche M, Lepiniec L, Rochat C (2005) The AtSUC5 sucrose transporter specifically expressed in the endosperm is involved in early seed development in *Arabidopsis*. Plant J 43:824–836

Berger F (1999) Endosperm development. Curr Opin Plant Biol 2:28–32

Berger F (2003) Endosperm: the crossroad of seed development. Curr Opin Plant Biol 6:42–50

Berger F (2004) Plant sciences. Imprinting–a green variation. Science 303:483–485

Berger F, Grini PE, Schnittger A (2006) Endosperm: an integrator of seed growth and development. Curr Opin Plant Biol 9:664–670

Boisnard-Lorig C, Colon-Carmona A, Bauch M, Hodge S, Doerner P, Bancharel E, Dumas C, Haseloff J, Berger F (2001) Dynamic analyses of the expression of the HISTONE::YFP fusion protein in *Arabidopsis* show that syncytial endosperm is divided in mitotic domains. Plant Cell 13:495–509

Brown RC, Lemmon BE, Nguyen H, Olsen O-A (1999) Development of endosperm in *Arabidopsis thaliana*. Sex Plant Reprod 12:32–42

Canales C, Bhatt AM, Scott R, Dickinson H (2002) EXS, a putative LRR receptor kinase, regulates male germline cell number and tapetal identity and promotes seed development in *Arabidopsis*. Curr Biol 12:1718–1727

Chanvivattana Y, Bishopp A, Schubert D, Stock C, Moon YH, Sung ZR, Goodrich J (2004) Interaction of Polycomb-group proteins controlling flowering in *Arabidopsis*. Development 131:5263–5276

Chaudhury AM, Berger F (2001) Maternal control of seed development. Semin Cell Dev Biol 12:381–386

Chaudhury AM, Ming L, Miller C, Craig S, Dennis ES, Peacock WJ (1997) Fertilization-independent seed development in *Arabidopsis thaliana*. Proc Natl Acad Sci USA 94:4223–4228

Choi Y, Gehring M, Johnson L, Hannon M, Harada JJ, Goldberg RB, Jacobsen SE, Fischer RL (2002) DEMETER, a DNA Glycosylase Domain Protein, Is Required for Endosperm Gene Imprinting and Seed Viability in *Arabidopsis*. Cell 110:33–42

Cross JC (2005) How to make a placenta: mechanisms of trophoblast cell differentiation in mice–a review. Placenta 26(A):3-9

Day RC, Grossniklaus U, Macknight RC (2005) Be more specific! Laser-assisted microdissection of plant cells. Trends Plant Sci 10:397–406

Debeaujon I, Nesi N, Perez P, Devic M, Grandjean O, Caboche M, Lepiniec L (2003) Proanthocyanidin-accumulating cells in *Arabidopsis* testa: regulation of differentiation and role in seed development. Plant Cell 15:2514–2531

Dickinson H (2003) Plant cell cycle: cellularisation of the endosperm needs spatzle. Curr Biol 13:146–148

Drea S, Leader DJ, Arnold BC, Shaw P, Dolan L, Doonan JH (2005) Systematic spatial analysis of gene expression during wheat caryopsis development. Plant Cell 17:2172–2185

Dubreucq B, Berger N, Vincent E, Boisson M, Pelletier G, Caboche M, Lepiniec L (2000) The *Arabidopsis* AtEPR1 extensin-like gene is specifically expressed in endosperm during seed germination. Plant J 23:643–652

Ebel C, Mariconti L, Gruissem W (2004) Plant retinoblastoma homologues control nuclear proliferation in the female gametophyte. Nature 429:776–780

Edgar BA, Kiehle CP, Schubiger G (1986) Cell cycle control by the nucleo-cytoplasmic ratio in early Drosophila development. Cell 44:365–372

Edgar BA, O'Farrell PH (1989) Genetic control of cell division patterns in the Drosophila embryo. Cell 57:177–187

Favery B, Chelysheva LA, Lebris M, Jammes F, Marmagne A, De Almeida-Engler J, Lecomte P, Vaury C, Arkowitz RA, Abad P (2004) *Arabidopsis* formin AtFH6 is a plasma membrane-associated protein upregulated in giant cells induced by parasitic nematodes. Plant Cell 16:2529–2540

Friedman W (1998) The evolution of double fertilization and endosperm: an "historical" perspective. Sex Plant Reprod 11:1–16

Garcia D, Fitz Gerald JN, Berger F (2005) Maternal Control of Integument Cell Elongation and Zygotic Control of Endosperm Growth Are Coordinated to Determine Seed Size in *Arabidopsis*. Plant Cell 17:52–60

Garcia D, Saingery V, Chambrier P, Mayer U, Jürgens G, Berger F (2003) *Arabidopsis* haiku mutants reveal new controls of seed size by endosperm. Plant Phys 131:1661–1670

Ge L, Yong JW, Goh NK, Chia LS, Tan SN, Ong ES (2005) Identification of kinetin and kinetin riboside in coconut (Cocos nucifera L.) water using a combined approach of liquid chromatography-tandem mass spectrometry, high performance liquid chromatography and capillary electrophoresis. J Chromatogr B Analyt Technol Biomed Life Sci 829:26–34

Gehring M, Choi Y, Fischer RL (2004) Imprinting and seed development. Plant Cell 16(Suppl):203–213

Glotzer M (2003) Cytokinesis: progress on all fronts. Curr Opin Cell Biol 15:684–90

Gross-Hardt R, Lenhard M, Laux T (2002) WUSCHEL signaling functions in interregional communication during *Arabidopsis* ovule development. Genes Dev 16:1129–1138

Grossniklaus U, Vielle-Calzada JP, Hoeppner MA, Gagliano WB (1998) Maternal control of embryogenesis by *MEDEA*, a polycomb group gene in *Arabidopsis*. Science 280:446–450

Guitton AE, Berger F (2005) Loss of function of MULTICOPY SUPPRESSOR OF IRA 1 produces nonviable parthenogenetic embryos in *Arabidopsis*. Curr Biol 15:750–754

Guitton AE, Page DR, Chambrier P, Lionnet C, Faure JE, Grossniklaus U, Berger F (2004) Identification of new members of Fertilisation Independent Seed Polycomb Group pathway involved in the control of seed development in *Arabidopsis thaliana*. Development 131:2971–2981

Hattori T, Vasil V, Rosenkrans L, Hannah LC, McCarty DR, Vasil IK (1992) The Viviparous-1 gene and abscisic acid activate the C1 regulatory gene for anthocyanin biosynthesis during seed maturation in maize. Genes Dev 6:609–618

Haughn G, Chaudhury A (2005) Genetic analysis of seed coat development in *Arabidopsis*. Trends Plant Sci 10:472–477

Hennig L, Bouveret R, Gruissem W (2005) MSI1-like proteins: an escort service for chromatin assembly and remodeling complexes. Trends Cell Biol 15:295–302

Hennig L, Taranto P, Walser M, Schonrock N, Gruissem W (2003) *Arabidopsis* MSI1 is required for epigenetic maintenance of reproductive development. Development 130:2555–2565

Ingouff M, Fitz Gerald JN, Guerin C, Robert H, Sørensen MB, Van Damme D, Geelen D, Blanchoin L, Berger F (2005a) Plant formin AtFH5 is an evolutionarily conserved actin nucleator involved in cytokinesis. Nat Cell Biol 7:374–380

Ingouff M, Haseloff J, Berger F (2005b) Polycomb group genes control developmental timing of endosperm. Plant J 42:663–674

Ingouff M, Jullien J, Berger F (2006) The female gametophyte and the endosperm control cell proliferation and differentiation of the seed coat in *Arabidopsis*. Plant Cell 18:3491–3501

Jofuku KD, Omidyar PK, Gee Z, Okamuro JK (2005) Control of seed mass and seed yield by the floral homeotic gene APETALA2. Proc Natl Acad Sci USA 102:3117–3122

Jullien PE, Katz A, Oliva M, Ohad N, Berger F (2006a) Polycomb Group Complexes Self-Regulate Imprinting of the Polycomb Group Gene MEDEA in *Arabidopsis*. Curr Biol 16:486–942

Jullien PE, Kinoshita T, Ohad N, Berger F (2006b) Maintenance of DNA Methylation during the *Arabidopsis* Life Cycle Is Essential for Parental Imprinting. Plant Cell 18:1360–1372

Jürgens G (2001) Apical-basal pattern formation in *Arabidopsis* embryogenesis. Embo J 20:3609–3616

Jürgens GMU (1994) *Arabidopsis*. In: Bard JBL (ed) Embryos, Colour Atlas of Development. Wolfe Publishing, London, pp 7–21

Katz A, Oliva M, Mosquna A, Hakim O, Ohad N (2004) FIE and CURLY LEAF polycomb proteins interact in the regulation of homeobox gene expression during sporophyte development. Plant J 37:707–719

Kiyosue T, Ohad N, Yadegari R, Hannon M, Dinneny J, Wells D, Katz A, Margossian L, Harada JJ, Goldberg RB, Fischer RL (1999) Control of fertilization-independent endosperm development by the MEDEA polycomb gene in *Arabidopsis*. Proc Natl Acad Sci USA 96:4186–4191

Köhler C, Grossniklaus U (2005) Seed development and genomic imprinting in plants. Prog Mol Subcell Biol 38:237–262

Köhler C, Hennig L, Bouveret R, Gheyselinck J, Grossniklaus U, Gruissem W (2003a) *Arabidopsis* MSI1 is a component of the MEA/FIE Polycomb group complex and required for seed development. Embo J 22:4804–4814

Köhler C, Hennig L, Spillane C, Pien S, Gruissem W, Grossniklaus U (2003b) The Polycomb-group protein MEDEA regulates seed development by controlling expression of the MADS-box gene PHERES1. Gene Dev 17:1540–1553

Lefebvre V, North H, Frey A, Sotta B, Seo M, Okamoto M, Nambara E, Marion-Poll A (2006) Functional analysis of *Arabidopsis* NCED6 and NCED9 genes indicates that ABA synthesized in the endosperm is involved in the induction of seed dormancy. Plant J 45:309–319

Lepiniec L, Debeaujon I, Routaboul JM, Baudry A, Pourcel L, Nesi N, Caboche M (2006) Genetics and Biochemistry of Seed Flavonoids. Annu Rev Plant Biol 57:405–430

Lid SE, Olsen L, Nestestog R, Aukerman M, Brown RC, Lemmon B, Mucha M, Opsahl-Sorteberg HG, Olsen OA (2005) Mutation in the Arabidopisis *thaliana* DEK1 calpain gene perturbs endosperm and embryo development while over-expression affects organ development globally. Planta 221:339–351

Liu PP, Koizuka N, Homrichhausen TM, Hewitt JR, Martin RC, Nonogaki H (2005) Large-scale screening of *Arabidopsis* enhancer-trap lines for seed germination-associated genes. Plant J 41:936–944

Lopes MA, Larkins BA (1993) Endosperm origin, development, and function. Plant Cell 5:1383–1399

Lukowitz W, Mayer U, Jürgens G (1996) Cytokinesis in the *Arabidopsis* embryo involves the syntaxin-related KNOLLE gene product. Cell 84:61–71

Luo M, Bilodeau P, Dennis ES, Peacock WJ, Chaudhury A (2000) Expression and parent-of-origin effects for *FIS2*, *MEA*, and *FIE* in the endosperm and embryo of developing *Arabidopsis* seeds. Proc Natl Acad Sci USA 97:10637–10642

Luo M, Dennis ES, Berger F, Peacock WJ, Chaudhury A (2005) MINISEED3 (MINI3), a WRKY family gene, and HAIKU2 (IKU2), a leucine-rich repeat (LRR) KINASE gene, are regulators of seed size in *Arabidopsis*. Proc Natl Acad Sci USA 102:17531–17536

Lur HS, Setter TL (1993) Role of Auxin in Maize Endosperm Development (Timing of Nuclear DNA Endoreduplication, Zein Expression, and Cytokinin). Plant Physiol 103:273–280

Maheshwari P (1950) An Introduction to the Embryology of Angiosperms. McGraw-Hill, New-York

Mansfield SG, Briarty LG (1990)a Endosperm cellularization in *Arabidopsis thaliana*. Arab Inf Serv 27:65–72

Mansfield SG, Briarty LG (1990)b Development of the free-nuclear endosperm in *Arabidopsis thaliana*. Arab Inf Serv 27:53–64

Mayer U, Herzog U, Berger F, Inze D, Jürgens G (1999) Mutations in the pilz group genes disrupt the microtubule cytoskeleton and uncouple cell cycle progression from cell division in *Arabidopsis* embryo and endosperm. Eur J Cell Biol 78:100–108

McAbee JM, Hill TA, Skinner DJ, Izhaki A, Hauser BA, Meister RJ, Venugopala Reddy G, Meyerowitz EM, Bowman JL, Gasser CS (2006) ABERRANT TESTA SHAPE encodes a KANADI family member, linking polarity determination to separation and growth of *Arabidopsis* ovule integuments. Plant J 46:522–531

Meister RJ, Kotow LM, Gasser CS (2002) SUPERMAN attenuates positive INNER NO OUTER autoregulation to maintain polar development of *Arabidopsis* ovule outer integuments. Development 129:4281–4289

Miyawaki K, Matsumoto-Kitano M, Kakimoto T (2004) Expression of cytokinin biosynthetic isopentenyltransferase genes in *Arabidopsis*: tissue specificity and regulation by auxin, cytokinin, and nitrate. Plant J 37:128–138

Nguyen H, Brown RC, Lemmon BE (2002) Cytoskeletal organization of the micropylar endosperm in Coronopus didymus L. (Brassicaceae). Protoplasma 219:210–220

Ohto MA, Fischer RL, Goldberg RB, Nakamura K, Harada JJ (2005) Control of seed mass by APETALA2. Proc Natl Acad Sci USA 102:3123–3128

Olsen OA (2001) Endosperm development: Cellularization and cell fate specification. Ann Rev Plant Physiol Plant Mol Biol 52:241–267

Olsen OA (2004) Nuclear endosperm development in cereals and *Arabidopsis thaliana*. Plant Cell 16(Suppl):214-227

Otegui M, Staehelin LA (2000a) Cytokinesis in flowering plants: more than one way to divide a cell. Curr Opin Plant Biol 3:493–502

Otegui M, Staehelin LA (2000b) Syncytial-type cell plates: a novel kind of cell plate involved in endosperm cellularization of *Arabidopsis*. Plant Cell 12:933–947

Otegui MS, Mastronarde DN, Kang BH, Bednarek SY, Staehelin LA (2001) Three-dimensional analysis of syncytial-type cell plates during endosperm cellularization visualized by high resolution electron tomography. Plant Cell 13:2033–2051

Penfield S, Li Y, Gilday AD, Graham S, Graham IA (2006) Arabidopsis ABA INSENSITIVE 4 regulates lipid mobilization in the embryo and reveals repression of seed germination by the endosperm. Plant Cell 18:1887–1899

Raz V, Bergervoet JH, Koornneef M (2001) Sequential steps for developmental arrest in *Arabidopsis* seeds. Development 128:243–252

Riefler M, Novak O, Strnad M, Schmulling T (2006) *Arabidopsis* cytokinin receptor mutants reveal functions in shoot growth, leaf senescence, seed size, germination, root development, and cytokinin metabolism. Plant Cell 18:40–54

Rotman N, Durbarry A, Wardle A, Yang WC, Chaboud A, Faure JE, Berger F, Twell D (2005) A Novel Class of MYB Factors Controls Sperm-Cell Formation in Plants. Curr Biol 15:244–248

Roy PJ, Stuart JM, Lund J, Kim SK (2002) Chromosomal clustering of muscle-expressed genes in Caenorhabditis elegans. Nature 418:975–979

Schruff MC, Spielman M, Tiwari S, Adams S, Fenby N, Scott RJ (2006) The AUXIN RESPONSE FACTOR 2 gene of *Arabidopsis* links auxin signalling, cell division, and the size of seeds and other organs. Development 133:251–261

Scott RJ, Spielman M, Bailey J, Dickinson HG (1998) Parent-of-origin effects on seed development in *Arabidopsis thaliana*. Development 125:3329–41

Sieber P, Gheyselinck J, Gross-Hardt R, Laux T, Grossniklaus U, Schneitz K (2004) Pattern formation during early ovule development in *Arabidopsis thaliana*. Dev Biol 273:321–334

Sørensen MB, Chaudhury AM, Robert H, Bancharel E, Berger F (2001) Polycomb group genes control pattern formation in plant seed. Curr Biol 11:277–281

Sørensen MB, Mayer U, Lukowitz W, Robert H, Chambrier P, Jürgens G, Somerville C, Lepiniec L, Berger F (2002) Cellularisation in the endosperm of *Arabidopsis thaliana* is coupled to mitosis and shares multiple components with cytokinesis. Development 129:5567–5576

Spielman M, Vinkenoog R, Dickinson HG, Scott RJ (2001) The epigenetic basis of gender in flowering plants and mammals. Trends Genet 17:705–711

Stangeland B, Salehian Z, Aalen R, Mandal A, Olsen OA (2003) Isolation of GUS marker lines for genes expressed in *Arabidopsis* endosperm, embryo and maternal tissues. J Exp Bot 54:279–290

Steinborn K, Maulbetsch C, Priester B, Trautmann S, Pacher T, Geiges B, Kuttner F, Lepiniec L, Stierhof YD, Schwarz H, Jürgens G, Mayer U (2002) The *Arabidopsis* PILZ group genes encode tubulin-folding cofactor orthologs required for cell division but not cell growth. Genes Dev 16:959–971

Vielle-Calzada JP, Baskar R, Grossniklaus U (2000) Delayed activation of the paternal genome during seed development. Nature 404:91–94

Yadegari R, Kinoshita T, Lotan O, Cohen G, Katz A, Choi Y, Nakashima K, Harada JJ, Goldberg RB, Fischer RL, Ohad N (2000) Mutations in the *FIE* and *MEA* genes that encode interacting polycomb proteins cause parent-of-origin effects on seed development by distinct mechanisms. Plant Cell 12:2367–2381

Yang J, Zhang J, Huang Z, Wang Z, Zhu Q, Liu L (2002) Correlation of cytokinin levels in the endosperms and roots with cell number and cell division activity during endosperm development in rice. Ann Bot (Lond) 90:369–377

Plant Cell Monogr (8)
O.-A. Olsen: Endosperm
DOI 10.1007/7089_2007_118/Published online: 31 March 2007

Endosperm Cell and Organ Culture

D. Gruis[1] (✉) · H. Guo[1] · Q. Tian[1,2] · O.-A. Olsen[1,2]

[1]Pioneer Hi-Bred International, A DuPont Company, Johnston, IA 50131, USA
fred.gruis@pioneer.com

[2]*Present address:*
Monsanto Company, 700 Chesterfield Pkwy West, Chesterfield, MO 63017, USA

Abstract This chapter describes efforts to culture maize endosperm to enable direct developmental and metabolic studies of the cereal endosperm organ without the effects of the maternal plant. First, we describe production of endosperm in culture through either central cell fertilization in vitro or through isolation of embryo sacs following fertilization in vivo. These techniques have been used to culture cereal endosperm but appear to rely on use of maternal tissue in the culture to provide growth factors to stimulate endosperm development. Second, we summarize efforts aimed at the establishment of immortalized cultures of maize endosperm. Evolution of the basic protocol first reported almost 60 years ago resulted in in vitro endosperm cultures shown to be comprised of a mixture of cell types including those retaining endosperm-like characteristics. These cultures lacked organ identity, with endosperm-like cell types intermingled with undifferentiated callus-like cells. These cultures were successfully used in studies of starch as well as storage protein accumulation. Finally, we describe a recently reported system for in vitro culture of maize endosperm that preserves organ identity. These cultures have provided conclusive evidence that starchy endosperm and aleurone cell identity is position dependent without the need of further signaling from maternal tissues. Furthermore, this comparatively simple culture technique appears broadly applicable to several maize genotypes allowing for use of the large mutant collection generated in maize. With the real-time continuous accessibility of a developing cereal endosperm to perturbation and observation, it is hoped that this culture technique will be broadly useful for researchers to answer several previously intractable questions.

1 Introduction

As the cereal endosperm is of vital importance to our food chain, efforts to understand and improve upon the biochemical mechanisms underlying its development and storage function(s) will continue indefinitely. The diversity of endosperm composition and type of grains used in agriculture today is likely reflective of the demands imposed by the domesticator(s) more than the demands of the plant itself, and speaks to the potential for continued specialization of cereal grains. As the technology to directly amend a plant's genetic makeup comes of age, the prospect of speeding up this process will become reality. Two questions become relevant: (1) how can the endosperm be changed to improve upon a desired trait and (2) what are the gene(s) that will result in the desired change? As several possible answers likely exist for each of these questions, a means by which to efficiently test hypotheses that is fast, flexible, cost effective, and permits direct studies of endosperm without the confounding effect of maternal tissues is desirable. In addition, a system for creating transgenic cereal endosperm for rapid gene testing without the need for stably transformed plants would be highly attractive. In this chapter we describe efforts to understand early cereal endosperm development in in vitro fertilized and in vitro grown central cells as well as previous and current efforts to culture, study, and genetically alter developing endosperm in vitro.

2 Endosperm Cultures Derived from Central Cell Fertilization In Vitro, and Embryo Sac Isolation Post-Fertilization

The process of angiosperm double fertilization (Nawaschin 1898; Guignard 1901) raises numerous questions about the evolutionary origin of the endosperm as well as the role of maternal signaling in early endosperm development (see chapter by Brown and Lemmon, in this volume). One hundred years after its discovery, Kranz and colleagues were able to re-create the second event in double fertilization that gives rise to the endosperm by fusing isolated maize male gametes and central cells (Kranz 1998). In these experiments, protoplasts of central cells were electrofused with male gametes. Karyogamy within the central cell was observed after 1 h. Culturing was achieved by cocultivation with somatic barley protoplast feeder cells. Under these conditions, fertilized central cells developed into a syncytium after rapid nuclear divisions without cell wall formation, followed by cellularization around 5 days after pollination (DAP) (Kranz 1998). While still a syncytium, the fertilized central cell developed into a bipolar structure that consisted of an oblong part and a globular part. Of the two, the globular part developed more rapidly than the oblong part. The characteristic cellularization process of the in planta endosperm coenocyte (see chapter by Brown and Lemmon, in this volume) appears to also occur in vitro (Kranz 1998). The globular part contained small cells with dense cytoplasm, whereas the oblong part contained fewer nuclei. Detailed structural analyses of the microtubular arrays in these endosperms have not been published, and it is therefore difficult to assess the extent to which the developmental pathway of in vitro fertilized endosperm reflects the in planta situation. Interestingly, development of in vitro fertilized central cells proceeds in the absence of maternal tissues. It is currently unclear what role, if any, the feeder cells used to support the growth of the in vitro endosperm play in the growth and differentiation of these structures. However, it is notable that fertilization and at least partial development of an endosperm occurs in the absence of the endogenous maternal tissues. While this observation was an exciting result, the technical hurdles of easily and efficiently generating in vitro endosperm using this technique appear high.

Another approach to obtaining cultured endosperm cells, which avoids the technical hurdles associated with central cell fertilization in vitro, is through isolation and culture of the embryo sac shortly after fertilization in vivo. In maize this has been most recently accomplished through isolation of ovule sections containing the embryo sac 1 DAP and culture on a modified Murashige and Skoog (MS) medium containing 6–15% sucrose (Laurie et al. 1999). The authors used these cultures to generate viable embryos and mature plants without somatic embryogenesis, and describe the production of an endosperm from these embryo sacs in culture as well. However, similar to the technique described above, maternal tissue is included in the culture and it is indicated that the adjacent nucellar tissue may be supplying growth factors essential for stimulating development. Additionally, the remaining maternal tissue appears to continue to at least partially limit direct accessibility of the endosperm to direct manipulation and examination during culture.

3 Maize Endosperm In Vitro Cultures that Retain some Degree of Cell Differentiation

The monitoring and visualization of cereal endosperm development is hampered by its localization within maternal tissues, and furthermore this localization makes efforts of direct manipulation difficult as well. Maize is an example of an economically important cereal which has had a depth of endosperm research and provides a good model of nuclear endosperm function. Therefore, attempts to culture the endosperm to examine aspects of assimilate accumulation have been ongoing since at least 1947 (LaRue 1947). In these early efforts, the goal was to develop immortal cultures. Efforts examining different media, endosperm isolation time (days after pollination), growth conditions, and genotypes of maize culminated in a successful methodology to continuously culture cells with some degree of endosperm traits (LaRue 1949; Straus 1954, 1959, 1960). This methodology utilized isolation of endosperms from kernels 8–18 DAP, most optimally 12 DAP with growth on an agar medium containing basal salts, a nitrogen source (optimally 2 g/L asparagine and 2 g/L yeast hydrolyzate), and sucrose (8% w/v). No growth substances or hormones were required. Subculturing was accomplished by transfer of the non-necrotic buds to fresh media at approximately 20-day intervals. One disadvantage was that only one genotype (Black Mexican Sweet, BMS) was described as being amenable to this culturing methodology. Immortal cultures of starchy and waxy genotypes of maize could not be established.

Based on this basic protocol, Shannon et al. continued to test a wide variety of genotypes and identified inbreds A636 and R168 as capable of producing viable continuously cultured endosperm for the purpose of examining starch accumulation (Shannon 1973). Carbon source requirements were further examined and it was confirmed that glucose, fructose, and combinations of glucose and fructose would all support culture growth and starch accumulation. While the cultured endosperm was capable of accumulating starch, the quantity was discordant with that of mature endosperm in planta and reflected an amount more characteristic of younger endosperms (Chu and Shannon 1975). Examining culture growth in detail, the cyclic nature of the culture growth and potential dependence upon carbon source/concentration in the media was first alluded to. Following subculture mitotic division rates of > 5% were observed from days 2 to 8 and decreased to half that rate by day 13. These rates could be altered by further supplementation of the carbon source, such that increases in concentration early in the culture cycle would retard growth while supplementation later in the cycle would promote both growth and starch accumulation. It was also noted that other experimenters could obtain higher amounts of amylose in endosperm explants harvested at 14 DAP than they could obtain in the continuous cultures, suggesting that an in vivo "switch" had to be activated prior to culturing the endosperm to result in appropriate differentiation to fully activate the necessary starch synthesis machinery (Chu and Shannon 1975).

Storage proteins (the significant zeins) and protein bodies were also observed in maize endosperm cultures. Endosperm explants harvested prior to protein body formation in planta at 10 DAP would continue to develop and accumulate protein bodies in cultures, indicating that continued maturation of the endosperm was occurring in vitro. Additionally, long-term continuous cultures (21 months) were also found to contain zeins and protein bodies (Shimamoto et al. 1983; Lyznick 1989). The existence of developmental gradients within the endosperm was used to explain how the long-term in vitro cultures could continue both dividing and producing zeins. Cultured endosperm explants were later used to examine the relationship of maternal and zygotic components of protein deposition in maize using the Illinois high protein and low protein materials, demonstrating the maternal aspects of controlling protein accumulation in the endosperm (Balconi 1991).

In addition to the compositional analysis described, cellular organization of the long-term continuous cultured endosperm was also examined (Felker 1987). A variety of cell types were identified in these cultures, some containing a high proportion of protein and lipid bodies characteristic of aleurone and subaleurone cells, others starch granules, while others exhibited cell wall ingrowths characteristic of basal endosperm cells. Lastly, an abundance of highly vacuolated cells was also observed. No spatially organized endosperm structure was observed; rather, cell clusters of a particular cell type described above were observed. These results may be interpreted to indicate that endosperm cell specification was no longer operating in the positionally dependent fashion described in planta (Becraft et al. 2000; see chapter by Becraft, in this volume).

4
A Maize Endosperm Organ Culture System that Retains Cell and Organ Identity Without Maternal Tissue Presence

The advent of modern biotechnological tools has enhanced the ability to access real-time cell differentiation and endosperm organ development through use of cell-type specific expression of fluorescent reporters. Recently, the conclusion that endosperm cell differentiation and organization fails to be maintained in the cultures described above (Felker 1987) was confirmed using a transgenic maize line expressing yellow fluorescent protein (YFP) in aleurone cells. In this line, aleurone cells were labeled yellow from YFP expression driven by the barley *Ltp2* promoter (Gruis et al. 2006). In contrast to in planta endosperm, which has a single layer of YFP positive aleurone cells on the surface (see chapter by Becraft, in this volume), the endosperm cultures contained YFP positive cells scattered internally in callus-like structures (Fig. 1a).

To further facilitate the study of cell and organ identity in in vitro grown maize endosperm in experiments aimed at identifying improved culture conditions, a transgenic maize line was developed in which aleurone cells are labeled with YFP as described above, starchy endosperm cells with cyan fluorescence protein (CFP) driven by the 27-kDa gamma zein promoter, and transfer cells with red fluorescent protein (RFP) driven by the END1 promoter (Gruis et al. 2006). When cultivating manually isolated 6-DAP embryo sacs from this marker line on a modified MS medium containing 15% w/v sucrose (US Patent Pub. No. US 2006/0123518), the endosperm grew into a body of starchy endosperm cells that accumulated starch granules (Fig. 1b and c) covered by a single surface layer of YFP positive aleurone cells (Fig. 1d). The starchy endosperm fluoresced cyan from CFP driven by the 27-kDa γ-zein promoter (Gruis et al. 2006). These observations, as well as others described below, confirm that the culture conditions used support endosperm development in vitro that is similar to in planta development. The improved maize endosperm culture system should open new opportunities to study the mechanisms underlying endosperm development, as well as biosynthesis of components including starch and storage proteins.

4.1
The Rate of Mitotic Divisions in In Vitro Organ Cultured Endosperm is Similar to that in In Planta Endosperm

In addition to the study of endosperm cell identity and organization, cell division rates of in vitro grown endosperm were investigated (Gruis et al. 2006). In planta, the rate of division of peripheral endosperm cells peaks around 10 DAP, followed by a rapid decline to a low level (Mangelsdorf 1926). The results of this investigation showed a similar rate of cell division in endosperm

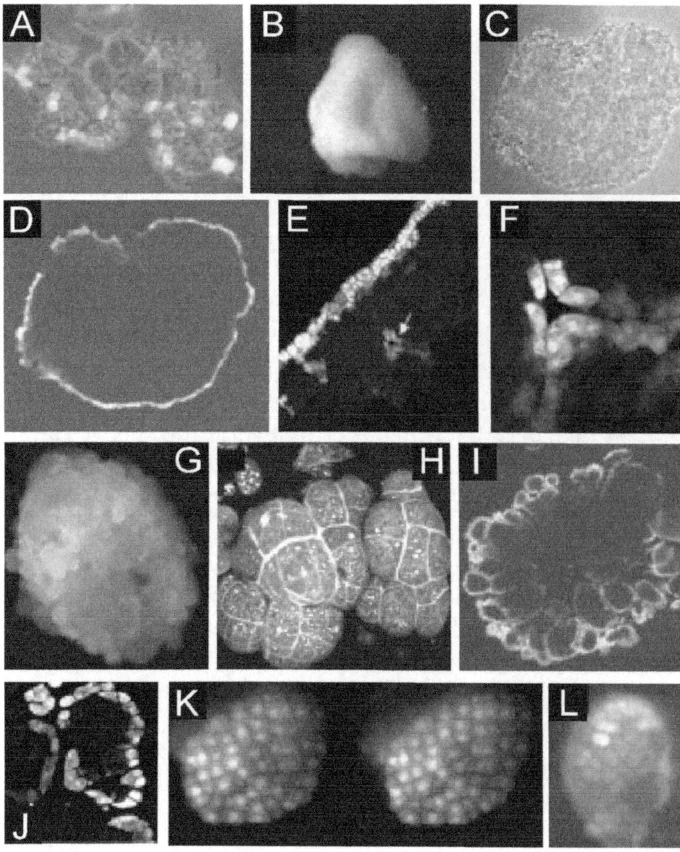

Fig. 1 **a** Endosperm suspension cultures grown according to Shannon (1973) contain a high proportion of nondifferentiated cells intermingled with cells that show features of differentiated *in planta* endosperm cells. The figure shows a transverse section of callus with an irregular pattern of YFP positive aleurone cells intermingled with nonfluorescing, highly vacuolated cells. **b** 6-DAP endosperm grown for 4 days in vitro. **c** Light micrograph of transverse section of 6-DAP endosperm grown for 10 days showing internal accumulation of starchy granules. **d** Epifluorescence micrograph of sections shown in c showing the surface layer of YFP positive aleurone cells. **e** Epifluorescence micrograph of transverse section of in vitro grown endosperm showing surface layer of aleurone cells and a group of internal YFP positive cells in the starchy endosperm (*arrow*). **f** Higher magnification of YFP positive cells identified by the *arrow* in **e**. **g** Surface of endosperm grown for 15 days in culture showing a high number of mini-endosperms on the surface. **h** Details of early bulges shown in confocal microscopy. **i** Epifluorescence microscopy of transverse sections of in vitro grown endosperm similar to **g**. Each globular structure on the surface represents one mini-endosperm. **j** Confocal microscopy image of one mini-endosperm with YFP positive cells on the surface. **k** 3-D representation of Ltp2:YFP positive sector after particle bombardment shown in side-by-side stereogram confocal image. **l** Complete mini-endosperm formed after bombardment shown in epifluorescence microscopy. Plates in this figure are from Gruis et al. 2006. Copyright American Society of Plant Biologists

cultivated in vitro, demonstrating that mitotic division control is endosperm autonomous, and not regulated by maternal tissue influences. These results support the conclusion that the in vitro endosperm system offers an opportunity to study molecular mechanisms of endosperm development in general and mitotic control in particular.

4.2 In Vitro Endosperm Offers the Possibility to Study Mechanisms of Aleurone and Starchy Endosperm Cell Fate Specification

Previous studies have suggested that endosperm form aleurone cells in response to positional information (i.e., a cell on the surface of the endosperm assumes the aleurone cell fate; see chapter by Becraft, in this volume). The study of in vitro grown endosperm described above conclusively demonstrated that aleurone cells develop in response to surface position (Gruis et al. 2006). The most striking demonstration of this phenomenon came from observations of YFP positive cells internally in the starchy endosperm (Fig. 1e). At closer inspection, it turned out that these YFP positive cells all surrounded voids that occasionally develop in in vitro endosperm and that they assumed aleurone cell identity in response to position (Fig. 1f). In addition to confirming the importance of positional signaling, these results demonstrated the developmental flexibility of endosperm cells, being able to convert between the starchy endosperm and the aleurone cell fates. Endosperm organ cultures thus also offer an opportunity for studies of the molecular mechanisms underlying aleurone and starchy endosperm cell fate specification and differentiation.

4.3 Basal Transfer Cells Fail to Develop in Endosperm In Vitro Cultures

In vivo, transfer cells are labeled with red fluorescence in the transgenic marker line described above. In contrast, red fluorescence is not observable in the basal part of in vitro grown endosperm, and neither are cells with typical transfer cell wall ingrowths (Gruis et al. 2006). This observation confirmed earlier indications that transfer cells require maternal tissue input to fully develop (see chapter by Royo et al., in this volume). Hopefully, in vitro endosperm may be useful to study maternal tissue influences, including the expression of genes suspected to represent maternal cell signals for transfer cell development, which has been shown to activate the promoter of several transfer-specific genes (see chapter by Royo et al., in this volume).

4.4 "Mini"-Endosperm of In Vitro Endosperm Cultures Represents a Transgenic Model for In Planta Endosperm

During the first 10 days in culture, the surface of the in vitro grown endosperm is relatively smooth (Fig. 1b). Subsequently, bulges (Fig. 1g–i) form on the surface that develop into spherical bodies with the same overall structure as in planta endosperm with a single surface layer of aleurone cells and an interior body of starchy endosperm cells. We termed these structures mini-endosperms (Fig. 1j). Interestingly, transformation experiments using particle bombardment suggested that single transformed cells have the ability to develop into a large sector of fluorescing cells (Fig. 1k) and also to a complete mini-endosperm (Fig. 1l). Provided that such transgenic mini-endosperms can be produced in sufficient numbers, they represent a potentially useful tool for gene testing in the endosperm, bypassing the need for the time-consuming process of developing stable transgenic corn lines.

5 Conclusions

The experiments in maize endosperm culture described herein present a methodology to culture endosperm without perturbing its development, and also speak to the autonomous nature of the developmental program of the endosperm storage tissue. This is of importance as it allows for manipulation and observation outside the confines of maternal tissues. Endosperm organ culture coupled with direct transgenic manipulation offers an unprecedented opportunity to test hypotheses as they relate to the development and therefore composition of endosperm. Ultimately it is hoped that further understanding of endosperm development will lead to further improvement of the use of this resource in food and industrial applications.

References

Balconi C, Rizzi E, Manzocchi L, Soave C, Motto M (1991) Analysis of in vivo and in vitro grown endosperms of high and low protein strains of maize. Plant Sci 73:1–9

Becraft PW (2007) Aleurone Cell Development (in this volume). Springer, Heidelberg

Becraft PW, Brown RC, Lemmon BE, Olsen O-A, Opsahl-Ferstad H-G (2000) Developmental biology of endosperm development. Kluwer, Dordrecht

Brown RC, Lemmon BE (2007) The Developmental Biology of Cereal Endosperm (in this volume). Springer, Heidelberg

Chu L, Shannon J (1975) In vitro cultures of maize endosperm—a model system for studying in vivo starch biosynthesis. Crop Sci 15:814–819

Felker FC (1987) Ultrastructure of maize endosperm suspension cultures. Am J Bot 74:1912–1920

Royo J, Gómez E, Hueros G (2007) Transfer Cells (in this volume). Springer, Heidelberg

Gruis DF, Guo H, Selinger D, Tian Q, Olsen O-A (2006) Surface position, and not signaling from surrounding maternal tissues, specifies aleurone epidermal cell fate in maize endosperm organ cultures. Plant Physiol 141:898–909

Guignard L (1901) La double fècondation dans le mais. J Bot 15:37–50

Kranz E, von Wiegen P, Quader H, Lorz H (1998) Endosperm development after fusion of isolated, single maize sperm and central cells in vitro. Plant Cell 10:511–524

LaRue CD (1947) Growth and regeneration of the endosperm of maize in culture. Am J Bot 34:585–586

LaRue CD (1949) Cultures of the endosperm of maize. Am J Bot 36:798

Laurie JD, Zhang G, McGann LE, Gordon Kamm WJ, Cass DD (1999) A novel technique for the partial isolation of maize embryo sacs and subsequent regeneration of plants. In Vitro Cell Dev Biol Plant 35:320–325

Lyznick LA, Tsai CY (1989) Protein synthesis in endosperm cell cultures of maize (*Zea mays* L.). Plant Sci 63:105–114

Mangelsdorf PC (1926) The genetics and morphology of some endosperm characters in corn. Conn Agric Exp Station Bull 279:513–614

Nawaschin S (1898) Resulte eine revison der Befruchtungsvorgange bei *Lilium martagon* und *Ritillaria tenella*. Bull Acad Imperial Sci (St Petersburg) 9:377–382

Shannon JC, Batey JW (1973) Inbred and hybrid effects on establishment of in vitro cultures of *Zea mays* L. endosperms. Crop Sci 15:491–493

Shimamoto K, Ackerman M, Dierks Ventling C (1983) Expression of zein in long-term endosperm cultures of maize (*Zea mays*). Plant Physiol 73:915–920

Straus J (1959) Anthrocyanin synthesis in corn endosperm tissue cultures. I. Identity of the pigments and general factors. Plant Physiol 34:536–541

Straus J (1960) Maize endosperm tissue grown in vitro III. Development of a synthetic medium. Am J Bot 47:641–647

Straus J, LaRue CD (1954) Maize endosperm tissue grown in vitro. I. Culture requirements. Am J Bot 41:687–694

Plant Cell Monogr (8)
O.-A. Olsen: Endosperm
DOI 10.1007/7089_2007_117/Published online: 15 March 2007

Anthocyanin Synthesis in Maize Aleurone Tissue

Karen C. Cone

Division of Biological Sciences, University of Missouri, Columbia, MO 65211, USA
conek@missouri.edu

Abstract Anthocyanin synthesis in maize has been the subject of intense genetic, biochemical, and molecular experimentation. These studies have identified the structural genes encoding enzymes and transporters of the biosynthetic pathway, as well as the key regulatory genes. The picture that emerges from these analyses is that pigmentation in the maize aleurone is orchestrated by a complex regulatory network involving interaction among tissue-specific, developmental and environmental factors.

1 Introduction

Anthocyanins are the red and purple flavonoid pigments made as products of secondary metabolism in plants. Anthocyanin synthesis has been studied extensively, through both biochemical and genetic analyses, and most of the enzymes and genes involved have been characterized. Moreover, because pigmentation levels reflect gene expression levels, the pathway has served as an outstanding model for studying gene regulation. Several recent reviews describing flavonoid and anthocyanin synthesis in different plant systems have been published (Grotewold 2006; Koes et al. 2005; Lepiniec et al. 2006; Springob et al. 2003). This chapter will focus on anthocyanin synthesis in maize aleurone (see also Becraft, in this volume), with emphasis on genetic analysis of the pathway and its regulation.

A diagram of the central anthocyanin biosynthetic pathway and the basic structure of an anthocyanin molecule are shown in Fig. 1. In maize kernels, the major anthocyanin constituents are pelargonidin and cyanidin (Abdel-Aal et al. 2006). The first committed step of the anthocyanin pathway is the condensation of three malonyl-CoA molecules with *p*-coumaroyl-CoA to produce a chalcone; this step is catalyzed by the enzyme chalcone synthase (CHS). The second step, isomerization of the chalcone to a flavanone, can occur without enzymatic activity, but production of the isomers that serve as exclusive substrates for downstream reactions occurs more efficiently through the action of chalcone-flavanone isomerase (CHI). Flavanones are converted to dihydroflavonols by hydroxylation of the carbon in position 3 of the C ring by the enzyme flavanone-3-hydroxylase (F3H). Both flavanones and dihydroflavonols can be hydroxylated on the 3′ position of the B ring by flavonoid

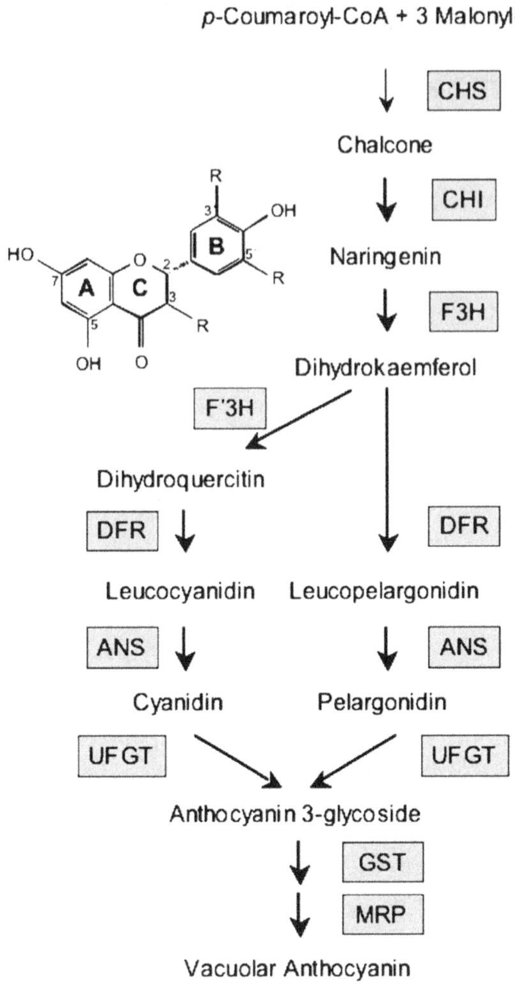

Fig. 1 Anthocyanin biosynthetic pathway in maize. Chemical structure shows basic skeleton of a flavonoid molecule. Enzyme abbreviations: *CHS* chalcone synthase, *CHI* chalcone isomerase, *F3H* flavanone 3-hydroxylase, *F3′H* flavonoid 3′-hydroxylase, *DFR* dihydroflavonol reductase, *ANS* anthocyanidin synthase, *UFGT* UDP flavonoid-3-*O*-glucosyl transferase, *GST* glutathione-*S*-transferase, *MRP* multidrug resistance-like transporter. Note that both flavanones (i.e., naringenin) and dihydroflavonols (i.e., dihydrokaemferol) are substrates for F3′H. Adapted from Koes et al. (2005)

3′-hydroxylase (F3′H). Dihydroflavonols are reduced to leucoanthocyanidins by dihydroflavonol 4-reductase (DFR), and the colorless leucoanthocyanidins are converted to colored anthocyanidins by anthocyanidin synthase (ANS), also known as leucoanthocyanidin dioxygenase/reductase (LDOX). Anthocyanidins are then glycosylated by UDP flavonoid 3-*O*-glucosyl transferase (UFGT), and the glycosylated molecules are transported to the vacuole by concerted action of a glutathione-*S*-transferase (GST) and a multidrug resistance-like transporter (MRP) localized in the vacuolar membrane.

2 Anthocyanin Pathway Proteins and Corresponding Structural Genes

The proteins responsible for anthocyanin biosynthesis and transport and the genes encoding those proteins are summerized in Table 1 and described in detail in the following sections. Gene–protein associations for most of the steps in the pathway were identified by biochemical analysis of mutants with altered pigmentation. However, for a few genes, no mutants exist; their roles in the pathway were deduced by extrapolating from information on flavonoid synthesis in other plants.

2.1 Chalcone Synthase (CHS)

The first enzyme in the pathway is encoded by duplicate genes, *colorless2* (*c2*) and *white pollen1* (*whp1*); Coe et al. 1988). The role of *c2* was established by assaying enzyme activity in colorless kernels. Aleurones from kernels that are

Table 1 Maize anthocyanin genes and their products

Gene	Gene product	Null alleles? [a]	Role [b]	Gene cloned? [a]
c2	Chalcone synthase	y	E	y
chi	Chalcone isomerase	n	E	y
pr1	Flavonoid 3′-hydroxylase	y	E	n
fht1 (F3H)	Flavanone 3-hydroxylase	n	E	y
a1	Dihydroflavonol 4-reductase	y	E	y
a2	Leucoanthocyanidin reductase/ anthocyanidin synthase	y	E	y
bz1	UDP glucose flavonol 3-*O*-glucosyl transferase	y	E	y
bz2	Glutathione-*S*-transferase	y	T	y
ZmMrp4	Multidrug resistance-like transporter	n	T	y
c1	R2 R3-MYB	y	R	y
pl1	R2 R3-MYB	y	R	y
r1	bHLH	y	R	y
b1	bHLH	y	R	y
in1	bHLH	y	R	y
pac1	WD40	y	R	y

[a] y yes, n no,
[b] E enzyme, T transport, R regulatory

homozygous for a recessive *c2* allele have low CHS activity, as do aleurones from pale or colorless kernels that carry the dominant inhibitory *C2-Idf* allele in heterozygous or homozygous state (Dooner 1983). The *c2* gene was cloned by transposon tagging using the element *Spm/En* (Wienand et al. 1986). Identity of the cloned gene was based on comparison of the sequence to that for the previously cloned CHS gene from parsley and demonstration that antibodies to parsley CHS could recognize the maize protein. RNA analysis showed that *c2* mRNA was present in pigmented tissues (kernels and tassels), but absent in tissues from *c2* mutants. Later, the duplicate *whp1* gene was cloned by virtue of its sequence similarity to *c2* (Franken et al. 1991). The presence of the dominant *Whp1* gene does not usually complement recessive *c2* in the aleurone. However, if *c2/c2; Whp1/whp1* kernels are also homozygous for a recessive *intensifier1* allele (*in1/in1*), anthocyanins are produced (Coe et al. 1988), and immunoblot analysis of proteins from such pigmented kernels with CHS-specific antiserum detects protein the size of CHS (Franken et al. 1991). This result indicates that under some circumstances, *whp1* expression can substitute for *c2* by producing CHS in the aleurone, thus leading to pigmented kernels.

2.2 Chalcone-Flavanone Isomerase (CHI)

A maize gene, *chalcone isomerase1* (*chi1*), for this enzyme was cloned using a PCR-based strategy employing primers that were complementary to conserved regions of CHI genes from petunia and Antirrhinum (Grotewold and Peterson 1994). RNA analysis showed that the gene was expressed in light-grown seedlings (which are pigmented) and in pigmented pericarps, but was not expressed in unpigmented pericarps. Expression in aleurone was not tested. Molecular analysis indicated that the maize genome contains more than one *chi* gene, mapping to three separate loci (Grotewold and Peterson 1994). It is not clear whether these additional genes are functional. If they are, then this genetic redundancy may explain why there are no mutations for this enzymatic step.

2.3 Flavanone 3-Hydroxylase (F3H)

The maize gene for F3H (*fht1*) was cloned by screening a cDNA library with a probe derived from the Antirrhinum F3H gene (Deboo et al. 1995). The gene appears to be single-copy in the maize genome. Its identity was confirmed by comparison of the deduced protein sequence to that for similar genes in a variety of other plants. RNA analysis showed that F3H expression correlated directly with pigmentation levels in kernels and flavonol levels in anthers. No mutants have been reported for this gene.

2.4 Flavonoid 3′-Hydroxylase (F3′H)

F3′H is encoded by the maize *red aleurone1* (*pr1*) gene. Assignment of *pr1* as the F3′H gene was made by measuring cyanidin and pelargonidin levels in aleurones and showing that higher cyanidin:pelargonidin ratios were directly correlated with increased doses of a dominant *Pr1* allele (Larson et al. 1986). Because these pigments differ in hydroxylation of the B ring, a modification that is associated with bluing of pigment color (Springob et al. 2003), this observation fits with the visual differences in kernel pigmentation, e.g., kernels with the dominant *Pr1* allele are purple, whereas homozygous recessive *pr1/pr1* kernels are red (Coe et al. 1988). Evidence for *Pr1* function was also seen in seedling and mature plant tissues. Detection of low-level F3′H activity in tissues of plants homozygous for recessive *pr1* suggested that there may be a duplicate gene (Larson et al. 1986). However, no genes for this enzyme have been cloned.

2.5 Dihydroflavonol 4-Reductase (DFR)

DFR is encoded by the *anthocyaninless1* (*a1*) gene, which was cloned from maize by transposon tagging using the *Spm/En* and *Mutator* transposable elements (O'Reilly et al. 1985). Two lines of evidence revealed that *a1* is the DFR structural gene: (i) NADPH-dependent reduction of dihydroquercitin to leucoanthocyanidin was detected in crude extracts of maize lines with functional *a1* alleles and not from lines where *a1* was inactive, and (ii) this same enzyme activity was demonstrated for the protein generated by in vitro translation of RNA from an *A1* cDNA (Reddy et al. 1987). A duplicate gene, which is detected on DNA gel blots hybridized with an *a1*-derived probe, was cloned and sequenced (Bernhardt et al. 1998). This gene, now called *anthocyaninless4* (*a4*), is predicted to code for a protein with 84% amino acid identity to the *a1*-encoded DFR. Although *a4* mRNA can be detected in many maize tissues (silks, leaves, leaf sheaths, and 20 DAP kernels), it is not clear whether an active enzyme is produced (Bernhardt et al. 1998). The fact that crosses of plants carrying a recessive *a1* allele produce anthocyaninless kernels in the ratios expected for a single-gene trait argues that *a4* does not functionally complement an *a1* defect in the aleurone.

2.6 Anthocyanidin Synthase (ANS)/Leucoanthocyanidin Dioxygenase (LDOX)

The enzymatic conversion of leucoanthocyanidins to anthocyanidins is controlled by the *anthocyaninless2* (*a2*) gene in maize, as demonstrated by intertissue complementation studies (Reddy and Coe 1962). The *a2* gene was cloned by transposon tagging using the elements *rcy* and *dSpm* (Menssen et al. 1990). Transient transformation of *a2* aleurones by particle bombardment with the *a2* cDNA led to production of pigmented cells (Menssen et al. 1990). The *a2* gene codes for a protein that shares sequence similarity to the family of 2-oxo-glutarate-dependent oxygenases like F3H (Menssen et al. 1990). Expression of the *a2* cDNA in bacterial cells produced an enzyme capable of catalyzing production of anthocyanidin by 2-oxoglutarate-dependent oxidation of leucoanthocyanidin (Nakajima et al. 2001)

2.7
UDP-Glucose Flavonoid 3-*O*-Glucosyl Transferase (UFGT)

Biochemical analysis of pigment accumulation and enzyme activity in the bronze aleurones of *bronze1* (*bz1*) mutant kernels established *bz1* as the structural gene for UFGT (Dooner and Nelson 1977a,b; Larson and Coe 1977). The *bz1* gene was cloned by transposon tagging using the element *Ac* (Fedoroff et al. 1984). Through DNA sequence analysis of different alleles, sequence polymorphisms were correlated with alterations in mRNA levels and/or UFGT activity (Furtek et al. 1988; Ralston et al. 1988).

2.8
Glutathione-S-Transferase (GST)

GST is encoded by the *bronze2* (*bz2*) gene, which was cloned by two groups, one using transposon tagging with the *Ds* element (Theres et al. 1987) and the other using an approach that combined *Mutator* transposon tagging with differential hybridization (McLaughlin and Walbot 1987). The protein encoded by *Bz2* was demonstrated to have GST activity in maize, transgenic Arabidopsis and bacteria. Anthocyanins extracted from maize protoplasts that had been transformed with *Bz2* were shown to be conjugated with glutathione, and treatment with vanadate, an inhibitor of the glutathione pump in vacuolar membranes, prevented accumulation of anthocyanins in the vacuole (Marrs et al. 1995).

2.9
Multidrug Resistance Transporter (MRP)

The involvement of a multidrug resistance-type transporter in sequestering anthocyanins in the vacuole was suggested by three observations: (i) vanadate, a specific inhibitor of ABC transporters, blocks vacuolar accumulation of anthocyanins (Marrs et al. 1995); (ii) the requirement of GST for vacuolar localization of anthocyanins is reminiscent of GST-mediated detoxification pathways for heterocyclic organic xenobiotic molecules (Marrs et al. 1995); and (iii) two genes for MRPs were upregulated in cells engineered to overexpress anthocyanin regulators (Bruce et al. 2000). ESTs corresponding to maize *mrp* genes were identified by their sequence similarity to MRPs from other organisms (Goodman et al. 2004; Swarbreck et al. 2003). Two of the *mrp* genes, *ZmMrp3* and *ZmMrp4*, which share 88% amino acid identity in the protein coding regions, are expressed in pigmented aleurones, but not in unpigmented aleurones that lack anthocyanin regulatory gene function. *ZmMrp3* is also expressed in pigmented plant tissues; an antisense mutant of *ZmMrp3* altered the localization of pigment in plant tissues, but had no effect on aleurone pigmentation (Goodman et al. 2004). These observations suggest that *ZmMrp4* is the key transporter in the aleurone.

3 Regulatory Genes

In the aleurone, the biosynthetic genes are coordinately activated at the transcriptional level by concerted action of three types of regulatory proteins: transcription factors with R2R3-MYB or basic helix–loop–helix (bHLH) domains and a WD40 protein. The R2R3-MYB proteins are encoded by duplicate genes *colorless1* (*c1*) and *purple plant1* (*pl1*); the bHLH proteins are encoded by duplicate genes *booster1* (*b1*) and *red plant color1* (*r1*) and by the more distantly related *intensifier1 (in1)* gene; and the WD40 protein is encoded by *pale aleurone color1* (*pac1*). The regulatory roles of all three classes of genes were suggested first by genetic and biochemical evidence and subsequently confirmed by cloning the genes and studying their regulatory functions.

3.1 *c1* and *pl1*

Production of anthocyanins in the kernel requires a dominant, functional *c1* allele (Emerson 1921). The idea that *c1* functions as a regulator of anthocyanin synthesis was suggested by several lines of evidence (as summarized in Chen and Coe 1977): Genetic analysis and allelism tests resulted in the recovery of an allelic series including dominant inhibitor (*C1-I*), functional colored (*C1*), conditional (*c1-p*) and null (*c1-n*) alleles (Coe 1962; Emerson et al. 1935). Analysis of kernel pigments from plants carrying various anthocyanin mutants showed that kernels homozygous for the null *c1-n* allele produced no flavonoids. This contrasted with the accumulation of flavonols in *a1* (Kirby and Styles 1970), leucoanthocyanidins in *a2* (Coe 1955), and brown pigments in *bz1* and *bz2*. Inter-tissue complementation studies placed *c1* at or near the beginning of the pathway (Reddy and Coe 1962). Finally, observations that *c1* affects pigmentation in the kernel (aleurone and scutellum), but not in other parts of the plant were consistent with the idea that *c1* regulates the tissue-specificity of anthocyanin synthesis.

This regulatory role was confirmed by assays of biosynthetic enzymes for the pathway. Activity of enzymes that act both early (CHS) and late (UFGT) in the pathway was dependent on the presence of functional *c1* and *r1* alleles (Dooner 1983; Dooner and Nelson 1977b). These results led to the hypothesis that all of the biosynthetic genes are coordinately regulated, a prediction that was borne out by analysis of cloned structural and regulatory genes.

The *c1* gene was cloned by transposon tagging from two different alleles tagged with the *Spm/En* element (Cone et al. 1986; Paz-Ares et al. 1986). To explore the putative regulatory role of *c1*, blots of RNAs from kernels with wild-type *C1* and dominant inhibitor *C1-I* alleles were hybridized with probes from the *a1* and *bz1* structural genes. Structural gene RNA was detected in pigmented *C1* kernels, but not in colorless *C1-I* kernels (Cone et al. 1986; Paz-Ares et al. 1986), confirming the pattern observed for detection of *bz1*-encoded UFGT enzyme activity (Dooner and Nelson 1977b). DNA sequencing of a wild-type *C1* allele revealed a predicted protein with features consistent with its regulatory function. The N-terminal domain is homologous to the DNA binding domain of mammalian *myb* proteins, and the C-terminal domain is rich in acidic amino acids, similar to the acidic transcriptional activation domains of mammalian and yeast transcription factors (Paz-Ares et al. 1987). The sequence of the dominant inhibitor *C1-I* allele showed a number of differences from *C1*, most notably an insertion that would truncate the protein, eliminating much of the C-terminal acidic domain (Paz-Ares et al. 1990). Analysis of three additional dominant inhibitor alleles derived from imprecise excision of an *En/Spm* element in the third exon of *C1* showed that these alleles all contained transposon footprints that would cause frameshifts, leading to production of truncated proteins lacking the C-terminal acidic region (Singer et al. 1998).

The *pl1* gene, a duplicate of *c1*, was cloned by virtue of its sequence homology to *c1* (Cone et al. 1993). Most *pl1* alleles control pigmentation in vegetative and floral parts of the plant, such as leaf sheath, culm, tassel, and husks. One allele, *Pl1-Blotched*, leads to a variegated, or blotched, pattern of pigmentation in both the plant and the kernel aleurone (Cocciolone and Cone 1993). This allele was originally described as a dominant gene (*Blotched*) responsible for blotched aleurone pigmentation in kernels that were homozygous recessive for *c1*. Thus, it was proposed to functionally substitute for *c1* in the aleurone (Rhoades 1948). Molecular studies established that *Blotched* is an allele of *pl1* (designated *Pl1-Blotched*) and that expression of anthocyanin structural gene mRNAs in blotched kernels is dependent on the presence of *Pl1-Blotched* (Cocciolone and Cone 1993). It is not clear why the pattern of pigmentation associated with *Pl1-Blotched* is variegated, although it may be due to some unusual feature of the epigenetically controlled chromatin state of this allele that sets it apart from other *pl1* alleles (Cocciolone and Cone 1993; Hoekenga et al. 2000).

3.2
r1 and *b1*

Members of the *r1/b1* gene family serve as regulatory partners of *c1/pl1* in activating anthocyanin synthesis. Evidence for this partnership came from genetic studies, which established the complementary action of *r1* and *c1* in the aleurone (Emerson 1921). Biochemical assays of CHS and UFGT enzyme activity demonstrated that functional alleles of both *r1* and *c1* were needed for wild-type activity of these enzymes in the aleurone (Dooner 1983; Dooner and Nelson 1977b). The discovery of "*R2*" (later designated *B-peru* and *B-Bolivia*) alleles of *b1* with aleurone and plant color, suggested that *r1* and *b1* were duplicate (Styles et al. 1973). The results of biochemical experiments confirmed this idea. In one study, UFGT activity in purple plant tissues, such as husks and leaf sheaths, was shown to be under the control of *b1* (Gerats et al. 1984). Qualitative comparison of the pigments produced in leaf sheaths, anthers, and aleurone showed that the same spectrum of anthocyanins was present in all three tissues (Styles et al. 1973), prompting the conclusion that alleles of *r1* and *b1* differ only in the tissue distribution of pigmentation under their control.

The spatial distribution of anthocyanin pigments, controlled by different *r1* and *b1* alleles, can be explained, in part, by the different genetic organization of these genes. Many alleles of *r1* have a complex organization and are composed of multiple units that are separable by recombination (Dooner and Kermicle 1971; Stadler and Emmerling 1956; Stadler and Nuffer 1953). The well-studied *R-r* allele contains two genetic components: (P), which leads to red pigmentation in plant tissues such as anthers, roots and coleoptiles, and (S), which leads to pigmentation in the aleurone layer of the seed. Each component can mutate independently, leading to production of alleles that have lost function of the (P) component or the (S) component (Stadler 1951). Not all *r1* alleles have this complex genetic organization; instead, some alleles have a single genetic component that confers pigmentation to both plant and seed tissues.

Mature tissues in the plant, such as culm, leaf sheaths, husks, and tassels, are pigmented by the action of *b1*. Although most *b1* alleles control pigmentation solely in plant tissues, some alleles, such as *B-peru* and *B-Bolivia*, also pigment the aleurone. This dual tissue-specificity prompted the speculation that these alleles might have a complex genetic organization like that of *R-r* (Coe 1979). However, genetic analysis of a group of transposable element-induced mutants of *B-peru* showed that each mutation altered pigmentation simultaneously in plant and aleurone tissues, arguing that *B-peru* constitutes a single genetic component (Patterson et al. 1991). Subsequent molecular comparisons of *B-peru* to other *b1* alleles showed that the sequences needed for aleurone pigmentation lie in the promoter region of the *B-peru* gene (Radicella et al. 1992; Selinger and Chandler 1999a; Selinger et al. 1998) and revealed that these sequences are shared by other *b1* alleles that pigment both plant and aleurone tissues (Selinger and Chandler 1999a).

The *r1* gene was cloned by transposon tagging with *Ac* (Dellaporta et al. 1988). Molecular characterization of the genetically complex *R-r* allele revealed that it actually contains four related gene segments: (P), two (S) genes and a fourth component, (Q), which has sequence homology to (S) and (P) (Dellaporta et al. 1988; Robbins et al. 1991; Walker et al. 1995). Analysis of the DNA sequences and organization of these segments provided molecular explanations for the meiotic instability and evolution of this allele (Robbins et al. 1991; Walker et al. 1995). Several other *r1* alleles and family members were cloned and sequenced, revealing that all of the genes encode predicted proteins with sequence similarity to the bHLH domain found in *myc*-type transcription factors (Ludwig et al. 1989; Perrot and Cone 1989; Petroni et al. 2000; Robbins et al. 1991; Consonni et al. 1992). Consistent with the proposed regulatory role for genes in the *r1* family members, functional alleles were shown to be required for the accumulation of structural gene transcripts (Ludwig et al. 1989; Tonelli et al. 1991).

Capitalizing on the idea that the functionally duplicate nature of *b1* and *r1* would be reflected by homology at the DNA sequence level, *r1* sequences were used as a hybridization probe to clone the *b1* gene (Chandler et al. 1989). It codes for a bHLH protein similar to those encoded by other *r1* family members (Consonni et al. 1993; Radicella et al. 1991). RNA analysis showed that structural gene expression was dependent on the presence of functional *b1* alleles (Chandler et al. 1989), thus confirming the regulatory role of *b1*.

3.3
in1

The recessive *in1* mutation intensifies the pigmentation of homozygous *pr1* aleurone from red to nearly black (Fraser 1924). Analysis of the pigments in the *in1* kernels indicated that this phenotypic change is due to quantitative increase of anthocyanins, but no qualitative change (Reddy and Peterson 1978). The *in1* mutation also leads to twofold higher levels of UFGT (Klein and Nelson 1983). The *in1* gene was cloned by transposon tagging using *Spm* (Burr et al. 1996). The gene codes for a bHLH protein with sequence similarity to the R1/B1 proteins. Expression of the gene is unusual in that most of the transcripts from the wild-type gene are misspliced and produce mRNAs that would lead to premature termination of translation of the encoded protein. Consequently, only a small proportion of the transcripts are predicted to be full-length and capable of being translated to make functional protein. In the recessive *in1* allele, expression is very low, due to reduction in the overall level of transcription and to missplicing of most of the transcripts produced (Burr et al. 1996).

Different theories have been proposed to account for the fact that recessive *in1* kernels are more pigmented than *In1* kernels. One idea, which was proposed to explain how recessive *in1/in1* leads to production of pigment and CHS protein in aleurones of *c2/c2; Whp1/Whp1* kernels, is that the wild-type IN1 protein inhibits translation of the *Whp1* mRNA either directly or by controlling expression of some other translational inhibitor (Franken et al. 1991). An alternate idea, consistent with the fact that *In1* codes for a bHLH protein similar to R1/B1, is that IN1 is a repressor that inhibits transcriptional activation of the anthocyanin structural genes, perhaps by competing with R1/B1 for interaction with C1 and thus preventing target gene activation (Burr et al. 1996). If this idea is correct, then it predicts that the higher the level of *In1* expression, the less pigment produced. This trend fits with the phenotypes associated with the *in1* allelic series: *In-D*, an allele for a dominant dilute factor, produces less pigment than the *In1* allele, and *In1* produces less pigment than recessive *in1/in1*. Transcript levels for *In-D* are higher than for *In1* (Rojek et al. 1997), and *In1* produces much more RNA than *in1* (Burr et al. 1996). In all three alleles, most of the transcripts are misspliced and thus do not have the potential to code for a full-length protein (Burr et al. 1996; Rojek et al. 1997). This may explain why even the *In-D* allele, with the highest RNA expression and presumably the highest level of IN1 repressor, does not lead to a completely colorless phenotype.

3.4 *pac1*

The *pac1* gene was identified in a mutational screen for new regulators of the maize anthocyanin pathway. The recessive mutation results in pale pigmentation of the aleurone and seedling roots. A regulatory role for this gene was suggested by the finding that in pale aleurones, structural gene mRNA levels are reduced (Selinger and Chandler 1999b). The *pac1* gene was cloned by transposon tagging with *Mutator* (Carey et al. 2004). It codes for a protein with high sequence similarity to WD40 repeat proteins that function as anthocyanin regulators in other plants, e.g., *Petunia hybrida* AN11 and *Arabidopsis thaliana* TTG1. The maize *Pac1* gene, expressed from the CaMV 35S promoter, complements *ttg1* function in transgenic Arabidopsis plants, indicating functional as well as sequence conservation between the maize and Arabidopsis proteins (Carey et al. 2004). The fact that *pac1* mutants have a pale, rather than colorless, phenotype suggested there may be a duplicate gene that provides residual function in the aleurone. This idea is reinforced by the observation of additional hybridizing bands on DNA blots hybridized with a *pac1* probe (Carey et al. 2004) and mapping of one of these bands to a locus distinct from *pac1* (T. Dinh and K. Cone, unpublished).

4
Action and Interaction of the Regulatory Genes

A host of experiments have been carried out to identify and functionally define the *cis*-acting elements in the promoters of the anthocyanin structural genes and the presumptive interaction domains of the regulatory proteins. Sequence alignments of structural gene promoters revealed sequences with similarity to binding sites for mammalian MYB proteins, as well as other consensus regions (Bodeau and Walbot 1992; Lesnick and Chandler 1998; Roth et al. 1991; Tuerck and Fromm 1994). Functional studies, which coupled deletion analysis and site-directed mutagenesis of the promoter regions with assays of binding and transcriptional activation by the C1 and R1/B1 regulatory proteins, defined two important types of *cis*-acting promoter elements: (i) a C1-binding site, whose consensus sequence differs from the mammalian MYB consensus; and (ii) an anthocyanin regulatory element (ARE), which is required for R1/B1 enhanced transcription (Lesnick and Chandler 1998; Sainz et al. 1997).

To characterize the functionally important domains of the C1 and B1 regulators, structural gene promoters were fused to reporter genes and used as targets in transient expression assays. The results confirmed genetic observations, e.g., activation of transcription from a structural gene promoter requires both C1 and B1; neither protein alone is sufficient (Goff et al. 1990). In C1, the MYB domain functions in DNA binding and the C-terminal acidic domain is required for transcriptional activation (Goff 1991). Substitution of the C-terminal acidic domain of C1 with the truncated C-terminal domain of *C1-I* abolishes the ability of the protein to function in transcriptional activation (Goff et al. 1991). There is no evidence that any regions of the B1 protein function independently as DNA binding or transcriptional activation domains (Goff et al. 1992). However, interaction of the N-terminal region of B1 and the R2R3-MYB domain of C1 is required to induce transcription of anthocyanin genes (Goff et al. 1992; Grotewold et al. 2000).

Identification of the *pac1*-encoded WD40 protein added another player to the maize anthocyanin regulatory complex. Similarity of *pac1* to WD40 regulators that function in dicot flavonoid synthesis (reviewed in Koes et al. 2005), together with the fact that *pac1* can functionally complement a *ttg1* mutation in Arabidopsis (Carey et al. 2004), predicted that PAC1 participates directly with R1/B1 and C1 in transcriptional activation.

A model for interaction of the anthocyanin regulatory factors with the promoters of their target genes incorporates features of two previously proposed models (Fig. 2; Hernandez et al. 2004; Koes et al. 2005). C1, R1 (or B1) and PAC1 interact with each other to form a complex. At a structural gene promoter, C1 binds via its R2R3-MYB domain to a conserved sequence motif (C1 binding site; CBS). The R1 protein does not directly bind DNA, but instead mediates binding of a hypothetical DNA binding protein (bHLH interacting

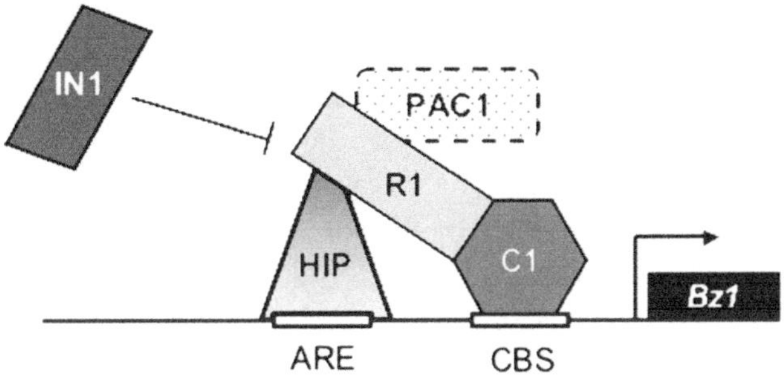

Fig. 2 Model for interaction among C1, R1 and PAC1 transcriptional regulators in activating transcription of an anthocyanin structural gene (*Bz1*). Interaction between C1, R1 and PAC1 is depicted. Uncertainty as to whether PAC1 remains associated with transcriptional complex is indicated by the *dashed outline*. HIP is a hypothetical bHLH interacting protein. IN1 acts as an inhibitor, perhaps by competing with R1 binding or activity. *CBS* C1 binding site, *ARE* anthocyanin regulatory element. Adapted from Hernandez et al. (2004) and Koes et al. (2005)

protein; HIP) to an anthocyanin regulatory element (ARE) that is adjacent to the CBS. It is not clear whether PAC1 remains associated with C1 and R1 at the promoter. IN1 may function as an inhibitor by interfering with the ability of R1 to form a productive transcription complex.

5 Regulation of the Regulators

5.1 Role of Development, Light, and Hormones

The dependence of anthocyanin synthesis on developmental, light, and hormonal cues can be traced in part to the regulatory effect of these signals on expression of *c1*. In most maize lines, pigmentation in the kernel is evident by 20–22 DAP. Prior to that time, little pigment is seen. Analysis of *c1* and *r1* RNA levels in *C1; R-sc* aleurones using a sensitive PCR-based assay showed that at 14 DAP only very low basal levels of *c1* mRNA levels are evident (Procissi et al. 1997). As development proceeds, *c1* mRNA levels increase, becoming maximal around 30 DAP when anthocyanin synthesis peaks (McCarty et al. 1989). In contrast, *r1* mRNA is readily detectable at 14 DAP and increases only slightly with development (Procissi et al. 1997).

The light requirement for anthocyanin synthesis is associated with light activation of different *c1* alleles. The recessive *c1-p* allele normally produces colorless kernels, but if the mature kernels are germinated in the light, they become pigmented (Chen and Coe 1977). Pigmentation of kernels with wild-type *C1* alleles also requires light. When light is excluded from developing *C1* ears, the seeds are colorless, but they develop pigment if subsequently exposed to light (Dooner and Ralston 1994). For both *c1-p* and wild-type alleles, light-induced pigmentation is correlated with light-induced increases in *c1* mRNA (Cone and Burr 1988; Procissi et al. 1997; Scheffler et al. 1994).

The *c1* gene is also regulated by an interaction between the maturation-regulating hormone abscisic acid (ABA) and *viviparous1* (*vp1*). Kernels homozygous for the recessive *vp1* mutation are colorless and germinate precociously on the ear (Coe et al. 1988). The lack of pigmentation in *vp1* kernels was correlated with absence of *c1* mRNA, suggesting that *vp1* might regulate *c1* (McCarty et al. 1989). The observation that *vp1* kernels, rescued before desiccation and germinated in the light, can produce anthocyanins suggested that light- and *vp1*-activated expression of *c1* might be separable (McCarty and Carson 1990). This idea was supported by experiments aimed at functionally dissecting the *cis* elements in the wild-type *C1* promoter (Hattori et al. 1992; Kao et al. 1996). Results indicated that there are three distinct promoter elements. One functions in light regulation and resembles light regulatory elements from other plants. Another element is important for both ABA and *vp1* regulation, and a third element is needed for regulation by ABA, but not *vp1*. Deletion of this last element recapitulates the promoter organization and phenotype of *c1-p*, i.e., expression during germination but not maturation. Together, the analyses indicated that expression of *c1* results from combinatorial interaction among protein factors that integrate developmental, light, and hormonal signals (Kao et al. 1996).

5.2 Epigenetic regulation

Chromatin-mediated (epigenetic) alterations in gene expression have been demonstrated for three of the anthocyanin regulatory loci: *pl1*, *r1*, and *b1*. The epigenetic state of *Pl1-Blotched* confers its unusual pattern of variegated pigmentation in both aleurone and plant tissues (Cocciolone and Cone 1993; Hoekenga et al. 2000). Paramutation at all three loci leads to allele-directed reduction of regulatory gene expression and hence pigmentation (reviewed in Chandler and Stam 2004). The recent finding that the gene silencing effects of paramutation can be relieved by mutation in a RNA-dependent RNA polymerase gene (Alleman et al. 2006) points to involvement of the small RNA pathway in controlling expression states of the anthocyanin regulatory genes. Another example of epigenetic regulation is the reduction of aleurone pigmentation due to imprinting at *r1* (Coe et al. 1988); the molecular basis for this parent-of-origin effect is not yet known. For a broad coverage of imprinting in endosperm development see Penterman et al., in this volume.

6
Conclusion

Genetic, biochemical, and molecular analyses of anthocyanin synthesis have proven extremely useful in identifying and characterizing gene regulatory mechanisms in maize. Moreover, because this pathway has been conserved in many plants, comparison among species will continue to provide important evolutionary insights into the complex regulatory networks that control plant pigmentation.

Acknowledgements The author is grateful to Ed Coe and Kyungju Shin for helpful comments on the manuscript. Current research on anthocyanin regulation in her laboratory is supported by the National Science Foundation Plant Genome Research Program (DBI-0421619) and the Genes and Genome Systems Cluster (MCB-0549447).

References

Abdel-Aal E-S, Young J, Rabalski I (2006) Anthocyanin composition in black, blue, pink, purple, and red cereal grains. J Agric Food Chem 54:4696–4704

Alleman M, Sidorenko L, McGinnis K, Seshadri V, Dorweiler J, White J, Sikkink K, Chandler VL (2006) An RNA-dependent RNA polymerase is required for paramutation in maize. Nature 442:295–298

Bernhardt J, Stich K, Schwarz-Sommer Z, Saedler H, Wienand U (1998) Molecular analysis of a second functional *A1* gene (*dihydroflavonol 40reductase*) in *Zea mays*. Plant J 14:483–488

Bodeau J, Walbot V (1992) Regulated transcription of the maize *Bronze-2* promoter in electroporated protoplasts requires the *C1* and *R* products. Mol Gen Genet 233:379–387

Bruce W, Folkerts O, Garnaat C, Crasta O, Roth B, Bowen B (2000) Expression profiling of the maize flavonoid pathway genes controlled by estradiol-inducible transcription factors CRC and P. Plant Cell 12:65–80

Burr F, Burr B, Scheffler B, Blewitt M, Wienand U, Matz E (1996) The maize repressor-like gene *intensifier1* shares homology with the *r1/b1* multigene family of transcription factors and exhibits missplicing. Plant Cell 8:1249–1259

Carey C, Strahle J, Selinger DA, Chandler VL (2004) Mutations in the *pale aleurone color1* regulatory gene of the *Zea mays* anthocyanin pathway have distinct phenotypes relative to the functionally similar *TRANSPARENT TESTA GLABRA1* gene in *Arabidopsis thaliana*. Plant Cell 16:450–464

Chandler VL, Radicella JP, Robbins TP, Chen J, Turks D (1989) Two regulatory genes of the maize anthocyanin pathway are homologous: Isolation of *B* utilizing *R* genomic sequences. Plant Cell 1:1175–1183

Chandler VL, Stam M (2004) Chromatin conversations: mechanisms and implications of paramutation. Nat Rev Genet 5:532–544

Chen S-M, Coe EH (1977) Control of anthocyanin synthesis by the *C* locus in maize. Biochem Genet 15:333–346

Cocciolone SM, Cone KCC (1993) *Pl-Bh*, an anthocyanin regulatory gene of maize that leads to variegated pigmentation. Genetics 135:575–588

Coe EH (1955) Anthocyanin synthesis in maize, the interaction of *A2* and *Pr* in leucoanthocyanin accumulation. Genetics 40:568

Coe EH (1962) Spontaneous mutation of the aleurone color inhibitor in maize. Genetics 47:779
Coe EH (1979) Specification of the anthocyanin biosynthetic function by *B* and *R* in maize. Maydica 24:49–58
Coe EH, Neuffer MG, Hoisington DA (1988) The genetics of corn. In: Sprague GF, Dudley JW (eds) Corn and corn improvement. ASA, Madison, WI, pp 81–258
Cone KC, Burr B (1988) Molecular and genetic analyses of the light requirement for anthocyanin synthesis in maize. In: Styles DE, Gavazzi GA, Racchi ML (eds) The genetics of flavonoids. Edizioni Unicopli, Milan, pp 143–145
Cone KC, Burr FA, Burr B (1986) Molecular analysis of the maize anthocyanin regulatory locus *C1*. Genetics 83:9631–9635
Cone KC, Cocciolone SM, Burr FA, Burr B (1993) The maize anthocyanin regulatory gene *pl* is a duplicate of *c1* that functions in the plant. Plant Cell 5:1795–1805
Consonni G, Geuna F, Gavazzi G, Tonelli C (1993) Molecular homology among members of the *R* gene family in maize. Plant J 3:335–346
Consonni G, Viotti A, Dellaporta SL, Tonelli C (1992) cDNA nucleotide sequence of *Sn*, a regulatory gene in maize. Nucleic Acids Res 20:373
Deboo GB, Albertsen MC, Taylor LP (1995) Flavanone 3-hydroxylase transcripts and flavonol accumulation are temporally coordinate in maize anthers. Plant J 7:703–713
Dellaporta SL, Greenblatt I, Kermicle JL, Hicks JB, Wessler SR (1988) Molecular cloning of the maize *R-nj* allele by transposon tagging with *Ac*. In: Gustafson P, Appels R (eds) 18th Stadler Genet. Symp. Plenum, New York, pp 263–282
Dooner HK (1983) Coordinate genetic regulation of flavonoid biosynthetic enzymes in maize. Mol Gen Genet 189:136–141
Dooner HK, Kermicle J (1971) Structure of the R^r tandem duplication in maize. Genetics 67:427–436
Dooner HK, Nelson OE (1977a) Controlling element-induced alterations in UDP glucose: flavonoid glucosyltransferase, the enzyme specified by the *bronze* locus in maize. Proc Natl Acad Sci USA 74:5623–5627
Dooner HK, Nelson OE (1977b) Genetic control of UDP glucose: flavonol 3-*O*-glucosyltransferase in the endosperm of maize. Biochem Genet 15:509–515
Dooner HK, Ralston E (1994) Light requirement for anthocyanin pigmentation of *C* aleurones. Maize Genet Newslett 65:74–75
Emerson RA (1921) The genetic relations of plant colors in maize. Cornell Univ Agr Exp Sta Mem 39:1–156
Emerson RA, Beadle G, Fraser A (1935) A summary of linkage studies in maize. Cornell Univ Agr Exp Sta Mem 180:1
Fedoroff NV, Furtek DB, Nelson OE Jr (1984) Cloning of the *bronze* locus in maize by a simple and generalizable procedure using the transposable controlling element *Activator (Ac)*. Proc Natl Acad Sci USA 81:3825–3829
Franken P, Niesbach-Klösgen U, Weydemann U, Maréchel-Drouard L, Saedler H, Wienand U (1991) The duplicated chalcone synthase genes *C2* and *Whp* (*white pollen*) of *Zea mays* are independently regulated; evidence for translational control of *Whp* expression by the anthocyanin intensifying gene *in*. EMBO J 10:2605–2612
Fraser A (1924) Heritable characters of maize XVII–Intensified red and purple aleurone color. J Hered 15:119–123
Furtek D, Schiefelbein JW, Johnston F, Nelson OEJ (1988) Sequence comparisons of three wild-type *Bronze-1* alleles from *Zea mays*. Plant Mol Biol 11:473–481
Gerats AGM, Bussard J, Coe EH Jr, Larson R (1984) Influence of *B* and *Pl* on UDPG:flavonoid-3-*O*-glucosyltransferase in *Zea mays* L. Biochem Genet 22:1161–1169

Goff SA, Cone KC, Chandler VL (1992) Functional analysis of the transcriptional activator encoded by the maize *B* gene: Evidence for a direct functional interaction between two classes of regulatory proteins. Genes Dev 6:864–875

Goff SA, Cone KC, Fromm ME (1991) Identification of functional domains in the maize transcriptional activator C1: comparison of wild-type and dominant inhibitor proteins. Genes Dev 5:298–309

Goff SA, Klein TM, Roth BA, Fromm ME, Cone KC, Radicella JP, Chandler VL (1990) Transactivation of anthocyanin biosynthetic genes following transfer of *B* regulatory genes into maize tissues. EMBO J 9:2517–2522

Goodman C, Casati P, Walbot V (2004) A multi-drug resistance-associated protein involved in anthocyanin transport in *Zea mays*. Plant Cell 16:1812–1826

Grotewold E (2006) The genetics and biochemistry floral pigments. Ann Rev Plant Biol 57:761–780

Grotewold E, Peterson T (1994) Isolation and characterization of a maize gene encoding chalcone flavanone isomerase. Mol Gen Genet 242:1–8

Grotewold E, Sainz MB, Tagliani LA, Hernandez J, Bowen BA, Chandler VL (2000) Identification of the residues in the Myb domain of maize C1 that specify the interaction with the bHLH cofactor R. Proc Natl Acad Sci USA 97:13579–13584

Hattori T, Vasil V, Rosenkrans L, Hannah LC, McCarty DR, Vasil IK (1992) The *Viviparous-1* gene and abscisic acid activate the *C1* regulatory gene for anthocyanin biosynthesis during seed maturation in maize. Genes Dev 6:609–618

Hernandez J, Heine G, Irani N, Feller A, Kim M-G, Matulnik R, Chandler VL, Grotewold E (2004) Different mechanisms participate in the R-dependent activity of the R2R3 MYB transcription factor C1. J Biol Chem 46:48205–48213

Hoekenga O, Muszynski M, Cone K (2000) Developmental patterns of chromatin structure and DNA methylation responsible for epigenetic expression of a maize regulatory gene. Genetics 155:1889–1902

Kao C-Y, Cocciolone SM, Vasil IK, McCarty DR (1996) Localization and interaction of the *cis*-acting elements for abscisic acid, VIVIPAROUS1, and light activation of the *C1* gene of maize. Plant Cell 8:1171–1179

Kirby LT, Styles ED (1970) Flavonoids associated with specific gene action in maize aleurone, and the role of light in substituting for the action of a gene. Can J Genet Cytol 12:934–940

Klein AS, Nelson OE (1983) The allelic state at *intensifier* influences the accumulation of UDP glucose:flavonoid 3-*O*-glucosyltransferase in maize. Phytochemistry 22:2634–2635

Koes R, Verweij W, Quattrocchio F (2005) Flavonoids: a colorful model for the regulation and evolution of biochemical pathways. Trends Plant Sci 10:236–242

Larson R, Bussard J, Coe EH (1986) Gene-dependent flavonoid 3′-hydroxylation in maize. Biochem Genet 24:615–624

Larson R, Coe EH (1977) Gene-dependent flavonoid glucosyltransferase in maize. Biochem Genet 15:153–156

Lepiniec L, Debeaujon I, Routaboul J-M, Baudry A, Pourcel L, Nesi N, Caboche M (2006) Genetics and biochemistry of seed flavonoids. Ann Rev Plant Biol 57:405–430

Lesnick M, Chandler VL (1998) Activation of the maize anthocyanin gene *a2* is mediated by an element conserved in many anthocyanin promoters. Plant Physiol 117:437–445

Ludwig SR, Habera LF, Dellaporta SL, Wessler SR (1989) *Lc*, a member of the maize *R* gene family responsible for tissue-specific anthocyanin production, encodes a protein similar to transcriptional activators and contains the *myc* homology region. Proc Natl Acad Sci USA 86:7092–7096

Marrs K, Alfenito M, Lloyd A, Walbot V (1995) A glutathione *S*-transferase involved in vacuolar transfer encoded by the maize gene *Bronze-2*. Nature 375:397–400

McCarty DR, Carson C (1990) The molecular genetics of seed maturation in maize. Physiol Plant 81:267–272

McCarty DR, Carson C, Stinard P, Robertson D (1989) Molecular analysis of *viviparous-1*: an abscisic acid-insensitive mutant of maize. Plant Cell 1:523–532

McLaughlin M, Walbot V (1987) Cloning of a mutable *bz2* allele of maize by transposon tagging and differential hybridization. Genetics 117:771–776

Menssen A, Hohmann S, Martin W, Schnable P, Peterson PA, Saedler H, Gierl A (1990) The En/Spm transposable element of Zea mays contains splice sites at the termini generating a novel intron from a dSpm element in the A2 gene. EMBO J 9:3051–3057

Nakajima J, Tanaka Y, Yamazaki M, Saito K (2001) Reaction mechanism from leucoanthocyanidin to anthocyanidin 3-glucoside, a key reaction for coloring in anthocyanin biosynthesis. J Biol Chem 276:25797–25803

O'Reilly C, Shepherd N, Pereira A, Schwarz-Sommer Z, Bertram I, Peterson PA, Saedler H (1985) Molecular cloning of the *a1* locus in *Zea mays* using the transposable elements *En* and *Mu1*. EMBO J 4:877–882

Patterson GI, Harris LJ, Walbot V, Chandler VL (1991) Genetic analysis of *B-Peru*, a regulatory gene in maize. Genetics 126:205–220

Paz-Ares J, Ghosal D, Saedler H (1990) Molecular analysis of the *C1-I* allele from *Zea mays*: a dominant mutant of the regulatory C1 locus. EMBO J 9:315–321

Paz-Ares J, Ghosal D, Wienand U, Peterson PA, Saedler H (1987) The regulatory *c1* locus of *Zea mays* encodes a protein with homology to myb protooncogene products and with structural similarities to transcriptional activators. EMBO J 6:3553–3558

Paz-Ares J, Wienand U, Peterson PA, Saedler H (1986) Molecular cloning of the *c* locus of *Zea mays*: a locus regulating the anthocyanin pathway. EMBO J 5:829–833

Perrot GH, Cone KC (1989) Nucleotide sequence of the maize *R – S* gene. Nucleic Acids Res 17:8003

Petroni K, Cominelli E, Consonni G, Gusmaroli G, Gavazzi G, Tonelli C (2000) The developmental expression of the maize regulatory gene *Hopi* determines germination-dependent anthocyanin accumulation. Genetics 155:323–336

Procissi A, Dolfini SF, Ronchi A, Tonelli C (1997) Light-dependent spatial and temporal expression of pigment regulatory genes in developing maize seeds. Plant Cell 9:1547–1557

Radicella JP, Brown D, Tolar LA, Chandler VL (1992) Allelic diversity of the maize B regulatory gene: Different leader and promoter sequences of two B alleles determine distinct tissue-specificities of anthocyanin production. Genes Dev 6:2152–2164

Radicella JP, Turks D, Chandler VL (1991) Cloning and nucleotide sequence of a cDNA encoding *B-Peru*, a regulatory protein of the anthocyanin pathway in maize. Plant Mol Biol 17:127–130

Ralston EJ, English JJ, Dooner HK (1988) Sequence of three *bronze* alleles of maize and correlation with the genetic fine structure. Genetics 119:185–197

Reddy A, Britsch L, Salamini F, Saedler H, Rohde W (1987) The *A1* (*anthocyanin-1*) locus in *Zea mays* encodes dihydroquercetin reductase. Plant Sci 42:7–13

Reddy A, Peterson PA (1978) The action of the *intensifier* (*In*) gene in flavonoid production in aleurone tissue of maize. Can J Genet Cytol 20:337–347

Reddy G, Coe EH (1962) Genetic control of leucoanthocyanidin formation in maize. Genetics 50:485–489

Rhoades MM (1948) Interaction of *Bh* with recessive *c*. Maize Genet Newslett 22:9

Robbins TP, Walker EL, Kermicle JL, Alleman M, Dellaporta SL (1991) Meiotic instability of the *R-r* complex arising from displaced intragenic exchange and intrachromosomal rearrangement. Genetics 129:271–283

Rojek R, Scheffler B, Wienand U (1997) Molecular analysis of *In-D*, a semi-dominant allele of the anthocyanin regulatory gene intensifier. Abstracts of 39th Annual Maize Genetics Conference, 13–16 March, Clearwater Beach, FL

Roth BA, Goff SA, Klein TM, Fromm ME (1991) *C1*- and *R*-dependent expression of the maize *Bz1* gene requires sequences with homology to mammalian *myb* and *myc* binding sites. Plant Cell 3:317–325

Sainz MB, Grotewold E, Chandler VL (1997) Evidence for direct activation of an anthocyanin promoter by the maize C1 protein and comparison of DNA binding by related Myb domain proteins. Plant Cell 9:611–625

Scheffler B, Franken P, Schütt E, Schrell A, Saedler H, Wienand U (1994) Molecular analysis of *C1* alleles in *Zea mays* defines regions involved in the expression of this regulatory gene. Mol Gen Genet 242:40–48

Selinger DA, Chandler VL (1999a) Major recent and independent changes in levels and patterns of expression have occurred at the *b* gene, a regulatory locus in maize. Proc Natl Acad Sci USA 21:15007–15012

Selinger DA, Chandler VL (1999b) A mutation in the *pale aleurone color1* gene identifies a novel regulator of the maize anthocyanin pathway. Plant Cell 11:5–14

Selinger DA, Lisch D, Chandler VL (1998) The maize regulatory gene *B-Peru* contains a DNA rearrangement that specifies tissue-specific expression through both positive and negative promoter elements. Genetics 149:1125–1138

Singer T, Gierl A, Peterson PA (1998) Three new dominant C1 suppressor alleles in *Zea mays*. Genet Res 71:127–132

Springob K, Nakajima J, Yamazaki M, Saito K (2003) Recent advances in the biosynthesis and accumulation of anthocyanins. Nat Prod Rep 20:288–303

Stadler LJ (1951) Spontaneous mutation in maize. Cold Spring Harbor Symp Quant Biol 16:49–63

Stadler LJ, Emmerling M (1956) Relation of unequal crossing over to the interdependence of R^r elements. Genetics 41:124–137

Stadler LJ, Nuffer MB (1953) Problems of gene structure II. Separation of *R-r* elements (S) and (P) by unequal crossing over. Science 117:471–472

Styles ED, Ceska O, Seah K-T (1973) Developmental differences in action of *R* and *B* alleles in maize. Can J Genet Cytol 15:59–72

Swarbreck D, Ripoll P-J, Brown D, Edwards K, Theodoulou F (2003) Isolation and characterisation of two multi-drug resistance associated protein genes from maize. Gene 215:153–164

Theres N, Scheele T, Starlinger P (1987) Cloning of the *Bz2* locus of *Zea mays* using the transposable element *Ds* as a gene tag. Mol Gen Genet 209:193–197

Tonelli C, Consonni G, Dolfini SF, Dellaporta SL, Viotti A, Gavazzi G (1991) Genetic and molecular analysis of *Sn*, a light-inducible, tissue-specific regulatory gene in maize. Mol Gen Genet 225:401–410

Tuerck J, Fromm ME (1994) Elements of the maize *A1* promoter required for transactivation by the anthocyanin *B/C1* or phlobaphene *P* regulatory genes. Plant Cell 6:1655–1663

Walker EL, Robbins TP, Bureau TE, Kermicle J, Dellaporta SL (1995) Transposon-mediated chromosomal rearrangements and gene duplications in the formation of the maize *R-r* complex. EMBO J 14:2350–2363

Wienand U, Weydemann U, Niesbach-Klösgen U, Peterson PA, Saedler H (1986) Molecular cloning of the *c2* locus of *Zea mays*, the gene coding for chalcone synthase. Mol Gen Genet 203:202–207

Plant Cell Monogr (8)
O.-A. Olsen: Endosperm
DOI 10.1007/7089_2007_115/Published online: 5 April 2007

Protein Body Biogenesis in Cereal Endosperms

T. Kumamaru[1] · M. Ogawa[2] · H. Satoh[1] · T. W. Okita[3] (✉)

[1]Laboratory of Plant Genetic Resources, Faculty of Agriculture, Kyushu University, Hakozaki, 812-8581 Fukuoka, Japan

[2]Department of Life Science, Yamaguchi Prefectural University, Sakurabatake, 753-8502 Yamaguchi, Japan

[3]Institute of Biological Chemistry, Washington State University, Pullman, WA 99164-6340, USA
okita@wsu.edu

Abstract Developing cereal endosperms accumulate storage proteins as a reserve to support the early postgerminating phase of plant growth and development. These storage proteins are concentrated and packaged into a specialized organelle, the protein body, which are derived from the endoplasmic reticulum lumen or vacuole. In this chapter, we describe the biochemical and cellular events that underlies the basis for the use of these two intracellular sites for storage protein deposition and accumulation.

1 Introduction

Storage proteins, which serve as a reservoir of carbon, nitrogen, and sulfur for the postgerminating seedling, constitute up to 90% of the total protein fraction in developing seeds. Because of their large quantities, plants have utilized the endomembrane system for sequestering the storage proteins into a highly compact form, the protein body (PB). Most plants, including several of the major cereals, use the protein storage vacuole (PSV) as the site of deposition. Other cereals have exploited the lumen of the endoplasmic reticulum (ER) as an alternative storage site. In this chapter we describe the processes involved in PB formation in the major cereals wheat, barley, maize, and rice, with particular emphasis on how each of these cereals has contributed to our knowledge in this topic area. The study of each of these cereals has shed light on specific aspects of PB formation which, when viewed in total, have resulted in a better understanding of how the storage proteins are packaged in the cell. The reader should refer to two monographs on the subject (Larkins and Vasil 1997; Shewry and Casey 1999) for other topics not covered here.

2 PB Formation in Wheat and Barley

The storage proteins of the *Triticeae* are homologous in structure (Shewry and Casey 1999). The wheat prolamines consist mainly of two types: monomeric gliadins solubilized by alcohol solution and polymeric glutenins solubilized by alcohol solution in the presence of a reducing agent. The barley prolamines are classified by their electrophoretic mobility into three groups: B hordein, C hordein, and D hordein. The B hordein and D hordein subunits are rich in cysteine residues and are present as monomeric and polymeric species, while sulfur-poor C hordein subunits are present as monomers.

The prolamines of wheat (Campbell et al. 1981) and barley (Cameron-Mills and von Wettstein 1980) are deposited within the PSV (Fig. 1a,d). The earliest event of PB formation in these cereal grains is the synthesis of the storage protein on the ER and their cotranslational import into the ER lumen. Despite their close phylogenetic relationship, different mechanisms have been suggested for how prolamines are exported from the ER lumen to the PSV and the involvement of the Golgi apparatus in PB formation in these plant species.

2.1 Golgi-Dependent Pathway in Wheat PB Formation

Consistent with the secretory pathway unraveled in animal cells, the Golgi apparatus was suggested to play a prominent role in mediating the transport of wheat storage proteins from the ER to the PSV (Buttrose 1963). Evidence for a direct role of the Golgi apparatus in wheat PB formation was obtained by immunocytochemical studies of the gliadins during endosperm development (Kim et al. 1988). When viewed by electron microscopy the earliest indication of PB formation was the packaging of gliadins within small (0.2–0.3 μm in diameter) electron-dense vesicles connected to the Golgi cisternae (Fig. 1c). The absence of antigen signals within the ER lumen and their presence in (protein accumulating) vesicles suggested that the gliadins were rapidly exported from the ER lumen and concentrated within dense vesicles. The Golgi observed during the mid-stage of grain filling were atypical from the four to seven closely stacked cisternae structure commonly observed in other plant tissues. Instead the Golgi apparati were highly modified in containing two or three highly curved cisternae with associated electron-dense and electron-lucent vesicles. Dense vesicles in the cytoplasm were commonly observed associated with larger electron-lucent vesicles or vacuoles. Also distributed in the cytoplasm were small to medium size vacuoles containing one or more gliadin inclusion bodies, suggesting the coalescence of smaller inclusion granules to form larger ones which, in turn, merged with the larger PSV (Bechtel et al. 1982; Kim et al. 1988).

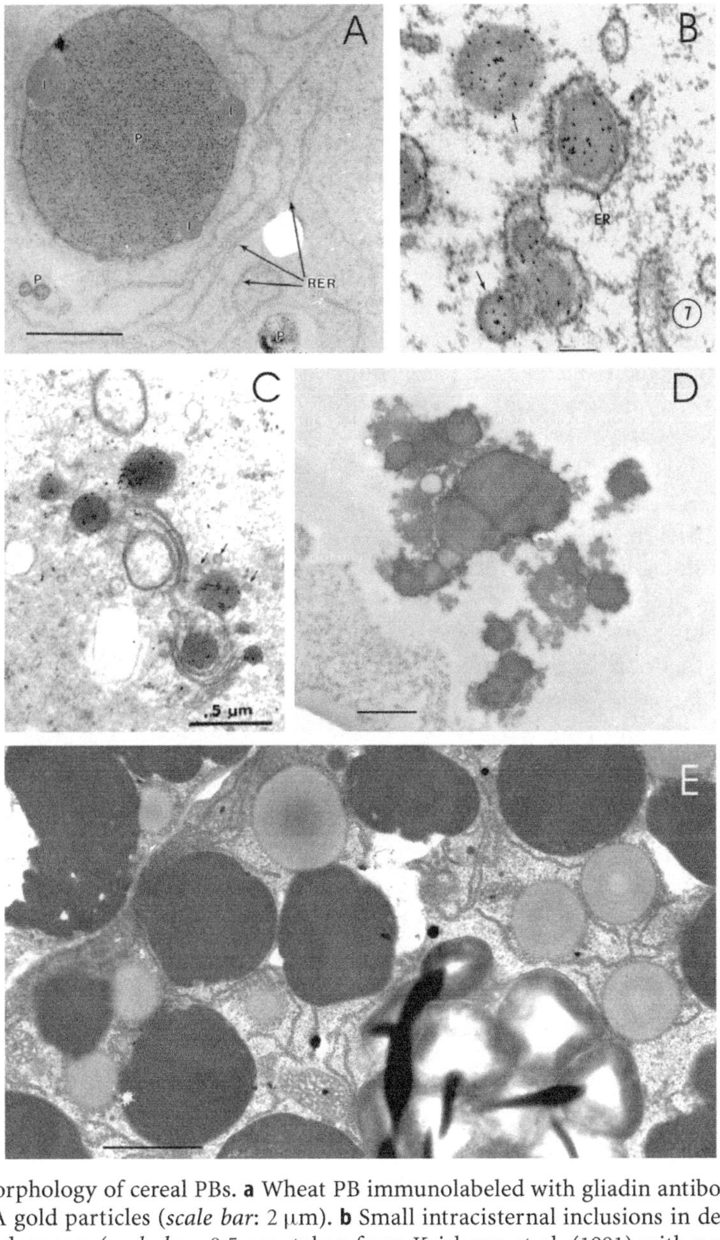

Fig. 1 Morphology of cereal PBs. **a** Wheat PB immunolabeled with gliadin antibodies and protein A gold particles (*scale bar*: 2 µm). **b** Small intracisternal inclusions in developing wheat endosperm (*scale bar*: 0.5 µm; taken from Krishnan et al. (1991) with permission from the author and NRC Research Press). **c** Unique Golgi structures containing highly curved cisternae with associated dense vesicles and electron-lucent vesicles in developing wheat endosperm. **d** Barley PB containing electron-light and -dense staining globules as well as a fibrillar matrix (*scale bar*: 1 µm). **e** Rice electron-lucent spherical prolamine-containing PBs and irregularly shaped, electron-dense glutelin-containing PSVs (*scale bar*: 2 µm)

2.2
Golgi-Independent Pathway in Wheat PB Formation

The role of the Golgi apparatus in wheat PB formation has been a subject of controversy ever since the first electron micrographic studies were undertaken in the early 1960s (for a review, see Parker 1982). The principal reason for dismissing the involvement of this organelle in wheat PB formation was that it was rarely detected throughout seed development or during the biosynthetic period when protein accumulation rates were at their highest (Bechtel et al. 1982; Campbell et al. 1981; Parker 1982). Moreover, in several studies (Bechtel et al. 1982; Campbell et al. 1981; Krishnan et al. 1991; Levanony et al. 1992; Parker 1982) but not all, large intracisternal inclusion granules bounded by rough ER were observed (Fig. 1b). These inclusion granules were thought to be too large to be exported from the ER to the Golgi via transport vesicles and were assumed to be transported to the PSV (Rubin et al. 1992). Lastly, the lumenal chaperone binding protein BiP was observed trapped within the PB inclusions (Levanony et al. 1992). As this protein normally resides in the ER lumen, its presence in the PB inferred the initial assembly of the inclusion granules in the ER lumen. Collectively, these three reasons suggested that the Golgi played, at best, a minor role in wheat PB formation. In turn, various novel transport processes from the ER to the PSV were proposed.

The latest and generally accepted rendition of a Golgi-independent pathway for wheat PB formation was proposed by Galili's laboratory in a series of studies (Galili 1997; Galili et al. 1995; Levanony et al. 1992). Although a Golgi-dependent pathway was acknowledged to exist, only small amounts of prolamines were transported from the ER to the storage vacuole by this pathway. The bulk of the proteins were assembled to form inclusion granules within the ER lumen which was then released into the cytoplasm, a proposal first suggested by Miflin et al. (1981). Once in the cytoplasm the inclusion granules were seen to be engulfed by small vesicles through a process akin to autophagy to form small vacuoles containing PB, which then merged with each other and eventually the large PSV. In addition to electron micrographs suggesting the presence of ER inclusion granules and remnant rough ER membranes within the PSV, the PSV contained the lumenal chaperone BiP, a marker of the ER lumen.

The existence of two pathways leading to PB formation in wheat was also inferred by biochemical studies. Wheat PBs consist of a low-density type (light PB) and a high-density type (dense PB) (Rubin et al. 1992). The gliadins were present in both light PBs and dense PBs, while the high molecular weight (HMW) glutenins were localized only in the dense PBs, suggesting that both prolamines are transported and deposited within PSV by different processes, one Golgi-dependent and the other Golgi-independent. Pulse–chase experiments showed that the dense PBs were formed from the light PBs, but the dense PBs were formed not by a gradual increase in density but by a quick process of storage protein aggregation.

2.3
PB Formation in Barley

The barley PSVs are highly complex as they contain electron-light and electron-dense staining globules, as well as a fibrillar matrix and vesicles (Fig. 1d) (Cameron-Mills and von Wettstein 1980). The electron-dense globules and the fibrillar matrix predominated in Risø 1508, a mutant line deficient in B and C hordeins, indicating that the electron-light staining globules were made up of B and C hordeins (Cameron-Mills and von Wettstein 1980), a conclusion later confirmed by immunogold labeling studies (Rechinger et al. 1993). Unlike wheat, where several studies demonstrated the existence of ER inclusion bodies, no such entities have been described in barley except in the Nevsky barley line, which is discussed below. Instead, dense vesicles containing hordeins associated with the Golgi were prominently observed (Galili and Herman 1997; Møgelsvang and Simpson 1998). In addition to those contained within the PSV, globules were also detected free in the cytoplasm. These cytoplasmic globules were not surrounded by a rough ER-like membrane. The cytoplasmic globules varied in size (0.34 μm on average) and were smaller than those present in the PSV (average size of 1.15 μm). Overall, these microscopic observations discount the existence of a direct ER to PSV pathway in developing barley endosperm. Instead, PB formation in barley is mediated by the Golgi, where the hordeins are initially concentrated into dense vesicles, and later in cytoplasmic globules which combine to form larger species and eventually with the PSV.

In the barley line Nevsky, which lacks γ_3-hordein (a minor B hordein), nearly all of the B and C hordeins are observed as globules contained within the ER lumen in the central starchy endosperm (Rechinger et al. 1993). These observations suggest a possible role for γ_3-hordein in conferring transport competence of B and C hordeins from the ER to the Golgi. Interestingly, subaleurone cells exhibited a normal PB phenotype of globules contained within the PSV and were nearly devoid of intracisternal inclusion granules in the ER. This apparent dichotomy depending on the cell type suggests that the extent of hordein assembly within the ER depends on relative rates of synthesis, of protein folding, and of their export from the ER (Tooze et al. 1989). In young subaleurone cells, the rate of hordein synthesis is lower than its export rate from the ER, thereby favoring hordein accumulation in the PSV. In the bulky endosperm the high rates of hordein synthesis exceed its capacity for export, conditions leading to elevated levels in the ER lumen and their assembly to form an inclusion granule.

3
PB Formation in Maize

The major storage proteins in maize kernels are the alcohol-soluble zeins (Coleman and Larkins 1999; Shotwell and Larkins 1989). Maize also contains smaller quantities of legumin-1, an 11S globulin (Yamagata et al. 2003), which is deposited in small PSVs. The zeins are composed of four major classes, α, β, γ, and δ, which are readily distinguished by their different molecular sizes and their dependence on reducing agents for their solubility in alcoholic solutions. The α-zeins are sulfur-deficient polypeptides of relative molecular mass (M_r) 19 000 and 22 000, and constitute the most abundant class (70%) of the prolamine fraction. The other zein classes are sulfur-rich and consist of the cysteine-rich γ-zeins, the second most abundant class, with major species at 16 and 27 kD as well as a third minor species of 50 kD identified by expressed sequence tag (EST) analysis of a maize developing endosperm library (Woo et al. 2001). The two remaining classes consist of β-zeins of M_r 15 000 and δ-zeins of M_r 10 000 (some inbred lines also have a δ-zein of M_r 18 000). These zein classes are sulfur-rich in containing a high mole percentage of cysteine and methionine residues. Sequence alignments indicate that the β-zeins belong to the γ-zein family in sharing the conserved N-terminal and C-terminal peptides (Woo et al. 2001). Both the γ-zeins and β-zeins require a reducing agent for their solubilization in alcoholic solution and exist as disulfide-linked protein complexes. The δ-zeins are the most hydrophobic and exist as monomeric species, readily extractable in the absence of reducing agents. Further details on the physical properties and protein gene structures can be found in Coleman and Larkins (1999) and Woo et al. (2001).

Zeins are packaged as accretions, spherical inclusion granules 1–2 μm in diameter, within the ER lumen (Lending et al. 1988). The zeins lack the canonical HDEL/KDEL retention/retrieval signal, and hence their retention and localization within the ER has been a subject of many studies over the past decade (Coleman and Larkins 1999). Although suggestions have been made for a peptide signal functional equivalent to HDEL/KDEL, the retention of zeins at the ER is likely due to their ability to assemble into a stable intracisternal inclusion granule (Geli et al. 1994; Kim et al. 2002) at a rate significantly higher than its capacity for ER export.

Based on its spherical shape and the lack of an unfolded protein response with accompanying high amounts of the lumenal chaperone BiP, zeins must be assembled and packaged in a highly ordered structure. Indeed, Lending and Larkins (1989) have shown in a series of elegant electron micrographs that the zein protein bodies undergo a series of morphological changes during their enlargement to a mature organelle. Young PBs, found in subaleurone cells, are small and electron dense and contain only γ- and β-zeins. During the maturation of the endosperm cell, the enlarging PB contains small, electron-lucent locules containing α- and δ-zeins (Esen and Stetler 1992; Lending and Larkins 1989). As additional α- and δ-zeins accumulate in older endosperm cells, the electron-lucent locules fuse to form a large central core of α- and δ-zeins surrounded by a thinner shell of β- and γ-zeins. Thus, the development of the zein PB occurs by a strict temporal expression pattern where the γ- and β-zeins are accumulated first, followed by the deposition and assembly of α- and δ-zeins to form the core region and, in turn, the displacement of γ- and β-zeins to the periphery. The spatial location of the various zeins within the inclusion granule is likely directed by their intrinsic physical properties.

3.1 The Role of Protein to Protein Interactions in Maize PB Formation

The biochemical basis for zein PB formation has been explored by expressing various zeins in a wide variety of heterologous hosts and plant tissues (Coleman and Larkins 1999). Despite the limitations of these systems, valuable information has been obtained from these studies, which have provided novel insights into how these proteins are retained and assembled to form a spherical intracisternal inclusion granule 1–2 μm in diameter.

The 27-kD γ-zeins form ER inclusion granules when expressed in *Arabidopsis* leaves (Geli et al. 1994). The γ-zein containing ER inclusions, however, were much smaller (< 0.5 μm) and somewhat amorphous in shape than those seen in maize. The 27-kD γ-zein has a modular structure containing an N-terminal domain with PPPVHL tandem repeats, and a conserved cysteine-rich C-terminal domain. Removal of the N-terminal proline-rich tandem repeat domain resulted in the export of the truncated γ-zein from the ER to the extracellular space, whereas removal of the conserved C-terminal domain transformed the small ER inclusions to large reticulated ER-bound structures. These observations indicate that the two domains have different roles in PB formation. The proline-rich tandem repeat domain is required for retention of the γ-zeins within the ER lumen by promoting protein to protein interactions and their assembly into a higher order protein conformation. The conserved C-terminal peptide is essential for the formation of a highly ordered spherical granular structure mediated by intermolecular disulfide bonds.

The β-zeins also have the capacity to form PBs in tobacco seeds and leaf tissues (Bagga et al. 1995), although the PB morphology was quite distinct from that seen for γ-zeins. In leaves, the PBs appeared as highly lobed, rosette-like structures. In seeds, the PBs resembled those found in maize endosperm, although multilobed PBs and multiple inclusion granules bounded by a single ER membrane were also apparent. In contrast to an earlier report (Hoffman et al. 1987), no significant levels of β-zein were observed in the crystalloid areas of the PSV although, in rare instances, a spherical β-zein PB was detected in the matrix region, an event suggested to occur by autophagy. A more recent study (Coleman et al. 2004), however, showed that the formation of ER-bound PBs is a relatively rare event and that the bulk of the β-zein is exported to the PSV (discussed below).

The β- and γ-zeins are each capable of forming PB-like structures, whose morphology reflects the intrinsic physical properties of each protein. In neither case were these PBs, formed by homotypic interactions of a single zein class, identical in morphology to the spherical compact accretions observed in maize. In light of their coordinate expression during the early endosperm development and colocalization in an early immature PB, it is likely that the interactions and assembly of γ- and β-zeins is a prerequisite for normal PB formation in maize.

Unlike the γ- and β-zeins, the α-zeins failed to accumulate to any appreciable extent when expressed in most heterologous hosts (Coleman and Larkins 1999). Coexpression with the 27-kD γ-zein, however, elevated the expression of a 22-kD α-zein by stabilizing protein levels (Coleman et al. 1996). Immunocytochemical analysis showed that both zein classes were located in spherical PBs similar in appearance to those observed during the early stages of maize development. The PBs were 0.1–0.5 μm in diameter and consisted of α-zein-associated electron-lucent locules contained within an electron-dense γ-zein matrix. The results support the view that γ-zeins help retain α-zeins within the ER by sequestering them and preventing their export from this compartment.

The accumulation of α-zein was also stabilized by coexpression with β-zein in tobacco seeds (Coleman et al. 2004). Closer examination showed that accumulation of α-zein but not β-zein was temporally regulated, suggesting that α-zein was more prone to proteolysis. Electron microscopic studies showed the presence of both multilobed and single spherical PBs. Interestingly, the multilobed PBs contained only β-zein while the single spherical PBs harbored both α- and β-zeins, although their spatial distribution was just the opposite from that seen in maize. The electron-lucent locules labeled with α-zein antibody were located on the periphery and surrounded the electron-dense β-zein core. Hence, β-zeins are capable of interacting with α-zeins and retaining them within the ER, but are not responsible for their spatial distribution within the protein body. This role is probably assumed by γ-zeins, which are capable of sequestering α-zeins to the core region.

Indirect immunofluorescence studies showed that although a few small PBs were observed, most of the α- and β-zeins were localized to the PSV (Coleman et al. 2004). The PSVs showed uniform fluorescence for both zein species, indicating that the zeins were randomly dispersed within this compartment. Therefore, the increase in α-zein accumulation in these transgenic plants was not due to their retention in the ER but, instead, to their protection by β-zeins from proteolysis within the PSV. In addition, these observations indicate that the α- and β-zeins are capable of ER export and at rates much higher than their capacity to form inclusion granules. The γ-zeins may also be exported from the ER efficiently when expressed in *Arabidopsis* and tobacco and only a small proportion may form ER-derived PBs.

It is unclear why the α- and β-zeins exit the ER so efficiently in tobacco. It is possible that the interaction of γ- and β-zeins and their assembly retards their export from the ER and that these two zein species work in unison to sequester α-zein within this compartment, as suggested by yeast two-hybrid analysis (discussed below). The existence of a receptor-like protein that retards the export of β-zein, however, cannot be excluded.

The 10-kD but not the 18-kD δ-zeins are capable of forming PBs in tobacco leaves and seeds (Bagga et al. 1997; Hinchliffe and Kemp 2002). These PBs, which differ considerably in structure from the γ-zein and β-zein types, contained amorphous inclusions surrounded by a thick layer of osmophilic material. Coexpression with β-zein resulted in a four- to five-fold increase in δ-zein accumulation. These results support the interaction between β- and δ-zeins where formation of the complex stabilizes δ-zein from proteolytic turnover. Interestingly, the mixed zein PBs were identical to that observed in the β-zein expressing plants, indicating that the structural properties of β-zein dictate the morphology of this inclusion granule.

The putative interactions between the various zein classes are supported by results obtained by yeast two-hybrid analysis. In particular, the protein binding specificity of the 16-kD γ-zein and 15-kD β-zein suggest that they play important roles in PB formation. These two zeins showed strong homotypic interaction with themselves, and strong heterotypic interactions with each other as well as with other zeins. The 16-kD γ-zein interacts with the 50- and 27-kD γ-zeins as well as with the 19- and 22-kD α-zeins and 10-kD δ-zein. Likewise, the 15-kD β-zein interacts with the 22-kD α-zein and 10-kD δ-zein. The interacting properties of the 16-kD γ-zein and 15-kD β-zein overlapped in specificity, suggesting that they play a dominant role in the assembly and formation of the PB.

3.2
RNA Localization in Maize

The formation of ER-bound PBs in rice has been suggested to be facilitated by the specific targeting of prolamine RNAs to the PB-ER (see Sect. 4.1). Using an in situ hybridization technique to visualize the density of RNAs on thin secretions of developing maize endosperm, Kim et al. (2002) showed that the 22-kD α-zein and 27-kD γ-zein RNAs were not concentrated but, instead, distributed uniformly on the cisternal ER and PB-ER membranes. This random distribution of RNAs was consistent with earlier biochemical fractionation studies (Larkins and Hurkman 1978), which showed that microsomal membranes enriched for cisternal ER and PB-ER membranes had about the same translational capacity for zein synthesis. These results led to the conclusion that zein protein interactions determine PB assembly and that RNA localization of zein RNAs played, at best, a minor role in this process.

Curiously, when expressed heterologously in developing rice endosperm, the 10-kD δ-zein RNA displayed asymmetric localization to the PB-ER (Washida and Okita, unpublished data). Further studies showed that this RNA contained zip code sequences that targeted green fluorescent protein (GFP) RNA to the PB-ER (Washida and Okita, unpublished data). To resolve these apparent differences in RNA localization in maize and rice, the localization of RNA coding for the various zein classes together with a legumin-like protein were reexamined in maize endosperm (Washida et al. 2004). Reverse transcriptase polymerase chain reaction (RT-PCR) analysis showed that the various zein RNAs had restricted localization to the PB-ER, whereas legumin RNAs were targeted to the cisternal ER. These results support a role for RNA localization in PB formation in maize.

The basis for the apparent differences in results between these two studies (Kim et al. 2002; Washida et al. 2004) remain to be resolved. Further studies must be conducted in maize to show whether these zein RNAs actually contain zip codes that direct RNA localization to the PB-ER, and whether PB formation is dependent on the targeting of zein transcripts to these membranes.

Overall, maize PB formation is dependent on the temporal expression patterns of the various zein classes, and their intrinsic properties to interact and assemble with each other and with other zein classes to form a higher ordered, stable inclusion granule. The initial deposition of γ- and β-zeins in early development and the propensity of these zeins to form intracisternal inclusion granules by homotypic protein to protein interactions support the dominant role of these proteins in initiating PB formation. Both zein types interact with α- and δ-zeins, which enables these zeins to be retained in the ER lumen so that they eventually interact with themselves to form locules and then later the central PB core.

4
PB Formation in Rice

Unlike other plants which store predominantly a single class of storage proteins, rice accumulates three major types (Muench et al. 1999). The most abundant on a weight basis are the glutelins, storage proteins homologous to 11S globulins, which constitute up to 60% of the total protein. Two other classes are the alcohol-soluble prolamines ($\sim$ 20%) and α-globulins ($\sim$ 10%). These storage proteins are transported to and accumulated in different sites of the endomembrane system (Krishnan et al. 1986; Tanaka et al. 1980). When viewed by electron microscopy, two types of PBs are evident in the bulky endosperm: irregularly shaped, electron-dense PSVs containing glutelins and globulins, and spherical ER-bound inclusion granules containing prolamines (Fig. 1e).

The rice prolamines, which lack repetitive sequences, are composed of sulfur-rich (10-, 13-, and 16-kD molecules) and sulfur-poor (13-kD molecule) species (Ogawa et al. 1987). The sulfur-rich species comprise about 60% of the total prolamine fraction and possess the ABC modular structure of the prolamine superfamily where each domain contains a single or a pair of conserved cysteine residues (Shewry et al. 1995). When viewed by electron microscopy, the spherical ER-derived PBs possess a high electron density core with a surrounding lamellar structure (Bechtel and Juliano 1980; Ogawa et al. 1987; Yamagata and Tanaka 1986). Immunocytochemical analysis indicates that the sulfur-rich prolamines reside in the electron-dense core region surrounded by sulfur-poor prolamines (Ogawa et al., unpublished data). Temporal expression studies (Ogawa, unpublished data) show that the sulfur-rich 10- and 13-kD prolamines are initially deposited within the ER lumen to form the electron-dense core followed by the accumulation of the other prolamine species which form the peripheral lamellar structure. This spatial relationship between sulfur-rich and sulfur-poor prolamines within the intracisternal inclusion granule is just the opposite to that seen in maize, where the sulfur-poor zeins constitute the core which, in turn, is surrounded by the sulfur-rich species (Lending et al. 1988). The PBs found in the endosperm of sorghum (Shull et al. 1992) and *Setaria* (Rost 1971) resemble the rice prolamine PBs, suggesting that they are formed by a similar process.

The mechanism by which prolamine polypeptides are assembled into a structurally ordered stratified inclusion granule remains largely unexplored. As suggested for maize zein PB formation, specific protein to protein interactions among the various rice prolamine classes are likely responsible for the ordered spatial organization of the PBs. In addition to homotypic and heterotypic interactions, there is compelling evidence that the lumenal chaperones are actively involved in this cellular process. Rice endosperm contains unusually high amounts of the lumenal chaperone BiP, which can be easily detected by sodium dodecyl sulfate polyacrylamide gel electrophoresis (SDS-PAGE) of PB extracts (Li et al. 1993a). This lumenal chaperone was found to be associated with nascent prolamine chains bound to polysomes, free prolamine polypeptides, and on the surface of the prolamine inclusion granule. The association with the latter was ATP dependent. Moreover, immunocytochemical studies showed that BiP was not uniformly distributed to the ER but, instead, concentrated on the periphery of the prolamine PBs (Muench et al. 1997). Overall, these observations indicate a role for BiP in prolamine polypeptide folding and assembly to form an intracisternal inclusion granule.

Glutelin synthesis is detected in 5-day-old developing seeds, whereas the onset of prolamine accumulation occurs 5 days later (Yamagata and Tanaka 1986). Glutelins are initially synthesized on ER-bound polysomes (Yamagata and Tanaka 1986) as a 57-kD precursor, which is then cleaved to acidic and basic subunits within the PSV (Yamagata et al. 1982). Transport from the ER to the PSV is through the Golgi apparatus, an organelle containing two highly curved cisternae with attached glutelin-containing dense vesicles (Krishnan et al. 1990). Larger dense vesicles were evident in the cytoplasm, suggesting that they were formed by fusion of smaller vesicles (Oparka and Harris 1982). The mature PSV is composed of glutelin crystalline inclusions separated by an electron-dense matrix (Bechtel and Juliano 1980; Krishnan et al. 1986).

Few studies have been directed at understanding the expression of the third most abundant rice storage protein, the α-globulins. These proteins are synthesized on the ER and exported to the Golgi apparatus where they form dense vesicles, which are then transported to the PSV (Krishnan et al. 1992). The globulins accumulate in the matrix located at the periphery and the border regions between the glutelin crystalline inclusions. This partitioning of α-globulins from glutelins within the PSV suggests that these proteins are not cotransported together to the PSV but, instead, use different pathways.

Despite its solubility in saline solutions, the 26-kD α-globulin is a member of the prolamine superfamily (Krishnan and Pueppke 1993). Kawagoe et al. (2005) showed in an elegant study that the conserved cysteine residues in the motifs LxxC (A domain), CCxQL (B domain), and PxxC (C domain) formed two intra-disulfide bonds, which were essential for proper polypeptide folding and intracellular localization. GFP fusions containing the AB domain were found to be mislocalized to the prolamine PB, while the same protein containing a C79S ($CC_{79}xQL$ of the B domain) replacement was properly sorted to the PSV. As cysteine-79 forms a disulfide bond with the conserved cysteine (PxxC) of the C domain, the unreacted residue interacts with a sulfur-rich prolamine resulting in its retention within the ER lumen. Replacement of cysteine-79 enables the GFP fusion to fold properly and gain competence for protein export.

4.1 Localization of Rice Storage Protein RNAs

The presence of a signal peptide and translation by membrane-bound polysomes indicates the synthesis of these storage proteins on the ER (Muench et al. 1999; Takaiwa et al. 1999). Early studies have suggested that the storage proteins were not randomly synthesized on this membrane complex but distributed to specific subdomains. In vitro translation of polyA$^+$-RNA isolated from a highly enriched prolamine PB fraction produced only prolamine, indicating that prolamine mRNA was highly enriched on the PB-ER (Yamagata et al. 1986). In contrast, glutelins were present at more than twofold greater levels than prolamine transcripts RNA in a microsomal fraction enriched for cisternal ER vesicles (Kim et al. 1993). The nonrandom distribution of these storage protein RNAs on the ER subdomains was unequivocally demonstrated by results obtained by biochemical and high-resolution in situ hybridization studies (Li et al. 1993b). Prolamine RNAs were highly enriched on the PB-ER while glutelin RNA predominated on the cisternal ER. By assessing the spatial location of RNA transcripts from synthetic prolamine gene constructs in transgenic developing endosperm tissue, targeting to PB-ER was dependent on signals contained within the prolamine RNA and not its primary sequence (Choi et al. 2000). Interestingly, protein synthesis was required for PB-ER localization, although this requirement could be provided by translatable reporter gene sequences such as β-glucuronidase (GUS) or GFP independent of prolamine sequences. Deletion analysis confirmed the presence of two partially redundant *cis* elements or zip codes, located in the prolamine RNA coding sequence and 3′ untranslated region, which were required for PB-ER targeting (Hamada et al. 2003a). RNA transport was shown to occur by the formation and movement of a particle, the latter dependent on intact actin filaments but not on intact microtubules (Hamada et al. 2003b).

4.2 Storage Protein Mutants

Three mutants *esp1*, *esp3*, and *Esp4*, generated by *N*-methyl-*N*-nitrosourea mutagenesis, contained significant alterations in prolamine accumulation. The *esp1* mutant was deficient in sulfur-poor 13-kD polypeptides, while *esp3* and *Esp4* contained depressed and elevated levels of the sulfur-rich prolamines, respectively (Kumamaru et al. 1987, 1988). The spherical PBs in *esp3* and *Esp4* mutations lacked the typical lamellar structure, although *Esp4* PBs showed an enlarged electron-dense core surrounded by a region of uniform electron density while PBs of *esp1* had the same structure as wild type (Ogawa et al. 1989), suggesting that sulfur-rich prolamines are essential for the core formation of ER-derived PBs.

Several *glup* mutants have been identified that contain abnormal elevated amounts of the 57-kD glutelin precursor, which is normally processed to acidic and basic subunits in the PSV compartment. One mutant, *esp2*, lacks a functional 60-kD protein disulfide isomerase (PDI) (Takemoto et al. 2002). Electron microscopic analysis showed that PSVs were normal in appearance but that the formation of prolamine-accumulating PBs was significantly altered where numerous small, low electron density ER-derived PBs of uniform size (0.5 μm in diameter) were readily observed in place of the normally larger spherical PBs possessing stratified regions of varying electron density. Immunocytochemical and biochemical studies showed that these abnormally small PBs contained glutelin as well as prolamine bound together by intermolecular disulfide bonds (Takemoto et al. 2002). The retention of glutelins within the ER lumen via their interaction with prolamine polypeptides indicates that this lumenal chaperone is required for intramolecular disulfide bond formation for these storage proteins. Glutelin is not absolutely dependent on PDI for intramolecular S – S bond formation as normal PSVs are readily evident, which may occur especially in 5- to 10-day-old developing seeds where glutelin synthesis occurs in the absence of prolamine. PDI may be essential at 10 days and later where synthesis and accumulation of these storage proteins are occurring at their maximum rates.

Immunocytochemistry of the *glup4* mutant showed a significant disruption in transport of glutelins to the PSV. Instead of normal PSVs 2–3 μm in size, numerous smaller cytoplasmic inclusions were present. Moreover, regions of the ER were highly distended and contained glutelins. Map-based cloning studies showed that the *glup4* gene encodes a Rab5a-like protein, which possesses GTPase activity and functions in vesicle traffic (Satoh, unpublished data). The accumulation of glutelin precursor within the ER in *glup4* endosperm may be due to a defect in the export of glutelins from the ER.

Analysis of the *glup3* mutant showed that the PSVs contained large quantities of unprocessed glutelin precursors, suggesting a defect in proteolytic processing (Kumamaru, unpublished data). Subsequent analysis revealed that the vacuolar processing enzyme (VPE) activity in the developing seeds from three independent *glup3* lines in rice was markedly reduced compared to wild type. DNA sequencing of the VPE gene in these *glup3* lines revealed single or multiple amino acid replacements or the formation of premature stop codons in the coding region. The *glup3* gene encodes a major VPE, which is responsible for the proteolytic processing of glutelin precursor within the PSV.

5 Conclusions

The major cereal grains deposit their reserve protein in the PSV or ER lumen. Irrespective of the endomembrane site used for storage, the processes responsible for the formation of the inclusion granule within the ER lumen or within Golgi-associated dense vesicles are likely to be functionally similar. In the ER lumen, the storage proteins must be correctly folded into a competent state, a process facilitated by chaperones, for subsequent protein to protein interactions or for their export from the ER. Another shared requirement for PB formation is that the storage proteins must be concentrated, a condition conducive for protein to protein interactions, at its initial site of inclusion granule formation, whether it is the ER lumen or Golgi-associated dense vesicles. Localization of RNAs to a defined ER subdomain, such as the PB-ER in rice endosperm, would effectively concentrate the levels of the newly synthesized prolamine polypeptide at this site. The assembly of storage proteins within Golgi-associated dense vesicles is not known, although one obvious mechanism would be the retrograde transport of vesicles from the *cis*-Golgi to the ER resulting in the concentration of the storage proteins within this compartment and their assembly into an inclusion within the dense vesicle. Lastly, the site of final deposition utilized by the different cereals may depend on the relative rates of storage protein synthesis, rates of polypeptide folding, and rates of export from the ER as inferred by the example seen for the barley Nevsky line.

The proposed Golgi-independent, direct ER to PSV pathway requires further study for its role in PB formation in wheat. Questions that remain to be addressed include the biochemical processes involved in the release of the intracisternal inclusion granule from the ER and the autophagic uptake of the intracisternal inclusion granule into a vacuole, the biochemical basis for the trapping of BiP within the inclusion granules, and the determination of the relative contribution of this pathway compared to the Golgi-dependent pathway in PB formation. These studies should provide critical insights into how closely related wheat and barley can exploit distinct cellular pathways for the packaging of nearly identical proteins to the PSV.

Acknowledgements Our ongoing collaborative research activities are supported by a Grant-in-Aid for Scientific Research from the Japanese Society for the Promotion of Science, No. 16830009, and grants from the National Research Initiative of the USDA Cooperative State Research, Education, and Extension Service, No. 2003-35301-13270, and the National Science Foundation IOB-0544469.

References

Bagga S, Adams H, Kemp JD, Sengupta-Gopalan C (1995) Accumulation of 15-kD zein in novel protein bodies in transgenic tobacco. Plant Physiol 107:13–23

Bagga S, Adams HP, Rodriguez FD, Kemp JD, Sengupta-Gopalan C (1997) Coexpression of the maize δ-zein and β-zein genes results in stable accumulation of δ-zein in endoplasmic reticulum-derived protein bodies formed by β-zein. Plant Cell 9:1683–1696

Bechtel DR, Juliano BO (1980) Formation of protein bodies in the starchy endosperm of rice (*Oryza sativa* L.): a reinvestigation. Ann Bot 65:684–691

Bechtel DB, Gaines RL, Pomeranz Y (1982) Protein secretion in wheat endosperm—formation of the matrix protein. Cereal Chem 59:336–343

Buttrose MS (1963) Ultrastructure of the developing wheat endosperm. Aust J Biol Sci 16:305–317

Cameron-Mills V, von Wettstein D (1980) Protein body formation in the developing barley endosperm. Carlsberg Res Commun 45:577–594

Campbell WP, Lee JW, O'Brien TP, Smart MG (1981) Endosperm morphology and protein body formation in developing wheat grains. Aust J Plant Physiol 8:5–19

Choi S-B, Wang C, Muench DG, Ozawa K, Franceschi VR, Wu Y, Okita T (2000) Messenger RNA targeting of rice seed storage proteins to specific ER subdomains. Nature 407:765–767

Coleman CE, Larkins BA (1999) The prolamins of maize. In: Shewry PR, Casey R (eds) Seed proteins. Kluwer, Dordrecht, pp 109–139

Coleman CE, Herman EM, Takasaki K, Larkins BA (1996) The maize γ-zein sequesters α-zein and stabilizes its accumulation in protein bodies of transgenic tobacco endosperm. Plant Cell 8:2335–2345

Coleman CE, Yoho PR, Escobar S, Ogawa M (2004) The accumulation of α-zein in transgenic tobacco endosperm is stabilized by co-expression of β-zein. Plant Cell Physiol 45:864–871

Esen A, Stetler DA (1992) Immunocytochemical localization of δ-zein in the protein bodies of maize endosperm cells. Am J Bot 79:243–248

Galili G (1997) The prolamin storage proteins of wheat and its relatives. In: Larkins BA, Vasil IK (eds) Cellular and molecular biology of plant seed development. Kluwer, Dordrecht, pp 221–256

Galili G, Herman EM (1997) Protein bodies: storage vacuoles in seeds. Adv Bot Res 25:113–140

Galili G, Altschuler Y, Levanony H, Giorini-Silfen S, Shimoni Y, Shani N, Karchi H (1995) Assembly and transport of wheat storage proteins. J Plant Physiol 145:626–631

Geli MI, Torrent M, Ludevid D (1994) Two structural domains mediate two sequential events in γ-zein targeting: protein endoplasmic reticulum retention and protein body formation. Plant Cell 6:1911–1922

Hamada S, Ishiyama K, Choi SB, Wang C, Singh S, Kawai N, Franceschi VR, Okita TW (2003a) The transport of prolamine RNAs to prolamine protein bodies in living rice endosperm cells. Plant Cell 15:2253–64

Hamada S, Ishiyama K, Sakulsingharoj C, Choi SB, Wu Y, Wang C, Singh S, Kawai N, Messing J, Okita TW (2003b) Dual regulated RNA transport pathways to the cortical region in developing rice endosperm. Plant Cell 15:2265–72

Hinchliffe DJ, Kemp JD (2002) β-Zein protein bodies sequester and protect the 18-kD δ-zein protein from degradation. Plant Sci 163:741–752

Hoffman LM, Donaldson DD, Bookland R, Rashka K, Herman EM (1987) Synthesis and protein deposition of maize 15-kD zein in transgenic tobacco seeds. EMBO J 6:3213–3221

Kawagoe Y, Suzuki K, Tasaki M, Yasuda H, Akagi K, Katoh E, Nishizawa NK, Ogawa M, Takaiwa F (2005) The critical role of disulfide bond formation in protein sorting in the endosperm of rice. Plant Cell 17:1141–1153

Kim CS, Woo Y-M, Clore AM, Burnett RJ, Carneiro NP, Larkins BA (2002) Zein protein interactions, rather than the asymmetric distribution of zein mRNAs on endoplasmic reticulum membranes, influence protein body formation in maize endosperm. Plant Cell 14:655–672

Kim WT, Franceschi VR, Krishnan HB, Okita TW (1988) Formation of wheat protein bodies: involvement of the Golgi apparatus in gliadin transport. Planta 173:173–182

Kim WT, Li X, Okita TW (1993) Expression of storage protein multigene families in developing rice endosperm. Plant Cell Physiol 34:595–603

Krishnan HB, Pueppke SG (1993) Nucleotide sequence of an abundant rice seed globulin: homology with the high molecular weight glutelins of wheat, rye and triticale. Biochem Biophys Res Commun 193:460–466

Krishnan HB, Franceschi VR, Okita TW (1986) Immunochemical studies on the role of the Golgi complex in protein body formation in rice seeds. Planta 169:471–480

Krishnan HB, White JA, Pueppke SG (1990) Immunocytochemcial evidence for the involvement of the Golgi apparatus in the transport of the vacuolar protein, gamma-secalin, in rye (*Secale cereale*) endosperm. Cereal Chem 67:360–366

Krishnan HB, White JA, Pueppke SG (1991) Immunocytochemical localization of wheat prolamins in the lumen of the rough endoplasmic reticulum. Can J Bot 69:2574–2577

Krishnan HB, White JA, Pueppke SG (1992) Characterization and localization of rice (*Oryza sativa* L.) seed globulins. Plant Sci 81:1–11

Kumamaru T, Satoh H, Iwata N, Omura T, Ogawa M (1987) Mutants for rice storage proteins. 3. Genetic analysis of mutants for storage proteins of protein bodies in the starchy endosperm. Theor Jpn J Genet 62:333–339

Kumamaru T, Satoh H, Iwata N, Omura T, Ogawa O, Tanaka K (1988) Mutants for rice storage proteins. 1. Screening of mutants for rice storage proteins of protein bodies in the starchy endosperm. Theor Appl Genet 76:11–16

Larkins BA, Hurkman WJ (1978) Synthesis and deposition of zein protein bodies of maize endosperm. Plant Physiol 62:256–263

Larkins BA, Vasil IK (1997) Cellular and molecular biology of plant seed development. Kluwer, Dordrecht

Lending CR, Larkins BA (1989) Changes in the zein composition of protein bodies during maize endosperm development. Plant Cell 1:1011–1023

Lending CR, Kriz AL, Larkins BA, Bracker CE (1988) Structure of maize protein bodies and immunocytochemical localization of zeins. Protoplasma 143:51–62

Levanony H, Rubin R, Altschuler Y, Galili G (1992) Evidence for a novel route of wheat storage proteins to vacuoles. J Cell Biol 119:1117–1128

Li X, Wu Y, Zhang D-Z, Gillikin JW, Boston R, Franceschi VR, Okita TW (1993a) Rice prolamine protein body biosynthesis: a BiP-mediated process. Science 262:1054–1056

Li X, Franceschi VR, Okita TW (1993b) Segregation of storage protein mRNAs on the rough endoplasmic reticulum membranes of rice endosperm cells. Cell 72:869–879

Miflin BJ, Burgess SR, Shewry PR (1981) The development of protein bodies in the storage tissues of seeds: subcellular separation of homogenates of barley, maize and wheat endosperm and pea cotyledon. J Exp Bot 32:199–219

Møgelsvang S, Simpson DJ (1998) Changes in the levels of seven proteins involved in polypeptide folding and transport during endosperm development of two barley genotypes differing in storage protein localisation. Plant Mol Biol 36:541–552

Muench DG, Wu Y, Zang Y, Li X, Boston R, Okita TW (1997) Molecular cloning, expression and subcellular localization of a BiP homolog from rice endosperm tissue. Plant Cell Physiol 38:404–412

Muench DG, Ogawa M, Okita TW (1999) The prolamins of rice. In: Shewry PR, Casey R (eds) Seed proteins. Kluwer, Dordrecht, pp 93–108

Ogawa M, Kumamaru T, Satoh H, Iwata N, Omura T, Kasai Z, Tanaka K (1987) Purification of protein body-I of rice seed and its polypeptide composition. Plant Cell Physiol 28:1517–1527

Ogawa M, Kumamaru T, Satoh H, Omura T, Park T, Shintaku K, Baba K (1989) Mutants for rice storage proteins. 2. Isolation and characterization of protein bodies from rice mutants. Theor Appl Genet 78:305–310

Oparka N, Harris N (1982) Rice protein body formation: all types are initiated by dilation of the endoplasmic reticulum. Planta 154:184–188

Parker ML (1982) Protein accumulation in developing endosperm of a high-protein line of *Triticum dicoccoides*. Plant Cell Environ 5:237–243

Rechinger KB, Simpson DJ, Svendsen I, Cameron-Mills V (1993) A role for γ_3-hordein in the transport and targeting of prolamin polypeptides to the vacuole of developing barley endosperm. Plant J 4:841–853

Rost TL (1971) Fine structure of endosperm protein bodies in *Setaria lutescens* (*Gramineae*). Protoplasma 73:475–479

Rubin R, Levanony H, Galili G (1992) Evidence for the presence of two different types of protein bodies in wheat endosperm. Plant Physiol 99:718–724

Shewry PR, Casey R (eds) (1999) Seed proteins. Kluwer, Dordrecht

Shewry PR, Napier NA, Tatham AS (1995) Seed storage proteins: structures and biosynthesis. Plant Cell 7:945–956

Shotwell M, Larkins BA (1989) The biochemistry and molecular biology of seed storage proteins. In: Marcus E (ed) The biochemistry of plants: a comprehensive treatise. Academic, Orlando, FL, pp 296–345

Shull JM, Watterson JJ, Kirleis AW (1992) Purification and immunocytochemical localization of karifins in *Sorghum bicolor* (L Moench) endosperm. Protoplasma 171:64–74

Takaiwa F, Ogawa M, Okita TW (1999) Rice glutelin. In: Casey R, Shewry PR (eds) Seed proteins. Chapman and Hall, London, pp 401–426

Takemoto Y, Coughlan SJ, Okita TW, Satoh H, Ogawa M, Kumamaru T (2002) Protein disulfide isomerase is essential for the localization of prolamin and glutelin storage proteins to separate intracellular compartments. Plant Physiol 128:1212–1222

Tanaka K, Sugimoto T, Ogawa M, Kasai Z (1980) Isolation and characterization of two types of protein bodies in the rice endosperm. Agric Biol Chem 44:1633–1639

Tooze J, Kern HF, Fuller SD, Howell KE (1989) Condensation-sorting events in the rough endoplasmic reticulum of exocrine pancreatic cells. J Cell Biol 109:35–50

Washida H, Sugino A, Messing J, Esen A, Okita TW (2004) Asymmetric localization of seed storage protein RNAs to distinct subdomains of the endoplasmic reticulum in developing maize endosperm cells. Plant Cell Physiol 45:1830–1837

Woo Y-M, Hu DW-N, Larkins BA, Jung R (2001) Genomics analysis of genes expressed in maize endosperm identifies novel seed proteins and clarifies patterns of zein gene expression. Plant Cell 13:2297–2317

Yamagata H, Tanaka K (1986) The site of synthesis and accumulation of rice storage proteins. Plant Cell Physiol 27:135–145

Yamagata H, Tanaka K, Kasai Z (1982) Evidence for a precursor form of rice glutelin subunits. Agric Biol Chem 46:321–322

Yamagata H, Tamura K, Tanaka K, Kasai Z (1986) Cell-free synthesis of rice prolamin. Plant Cell Physiol 27:1419–1422

Yamagata T, Kato H, Kuroda S, Abe S, Davies E (2003) Uncleaved legumin in developing maize endosperm: identification, accumulation and putative subcellular localization. J Exp Bot 54:913–922

Plant Cell Monogr (8)
O.-A. Olsen: Endosperm
DOI 10.1007/7089_2007_113/Published online: 15 March 2007

Endosperm Cell Walls: Formation, Composition, and Functions

Marisa S. Otegui

Department of Botany, University of Wisconsin, 430 Lincoln Drive, Madison, WI 53706, USA
otegui@wisc.edu

Abstract This chapter deals with the mechanisms by which endosperm cell walls arise, with the composition and synthesis of endosperm cell wall polysaccharides, and with their specific functions during seed development and germination. The endosperm cell walls differ from somatic cell walls in polysaccharide composition. Endosperm walls often contain massive amounts of storage hemicelluloses such as mannan, galactomannan, glucomannans, xyloglucans, (1,3;1,4)-β-glucans but very little of cellulose. These hemicelluloses act as carbohydrate stores for the germinating embryo and, depending on their specific physico-chemical properties, may also play additional roles in water retention during post-imbibition drought, mechanical protection of the embryo, and signaling. In recent years, many of the genes responsible for the synthesis of endosperm cell wall polysaccharides have been identified. Endosperm cell walls have also evolved special mechanisms of formation. In the nuclear-type of endosperm development, which is the most common type in all analyzed species, the endosperm becomes multinucleate (syncytial) before cellularizing. Cellularization proceeds by the simultaneous formation of syncytial-type cell plates between sister and non-sister nuclei. Syncytial-type cell plates form at the boundaries of nuclear cytoplasmic domains, which are defined by radial systems of microtubules that organize on the envelope of endosperm nuclei. These cell plates are associated with clusters of microtubules called mini-phragmoplasts that transport the cell plate-forming vesicles to the division plane. In contrast to phragmoplasts in dividing somatic cells, mini-phragmoplasts never form a coherent cytoskeletal structure around the expanding cell plate. Syncytial-type cell plates undergo several developmental stages before completion of the new cell wall: vesicle fusion intermediates, wide tubules and wide tubular networks, convoluted fenestrated sheets, and planar fenestrated sheets.

1 Introduction

Primary cell walls consist of polysaccharides, smaller proportions of glycoproteins and, in some specialized cell types, various non-carbohydrate substances such as lignin, suberin, cutin, or silica. Wall polysaccharides are traditionally classified into three categories: pectins, hemicelluloses, and cellulose. Pectins and hemicelluloses are components of the wall "matrix", within which the cellulosic microfibrils are embedded. Cell wall formation in plants is mediated by the phragmoplast, an array of microtubules, actin filaments, and associated molecules that arise from the mitotic spindle remnants and act as a framework for the growing cell plate. In most dividing cells, the site where the new cell wall will form is determined at the beginning of mitosis by a cortical array of microtubules called preprophase band (Mineyuki 1999). Endosperm cell walls are highly specialized and differ from typical somatic cell walls in their composition, functions, and in most cases, in their mechanism of formation (cytokinesis).

The endosperm cell walls are often thickened due to the presence of massive amounts of hemicelluloses. These specialized cell walls represent an important source of carbohydrates during germination and, in some cases, they play additional functions in water retention, signaling, mechanical protection of the embryo, and dormancy (Reid 1985; DeMason 1997; Buckeridge et al. 2000). In many seeds, the embryo cell walls are also specialized in polysaccharide storage. However, in general terms, the endosperm can store much larger amounts of cell wall storage polysaccharides. In fact, in some extreme cases like fenugreek (*Trigonella foenum-graecum*) seeds, the endosperm cell walls almost completely occlude the cell lumen and, at seed maturity, the endosperm is a dead tissue except for an outer layer of living aleurone cells (Meier and Reid 1977).

This chapter deals with the mechanisms by which endosperm cell walls arise, with the composition and synthesis of endosperm cell wall polysaccharides, and with their specific functions during seed development and germination.

2
Mechanisms of Cell Wall Formation in the Endosperm

The temporal and spatial pattern of cell wall formation has been one of the key features in defining the three traditional categories of endosperm development: cellular-, nuclear-, and helobial-types (see Brown and Lemmon, in this volume). Whereas in the cellular-type endosperm cytokinesis is coupled to mitosis, several cycles of nuclear division occur in the nuclear-type endosperm before cell walls are formed. The helobial-type is a special developmental variant in which the first mitotic division of the endosperm cell leads to the formation of two unequally sized uninucleate cells. Later in development, both cells become multinucleate but only the largest one located toward the micropylar region cellularizes. The nuclear-type is the most common pathway of endosperm development. The cytokinetic mechanism leading to the cellularization of the nuclear-type endosperm has been extensively studied in cereals and *Arabidopsis thaliana* (see Berger et al., in this volume). The focus of this chapter is the mechanism of cell plate formation in the *Arabidopsis* endosperm as revealed by studies that analyze cryofixed/freeze-substituted endosperm samples by conventional electron microscopy, immunolabeling, and electron tomography.

2.1
Syncytial-Type Cell Wall Formation in *Arabidopsis*

Nuclear-type endosperms undergo several cycles of nuclear division and repositioning of nuclei before cellularization. In *Arabidopsis*, the cellularization process, which involves the simultaneous formation of multiple cell plates between sister and non-sister nuclei, starts at the micropylar domain when the embryo reaches the globular stage of development (Fig. 1) (Otegui and Staehelin 2000b). This particular type of cytokinesis is called syncytial-type cytokinesis and differs from somatic-type cytokinesis in two important aspects: (1) there are no pre-prophase bands of microtubules determining the position of the new cell walls (Mineyuki 1999); and (2) the phragmoplasts associated with syncytial-type cell plate do not arise from mitotic spindles but from sets of overlapping microtubules located at the boundaries of nuclear cytoplasmic domains (Otegui and Staehelin 2000a,b; Brown and Lemmon 2001a).

Just before syncytial-type cytokinesis occurs, microtubules assemble into microtubule-organizing centers associated with the outer membrane of the nuclear envelope (Canaday et al. 2000) and radiate outwards, intersecting the microtubules extending from adjacent nuclei (Fig. 1b). The portion of cytoplasm encompassed by a radial array of microtubules is called a nuclear–cytoplasmic domain (Brown et al. 1994; Pickett-Heaps et al. 1999; Brown and Lemmon 2001b). At this stage, the endosperm nuclei are located in a peripheral cytoplasmic layer lining the plasma membrane and the central vacuole (or the developing embryo in the endosperm micropylar domain). Cellularization starts with the formation of phragmoplasts at the boundaries of adjacent nuclear cytoplasmic domains, where opposite sets of microtubules overlap (Fig. 1c). These phragmoplasts consist of clusters of approximately 20 microtubules and they are called mini-phragmoplasts (Otegui and Staehelin 2000b; Otegui et al. 2001). Although conventional phragmoplasts in somatic cytokinesis also arise as microtubule clusters or phragmoplast initials, in later stages of cytokinesis they tend to form a continuous structure around the developing cell plate (Segui-Simarro et al. 2004; Austin et al. 2005), whereas in the endosperm, mini-phragmoplasts never form a coherent structure.

The combination of high-pressure freezing/freeze substitution and electron tomography have revealed amazing details about the complex three-dimensional organization of membranes and cytoskeletal elements in the cytokinetic apparatus of syncytial-type cell plates in the endosperm of *Arabidopsis* (Otegui and Staehelin 2000a; Otegui et al. 2001; Otegui and Austin 2006). Golgi-derived vesicles travel along mini-phragmoplast microtubules and fuse with each other within a dense matrix called the cell plate assembly matrix (CPAM). The exact composition of the CPAM is not known, but it appears to mediate membrane fusion events and to stabilize microtubule ends during both somatic- and syncytial-type cytokinetic processes.

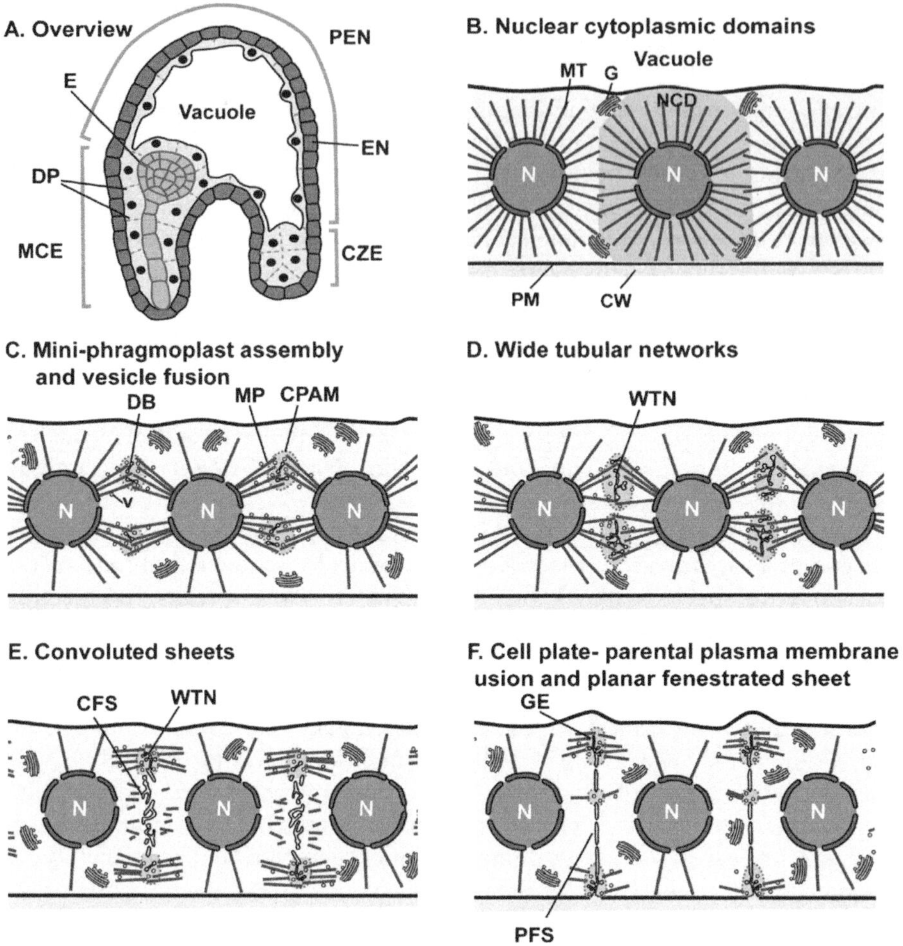

Fig. 1 Model depicting the main developmental stages in the assembly of syncytial-type cell plates. **a** Overview of a developing *Arabidopsis* seed just prior to endosperm cellularization. Cellularization starts at the micropylar endosperm domain (*MCE*) when the embryo (*E*) reaches the globular stage. **b** At this stage, the nuclear endosperm is organized into nuclear cytoplasmic domains (*NCD*) defined by radial systems of microtubules (*MT*). **c** Cellularization starts with the assembly of mini-phragmoplast (*MP*) at the boundaries of adjacent nuclear cytoplasmic domains and vesicle fusion at the division plane. **d** Assembly of multiple wide tubular networks (*WTN*) across the division plane. **e** Formation of a coherent cell plate by fusion of adjacent wide tubular networks and maturation of the central cell plate domains into convoluted fenestrated sheets (*CFS*). **f** Fusion of the cell plate with the parental plasma membrane and conversion of convoluted sheets into planar fenestrated sheets (*PFS*). *CPAM* cell plate assembly matrix, *CW* cell wall, *CZE* chalazal endosperm, *DB* dumbbell-shaped intermediate, *DP* division plane, *EN* endothelium, *G* Golgi stack, *GE* cell plate growing edge, *N* nucleus, *PEN* peripheral endosperm, *PM* plasma membrane, *V* Golgi-derived vesicle

As vesicles fuse, they first form dumbbell intermediates and later, they give rise to wide tubules of approximately 45–50 nm in diameter. These wide tubules become longer and branched, and fuse with each other originating a wide tubular network at every mini-phragmoplast site (Figs. 1d and 2a). These wide tubular structures appear to be stabilized by the presence of DRP1A (dynamin-related protein 1A) helical polymers, which are organized in rings and spirals that constrict the cell plate tubules (Figs. 2b and 2b′) (Otegui et al. 2001). As the wide tubular networks assemble in an expanding division plane, they fuse with each other forming a coherent cell plate between two adjacent endosperm nuclei. By this time, callose starts to be synthesized in the forming cell plate and the central, more mature wide tubular network domains convert into a system of highly convoluted fenestrated sheets, while the expanding cell plate edges keep on incorporating new vesicles and wide tubules (Figs. 1e and 3a).

As the cell plate reaches the parental plasma membrane, it fuses with it. This appears to trigger the massive removal of membrane by the budding off of clathrin-coated vesicles and the conversion of the convoluted fenestrated sheets into a more planar system of fenestrated sheets (Figs. 1f and 3b). In fact, up to 75% of the total membrane surface area is removed from the cell plate at this time. Also, the CPAM disappears from the planar fenestrated sheet domains, although it remains at the cell plate growing edges (Otegui et al. 2001).

Due to the asymmetric architecture of the *Arabidopsis* endosperm (see Berger et al., in this volume), the cell plate forming in different endosperm domains have different fates. Those cell plates growing in the micropylar domain, between the embryo and the parental endosperm cell wall facing the endothelium, fuse with one another and with the parental plasma membrane at both the embryo and the endothelium sides, and mature into cell walls. In contrast, cell plates assembling in the thin layer of cytoplasm located between the plasma membrane and the tonoplast of the central vacuole, fuse first with the parental plasma membrane and with the adjacent cell plates, whereas the cell plate edges facing the central vacuole keep on growing (Fig. 1f). This particular pattern of cellularization leads to the formation of open alveoli containing one nucleus each. Once these first set of anticlinal cell walls have formed, the following cycles of nuclear division are coupled to cytokinesis (Sorensen et al. 2002; Olsen 2004).

Although syncytial- and somatic-type cytokinetic mechanisms are overall similar and utilize many of the same enzymatic and regulatory molecules (Jürgens 2005), syncytial-type cell plates exhibit specific architectural differences that relate to the spatial and developmental features of the endosperm. For example, whereas syncytial-type cell plates develop wide tubular membrane network intermediates, somatic type cell plates in meristematic cells exhibit tubulo-vesicular network/fenestrated sheet geometries (Segui-Simarro et al. 2004). It is reasonable to believe that the tubular net-

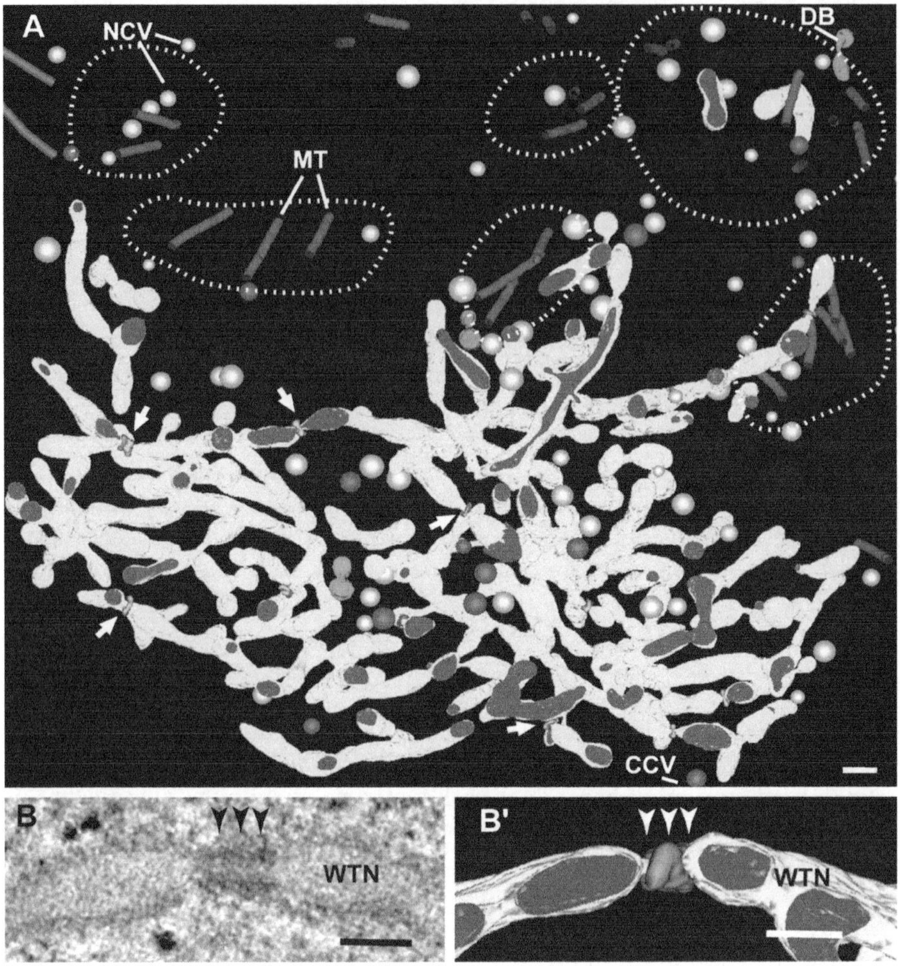

Fig. 2 Tomographic analysis of syncytial-type cell plate assembly in the *Arabidopsis* endosperm. **a** Tomographic reconstruction of a wide tubular network. The mini-phragmoplasts located at the edges of the cell plates are encircled by *dashed lines*. Numerous vesicles and dumbbell fusion intermediates (*DB*) are found in the vicinity of the cell plate. For clarity, the cell plate assembly matrix is not shown in this reconstruction but it encloses the totality of the wide tubular network. **b** Tomographic slice and **b′** corresponding tomographic model of a tubular domain of an endosperm cell plate. Note the presence of an electron dense collar (*arrowheads*) corresponding to a DRP1A polymer (Otegui et al. 2001) associated with a constricted membrane tubule. *CCV* clathrin-coated vesicle, *MT* microtubules, *NCV* non-coated vesicle, *WTN* wide tubular network. *Scales* = 100 nm in **a**; 50 nm in **b** and **b′**. Panel **a** reprinted from Otegui et al. (2001, © 2001 American Society of Plant Biologists)

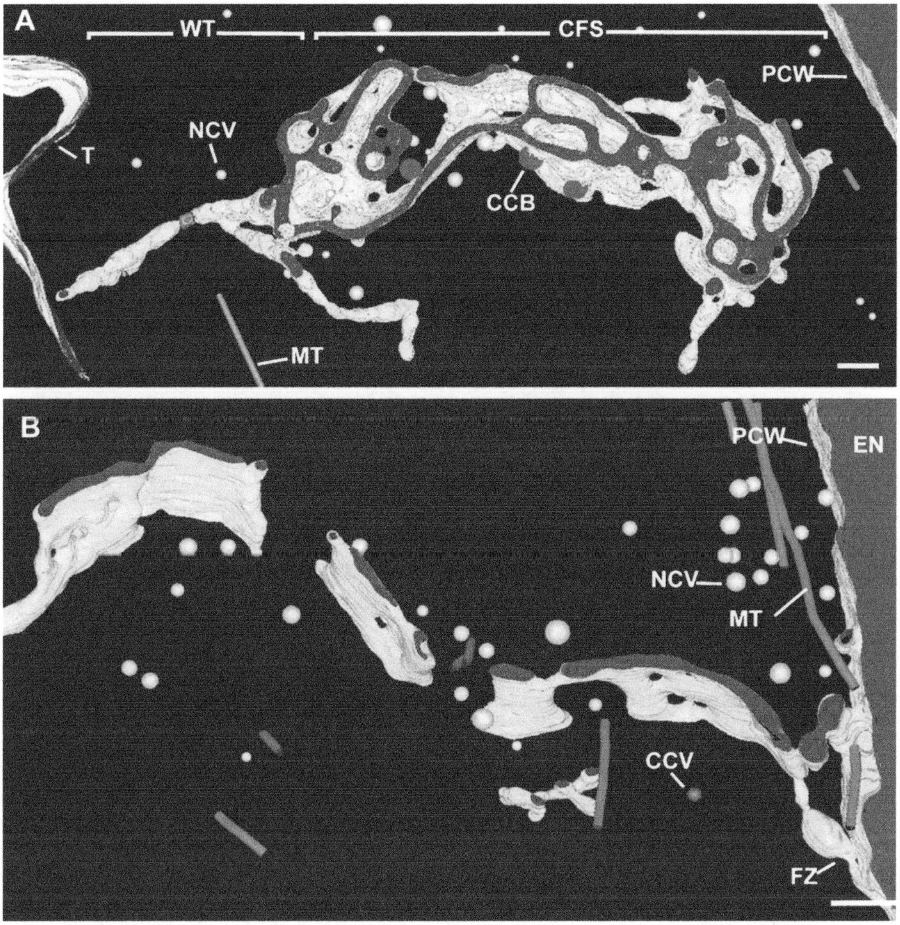

Fig. 3 Tomographic analysis of syncytial-type cell plates in the *Arabidopsis* endosperm. **a** Tomographic reconstruction of a cell plate consisting of a central convoluted fenestrated sheet domain (*CFS*) and peripheral wide tubules (*WT*) at the growing edges. At the *left*, the cell plate growing edge extends toward the tonoplast of the central vacuole, whereas at the *right*, the growing edge is close to the parental cell wall (*PCW*). **b** Portion of a planar fenestrated sheet already fused to the parental cell wall via two wide tubules. A cluster of mini-phragmoplast microtubules appears to be directing vesicles to the fusion zone (*FZ*). *CCB* clathrin-coated bud, *CCV* clathrin-coated vesicle, *EN* endothelium, *MT* microtubule, *NCV* non-coated vesicle, *T* tonoplast. *Scales* = 100 nm. Panels **a** and **b** reprinted from Otegui et al. (2001, © 2001 American Society of Plant Biologists)

work configuration provides more mechanical flexibility to the cell plates in the endosperm, which rapidly changes in size and shape as the embryo enlarges.

As in the somatic-type cytokinesis, syncytial-type cell plates also synthesize callose during maturation (Samuels et al. 1995; Otegui and Staehelin

2000b; Hong et al. 2001). However, in contrast to somatic-type cytokinesis, the callose deposits produced in the syncytial cell plates persist after completion of the cell walls. Once again, this peculiar variation in cell wall composition might relate to the specific functions of the endosperm. Callose-rich endosperm walls might be advantageous because: (1) in contrast to cellulose, callose can be readily broken down into glucose residues by specific cell wall enzymes (Verma and Hong 2001), and therefore, callose is suitable for use as a carbohydrate reserve for the embryos; (2) callose has gel-like properties that can confer high plasticity to the endosperm cell walls, allowing for the spatial remodeling of the endosperm around the enlarging embryo. Although in low quantities, cellulose is also synthesized in the endosperm cell plates in *Arabidopsis* (Otegui and Staehelin 2000b)

2.2 Polysaccharide Composition of Mature Endosperm Cell Walls

Mature endosperm cells usually exhibit thickened cell walls. In contrast to primary cell walls in other tissues, polysaccharide storage cell walls in the endosperm often contain little cellulose and are highly enriched in hemicellulosic polysaccharides. It is not well understood how these cell walls with very limited amounts of cellulose can stably assemble. However, this is clearly an adaptation to their specific function as carbohydrate-storing compartments since hemicellulosic polysaccharides are much easier to degrade than cellulose. The main cell wall storage polysaccharides in the endosperm belong to different groups of hemicellulosic polysaccharides such as mannans, galactomannans, glucomannans, xyloglucans, and (1,3;1,4)-β-glucans, depending on the species.

2.2.1 The Mannan Group: Pure Mannans, Glucomannans, and Galactomannans

These all contain backbones rich in (1,4)-linked-β-D-mannose residues, which may also contain D-glucose residues and/or a galactosyl side chain substitutions.

"Pure" mannans contain at least 90% mannose and few (1,6)-β-D-galactose side chain substitutions (see Fig. 1 and 4a); they are insoluble in water and, to some extent, crystalline in the cell wall (Buckeridge et al. 2000). Mannans confer rigidity and hardness to endosperm cell walls and their structural features have been studied in seeds of palm species, such as the date palm (*Phoenix dactylifera*) and the ivory nut tree (*Phytelephas macrocarpa*) (Reid 1985). In addition, mannans are abundant in the endosperm cell walls of *Carum carvi* (Hopf and Kandler 1977) and others umbelliferous (Apiaceae) species, as well as in coffee seeds (*Coffea arabica*, Rubiaceae) and some legumes such as *Schizolobium amazonicum* (Petkowicz et al. 2001).

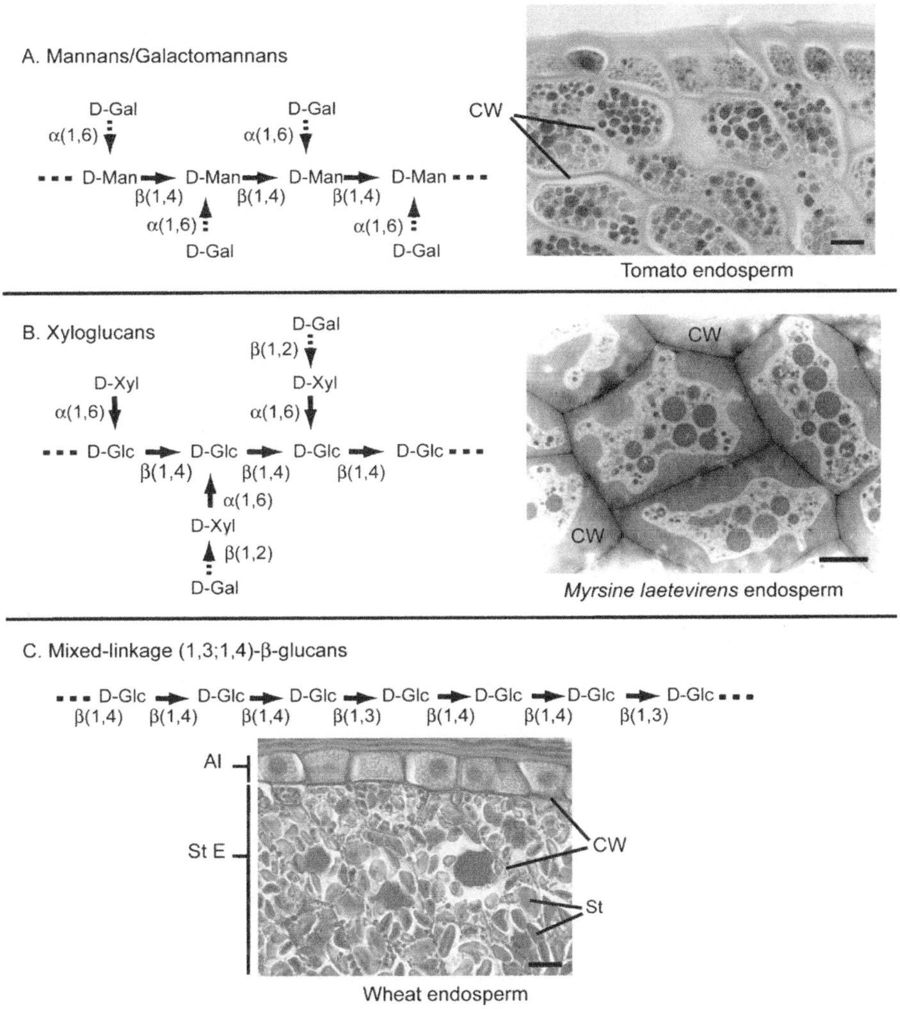

Fig. 4 Main groups of storage polysaccharides in the endosperm. Partial structures and micrographs of endosperm cell walls containing mannans and galactomannas (**a**), xyloglucans (**b**), and mixed-linkage (1,3;1,4)-β-glucans (**c**). *Solid arrows* indicate linkages that are always present whereas *dashed arrows* in mannans/galactomanns and in xyloglucans indicate variable substitution patterns. *Al* aleurone layer, *CW* cell wall, *Gal* galactose, *Glc* glucose, *Man* mannose, *St E* starchy endosperm, *St* starch granule, *Xyl* xylose

Galactomannans are related to mannans but they contain more (1,6)-β-D-galactosyl side chains (Fig. 4a). In contrast to mannans, galactomannans with a high degree of side chain substitutions are soft and highly hydrophilic. Galactomannans are found in the endosperm of endospermic leguminous seeds, such as fenugreek (*Trigonella foenum-graecum*), guar

(*Cyamopsis tetragonoloba*), and locust bean (*Ceratonia siliqua*); in the endosperm of tomato (*Solanum lycopersicum*) (Fig. 4a) and coffee seeds, and other groups such as Convolvulaceae, Asteraceae, and Arecaceae (Reid 1985; Buckeridge et al. 2000; Dhugga et al. 2004). The degree of galactose substitution in galactomannans varies from Man/Gal values of 1.1 anto about 3.5 (Reid et al. 2003) and it has a drastic effect on seed hardness and water retention capacities.

Galatomannans play multiple roles in seeds. Their hydrophilic properties enable the endosperm to retain water during post-imbibition drought, protecting the embryo from desiccation. In addition, they are mobilized as storage reserves during seed germination (Reid and Bewley 1979; Buckeridge and Dietrich 1996). Galactomannans from seeds are heavily used for industrial application. For example, guar and locust bean are used in food industry as viscosifiers, stabilizers, and gelling agents.

Glucomannans contains approximately equal amounts of (1,4)-β-D-mannose and D-glucose residues in their backbones, and sometimes galactosyl side chains. They are found in the endosperm of species of Liliaceae, such as asparagus (*Asparagus officinalis*) (Goldberg et al. 1992; Williams et al. 2001) and Iridaceae (Andrews et al. 1953; Reid 1985).

Thick endosperm cell walls rich in mannans and galactomannans with low level of side chain substitution are hard, and therefore, they may impose mechanical constraints on the germinating embryo. In some cases, such as coffee, asparagus, tobacco (*Nicotiana tabacum*), and tomato, the micropylar endosperm cells located close to the embryo radicle often differ from the remaining endosperm cells in having thinner cell walls. In addition, in some seeds, enzymatic activity is required for the weakening of the micropylar endosperm cell walls prior to germination. For example, in lettuce (*Lactuca sativa*) and *Datura ferox* seeds, there is an increase in endo-β-mannanase activity at the micropylar endosperm area prior to radicle emergence (Dutta et al. 1997; Sanchez and deMiguel 1997). In tomato seeds, both endo-β-mannanase and β-mannosidase activities increase prior to the completion of germination in the micropylar endosperm, and both increase in the lateral endosperm following germination (Mo and Bewley 2002, 2003).

2.2.2 Xyloglucans

Xyloglucans are one of the major hemicellulose polysaccharides in all primary cell walls of flowering plants. Xyloglucans consists of a (1,4)-β-D-glucose backbone with numerous β-D-xylose residues linked at regular sites to the O6 position of the glucose residues (Figs. 1 and 4b). Some of the xylosyl residues are substituted further with β-L-arabinose or β-D-galactose, depending on the species. In most primary cell walls, the side chains of xyloglucans bear a terminal β-L-fucosyl residue.

Biochemical studies have been conducted on seed storage xyloglucans from different species and tissues, such as endosperm cell walls of Myrsinaceae (Fig. 4b) (Otegui et al. 1998, 1999) and cotyledons of nasturtium (*Tropaeolum majus*, Tropaeolaceae), jojoba (*Simmondsia chinensis*, Simmondsiaceae), and several members of the legume family: tamarind (*Tamarindus indica*), jatoba (*Hymenaea courbaril*), *Afzelia africana*, *Detarium senegalense*, and *Copaifera langsdorfii* (Watanabe et al. 1980; Gidley et al. 1991; Buckeridge et al. 1992, 1997; York et al. 1993; Lima et al. 1995; Wang et al. 1996; Hantus et al. 1997; Yamagaki et al. 1998; Freitas et al. 2005; Ren et al. 2005). The main distinctive feature of storage xyloglucans in seeds is the low amount or complete absence of terminal fucosyl residues in the side chains. The significance of this structural peculiarity of seed xyloglucans is not clear. Fucosylation of the xyloglucan side chains does not appear to significantly affect the wall strength (Pena et al. 2004). In fact, *Arabidopsis* mutants with drastic decrease in xyloglucan fucosylation, such as *mur2* and *atfut1*, have no observable alterations in plant growth and cell wall strength (Vanzin et al. 2002; Perrin et al. 2003). However, in vitro studies suggest that the absence of fucose in storage xyloglucans may improve their self-associating properties, leading to a more efficient packing of xyloglucan molecules in the thick storage cell walls (Lima et al. 2004).

In addition to their role as storage polysaccharides during germination, seed xyloglucans may have important signaling functions (McDougall and Fry 1990). Interestingly, the biological responses triggered by xyloglucan oligosaccharides are highly dependent on the structural features of the xyloglucosyl oligomers and side chains. Fucosylated xyloglucan oligosaccharides from suspension cultures of *Rosa* sp. act as an anti-auxin 2,4-dichlorophenoxyacetic acid (2,4-D) growth promotor in etiolated pea stems (McDougall and Fry 1988) and are also able to inhibit gibberelic acid growth induction of pea segments (Yang et al. 1996). The anti-auxin activity depends on the presence of the β-L-fucopyranosyl β-(1 → 2)-D-galactopyranosyl β-(1 →) side chain of the xyloglucan (Vargas-Rechia et al. 1998). Interestingly, seed xyloglucan oligosaccharides from *Hymenaea courbaril* lack terminal fucosyl residues and promote growth of wheat (*Triticum aestivum*) coleoptiles independently of the presence of the auxin analog 2,4-D (Vargas-Rechia et al. 1998).

2.2.3 (1,3;1,4) *β*-glucans

Mixed-linkage (1,3;1,4)-β-D-glucans, commonly referred to as β-glucans, are abundant hemicelluloses in the order Poales, which includes cereals, but are not present in walls of other angiosperm species. They are one of the most abundant polysaccharide in the cereal endosperm cell walls (they constitute up to 70% of the endosperm walls) (Carpita 1996), although the main carbohydrate store in cereal grains is starch. These β-glucans are linear molecules with approximately 30% (1,3)- and 70% (1,4)-linkages. The composition of the walls of the starchy endosperm and the aleurone cells is qualitatively similar, but quantitatively different. In barley, the cell walls of the starchy endosperm consist of about 70% β-glucan and 20% arabinoxylan, whereas the aleurone cell walls contain 26% β-glucan and 67% arabinoxylan (Bacic and Stone 1981a,b; Woodward et al. 1983; Brennan and Cleary 2005). The (1,3)-linkages in the linear β-glucan backbone occur singly whereas the (1,4)-linkages are found mostly in sequences of two or three (Skendi et al. 2003) (Fig. 4c), but sequences of up to four have been reported (Cui et al. 2000). Hence the molecules may be regarded as being composed of (1,3)-β-linked cellotriosyl and cellotetraosyl units (Cui et al. 2000; Buckeridge et al. 2001; Tosh et al. 2004).

The composition of the endosperm cell walls has important effects on cereal grain uses, such as baking, malting and brewing, and animal and human nutrition (Philippe et al. 2006). As a major component of soluble fiber in the human diet, β-glucan has been implicated in reducing the risk of colorectal cancer and lowering serum cholesterol and glucose levels. However, the presence of (1,3;1,4)-β-glucans in the barley endosperm may have serious negative effects in the brewing process. The viscous β-D-glucans can interfere with fermentation and also inhibit the filtration process, leaving beer with an unappealing haze (Brennan and Cleary 2005; Keegstra and Walton 2006).

3
Synthesis of Endosperm Cell Wall Polysaccharides

The Arabidopsis genome contain almost 800 genes encoding putative glycosyltransferases or glycosyl hydrolases (approximately 3.3% of the genes), likely involved in cell wall synthesis, remodeling, and degradation (Henrissat et al. 2001; Coutinho et al. 2003; Somerville et al. 2004). Cell wall polysaccharides are synthesized in two organelles in plant cells. Callose and cellulose, both unsubstituted β-linked glucose polymers, are synthesized at the plasma membrane by the callose and cellulose synthase proteins, respectively, whereas the remaining non-cellulosic cell wall polysaccharides, such as pectins and hemicelluloses, are produced in the Golgi by glycan synthases and glycosyltransferases (Verma and Hong 2001; Scheible and Pauly 2004; Somerville et al. 2004; Cosgrove 2005; Liepman et al. 2005).

The main components of endosperm cell walls are hemicelluloses, and therefore, the enzymes responsible for their synthesis are assumed to be Golgi-localized. In addition, arabino-galactan proteins, which represent an abundant component in some endosperm cell walls, are also synthesized in the secretory pathway, that is, the endoplasmic reticulum and the Golgi.

Some of the putative golgi-localized glycan synthases are encoded by *cellulose-like synthase* (*CSL*) genes, which have been identified based on their sequence homology with the cellulose synthase catalytic subunit genes. The *CSL* genes have been divided into eight different subfamilies, *CSLA–H*. The *Arabidopsis* genome contains members of six *CSL* subfamilies, *CSLA–E* and *CSLG* (Richmond and Somerville 2000, 2001) whereas two additional subfamilies, CSLF and CSLH, have been identified in rice (Oryza sativa) (Hazen et al. 2002). Although only few *CSL* genes have been assigned a biological function, it is assumed that the *CSL* subfamilies are responsible for the synthesis of the backbones of all non-cellulosic β-linked polysaccharides (with the exception of callose) such as β-(1,4)-galactans, xylans, mannans, xyloglucans, the β-(1,3)- and β-(1,6)-galactans of arabino-galactans (Goubet et al. 2003), and the mixed-linkage (1,3;1,4)-β-glucans found in Poales.

Galactomannans are synthesized by the combined action of a GDP-mannose-dependent mannan synthase (ManS), which makes the β-(1,4)-linked mannan backbone, and a galactosyltransferase, which catalyzes the transfer of galactosyl residues from UDP-galactose to the mannan backbone to assemble the galactosyl side chains (Edwards et al. 1999, 2002, 2004; Reid et al. 2003). One member of the *CSLA* subfamily isolated from the developing endosperm of guar has ManS activity when expressed in embryonic soybean cultured cells. The protein is highly and specifically expressed in the guar endosperm. It is predicted to have five transmembrane domains and it localizes to the Golgi (Dhugga et al. 2004). In addition, at least three *CSLA* genes from *Arabidopsis*, *CSLA2*, *CSLA7*, and *CSLA9*, encode for ManS enzymes. In vitro activity assays of these enzymes indicate that they are capable of incorporating GDP-mannose or GDP-glucose into β-linked homo- or heteropolymers, depending on the presence or absence of the other nucleotide sugars in the incubation mixture (Liepman et al. 2005). Based on these studies, it has been postulated that all *CSLA* genes encode proteins with ManS activity.

At least four different types of enzymes are necessary for xyloglucan synthesis: a β-glucan synthase necessary for assembling the xyloglucan backbone, and xylosyl-, galactosyl-, and fucosyl-transferases for the synthesis of the xyloglucan side chains. The first gene identified as being involved in xyloglucan synthesis was *AtFUT1* (*fucosyltransferase 1*) (Perrin et al. 1999). In *Arabidopsis*, only the *AtFUT1* gene, appears to be responsible for xyloglucan fucosylation (Perrin et al. 2003; Scheible and Pauly 2004). However, as mentioned above, seed storage xyloglucans are typically non-fucosylated and therefore, xyloglucan fucosyltransferases are not expected to be highly expressed or active in endosperm cells. In *Arabidopsis*, a xylosyltransferase (AtXT1) that adds xylosyl residues to cello-oligosaccharides (Faik et al. 2002) and MUR3, a galactosyltransferease that transfers galactosyl residues to specific xylosyl residues on xyloglucan molecules (Madson et al. 2003) have been characterized. The β-glucan synthases necessary for assembling the xyloglucan backbone have not been identified yet but they are assumed to be encoded by one of the *CSL* gene subfamilies.

The mechanism of synthesis of mixed-linkage (1,3;1,4)-β-glucans has been extensively studied in excised maize coleoptiles (Buckeridge et al. 1999, 2001). In addition, the *CSLF* genes from rice have been shown to encode for (1,3;1,4)-β-D-glucan synthases (Burton et al. 2006).

4 Concluding Remarks

The endosperm is one of the most fascinating plant tissues. It is not only a unique system for studying plant development and cell differentiation but it is also one of the main sources of carbohydrate and proteins in the human diet worldwide. In addition, endosperm cell walls are used in multiple industrial processes. The polysaccharide storage endosperm cell walls offer a unique opportunity to study interactions between polysaccharide and cell wall assembly in the presence of limited amounts of cellulose. In recent years, many of the enzymes responsible for the synthesis of hemicelluloses present in the endosperm cell walls have been identified. These discoveries open the exciting possibility of designing molecular strategies to change the composition of the endosperm cell wall polysaccharides for both scientific research and industrial applications.

Acknowledgements I would like to thanks L. Andrews Staehelin (University of Colorado-Boulder) and Marek K. Sliwinski for the critical reading of this chapter. This work was supported by funds from the University of Wisconsin, Botany Department to MSO.

References

Andrews P, Hough L, Jones JKN (1953) Mannose-containing polysaccharides. Part 3. The polysaccharides in the seeds of *Iris ochroleuca* and *I. sibrica*. J Chem Soc 79:1186–1192

Austin JR 2nd, Segui-Simarro JM, Staehelin LA (2005) Quantitative analysis of changes in spatial distribution and plus-end geometry of microtubules involved in plant-cell cytokinesis. J Cell Sci 118:3895–3903

Bacic A, Stone BA (1981a) Isolation and ultrastructure of aleurone cell walls from wheat and barley. Aust J Plant Physiol 8:453–474

Bacic A, Stone BA (1981b) Chemistry and organisation of aleurone cell wall components from wheat and barley. Aust J Plant Physiol 8:475–495

Brennan CS, Cleary LJ (2005) The potential use of cereal (1 → 3, 1 → 4)-β-D-glucans as functional food ingredients. J Cereal Sci 42:1–13

Brown RC, Lemmon BE (2001a) The cytoskeleton and spatial control of cytokinesis in the plant life cycle. Protoplasma 215:35–49

Brown RC, Lemmon BE (2001b) Phragmoplasts in the abscence of nuclear division. J Plant Growth Regul 20:151–161

Brown RC, Lemmon BE, Olsen OA (1994) Endosperm development in barley: microtubule involvement in the morphogenetic pathway. Plant Cell 6:1241–1252

Buckeridge MS, Dietrich SMC (1996) Mobilisation of the raffinose family oligosaccharides and galactomannan in germinating seeds of Sesbania marginata Benth. (Leguminosae-Faboideae). Plant Sci 117:33–43

Buckeridge MS, Vergara CE, Carpita NC (1999) The mechanism of synthesis of a mixed-linkage (1 → 3),(1 → 4)-β-D-glucan in maize. Evidence for multiple sites of glucosyl transfer in the synthase complex. Plant Physiol 120:1105–1116

Buckeridge MS, Pessoa dos Santos H, Tine MAS (2000) Mobilisation of storage cell wall polysaccharides in seeds. Plant Physiol Biochem 38:141–156

Buckeridge MS, Vergara CE, Carpita NC (2001) Insight into multi-site mechanisms of glycosyl transfer in (1 → 4)-β-glycans provided by the cereal mixed-linkage (1 → 3),(1 → 4)-β-glucan synthase. Phytochemistry 57:1045–1053

Buckeridge MS, Rocha DC, Reid JSG, Dietrich SMC (1992) Xyloglucan structure and post-germinative metabolism in seeds of *Copaifera langsdorfii* from savanna and forest populations. Physiol Plant 86:145–151

Buckeridge MS, Crombie HJ, Mendes CJM, Reid JSG, Gidley MJ, Vieira CCJ (1997) A new family of oligosaccharides from the xyloglucan of *Hymenaea coubaril* L. (Leguminosae) cotyledons. Carbohydr Res 303:233–237

Burton RA, Wilson SM, Hrmova M, Harvey AJ, Shirley NJ, Medhurst A, Stone BA, Newbigin EJ, Bacic A, Fincher GB (2006) Cellulose synthase-like CslF genes mediate the synthesis of cell wall (1,3;1,4)-β-D-glucans. Science 311:1940–1942

Canaday J, Stoppin-Mellet V, Mutterer J, Lambert AM, Schmit AC (2000) Higher plant cells: Gamma-tubulin and microtubule nucleation in the absence of centrosomes. Microscopy Res Tech 49:487–495

Carpita NC (1996) Structure and biogenesis of the cell walls of grasses. Ann Rev Plant Physiol Plant Mol Biol 47:445–476

Cosgrove DJ (2005) Growth of the plant cell wall. Nat Rev Mol Cell Biol 6:850–861

Coutinho PM, Stam M, Blanc E, Henrissat B (2003) Why are there so many carbohydrate-active enzyme-related genes in plants? Trends Plant Sci 8:563–565

Cui W, Wood PJ, Blackwell B, Nikiforuk J (2000) Physicochemical properties and structural characterization by two-dimensional NMR spectroscopy of wheat β-D-glucan: Comparison with other cereal β-D-glucans. Carbohydr Polym 41:249–258

DeMason DA (1997) Endosperm structure and development. In: Larkins BA, Vasil IK (eds) Cellular and molecular biology of plant seed development. Kluwer Academic, Dordrecht, pp 73–115

Dhugga KS, Barreiro R, Whitten B, Stecca K, Hazebroek J, Randhawa GS, Dolan M, Kinney AJ, Tomes D, Nichols S, Anderson P (2004) Guar seed β-mannan synthase is a member of the cellulose synthase super gene family. Science 303:363–366

Dutta S, Bradford KJ, Nevins DJ (1997) Endo-β-mannanase activity present in cell wall extracts of lettuce endosperm prior to radicle emergence. Plant Physiol 113:155–161

Edwards ME, Marshall E, Gidley MJ, Reid JSG (2002) Transfer specificity of detergent-solubilized fenugreek galactomannan galactosyltransferase. Plant Physiol 129:1391–1397

Edwards ME, Dickson CA, Chengappa S, Sidebottom C, Gidley MJ, Reid JSG (1999) Molecular characterisation of a membrane-bound galactosyltransferase of plant cell wall matrix polysaccharide biosynthesis. Plant J 19:691–697

Edwards ME, Choo T-S, Dickson CA, Scott C, Gidley MJ, Reid JSG (2004) The seeds of *Lotus japonicus* lines transformed with sense, antisense, and sense/antisense galactomannan galactosyltransferase constructs have structurally altered galactomannans in their endosperm cell walls. Plant Physiol 134:1153–1162

Faik A, Price NJ, Raikhel NV, Keegstra K (2002) An Arabidopsis gene encoding an alpha-xylosyltransferase involved in xyloglucan biosynthesis. Proc Natl Acad Sci USA 99:7797–7802

Freitas RA, Martin S, Santos GL, Valenga F, Buckeridge MS, Reicher F, Sierakowski MR (2005) Physico-chemical properties of seed xyloglucans from different sources. Carbohydr Polym 60:507–514

Gidley MJ, Lillford PJ, Rowlands DW, Lang P, Dentini M, Crescenzi V, Edwards M, Fanutti C, Reid JSG (1991) Structure and solution properties of tamarind-seed polysaccharide. Carbohydr Res 214:299–314

Goldberg R, Guillou L, Prat R, Dubacq JP, Roland JC (1992) Structure and biochemistry of endosperm breakdown in *Asparagus officinalis* seedlings. Plant Physiol Biochem 30:477–485

Goubet F, Misrahi A, Park SK, Zhang Z, Twell D, Dupree P (2003) AtCSLA7, a cellulose synthase-like putative glycosyltransferase, is important for pollen tube growth and embryogenesis in Arabidopsis. Plant Physiol 131:547–557

Hantus S, Pauly M, Darvill AG, Albersheim P, York WS (1997) Structural characterization of novel L-galactose-containing oligosaccharide subunits of jojoba seed xyloglucans. Carbohydr Res 304:11–20

Hazen SP, Scott-Craig JS, Walton JD (2002) Cellulose synthase-like genes of rice. Plant Physiol 128:336–340

Henrissat B, Coutinho PM, Davies GJ (2001) A census of carbohydrate-active enzymes in the genome of Arabidopsis thaliana. Plant Mol Biol 47:55–72

Hong Z, Delauney AJ, Verma DPS (2001) A cell plate-specific callose synthase and its interaction with phragmoplastin. Plant Cell 13:755–768

Hopf H, Kandler O (1977) Characterization of the reserve cellulose of the endosperm of *Carum carvi* as a $\beta(1 \rightarrow 4)$-mannan. Phytochemistry 16:1715–1717

Jürgens G (2005) Cytokinesis in higher plants. Ann Rev Plant Biol 56:281–299

Keegstra K, Walton J (2006) β-Glucans – brewer's bane, dietician's delight. Science 311:1872–1873

Liepman AH, Wilkerson CG, Keegstra K (2005) Expression of cellulose synthase-like (Csl) genes in insect cells reveals that CslA family members encode mannan synthases. Proc Natl Acad Sci USA 102:2221–2226

Lima DU, Loh W, Buckeridge MS (2004) Xyloglucan–cellulose interaction depends on the sidechains and molecular weight of xyloglucan. Plant Physiol Biochem 42:389–394

Lima NN, Rechia CGV, Ganter J, Reicher F, Sierakowski MR (1995) Oligosaccharides derived from the xyloglucan isolated from the seeds of *Hymenaea courbaril* var stilbocarpa. Int J Biol Macromol 17:413–415

Madson M, Dunand C, Li X, Verma R, Vanzin GF, Caplan J, Shoue DA, Carpita NC, Reiter W-D (2003) The MUR3 gene of Arabidopsis encodes a xyloglucan galactosyltransferase that is evolutionarily related to animal exostosins. Plant Cell 15:1662–1670

McDougall GJ, Fry SC (1988) Inhibition of auxin-stimulated growth of pea stem segments by a specific nonasaccharide of xyloglucan. Planta 175:412–416

McDougall GJ, Fry SC (1990) Xyloglucan oligosaccharides promote growth and activate cellulase – Evidence for a role of cellulase in cell expansion. Plant Physiol 93:1042–1048

Meier H, Reid JSG (1977) Morphological aspects of galactomannan formation in the endosperm of *Trigonella foenum-graecum* L. (Leguminosae). Planta 133:243–248

Mineyuki Y (1999) The preprophase band of microtubules: its function as a cytokinetic apparatus in higher plants. Int Rev Cytol 187:1–49

Mo B, Bewley JD (2003) The relationship between β-mannosidase and endo-β-mannanase activities in tomato seeds during and following germination: a comparison of seed populations and individual seeds. J Exp Bot 54:2503–2510

Mo BX, Bewley JD (2002) β-Mannosidase (EC 3.2.1.25) activity during and following germination of tomato (*Lycopersicon esculentum* Mill.) seeds: purification, cloning and characterization. Planta 215:141–152

Olsen O-A (2004) Nuclear endosperm development in cereals and Arabidopsis thaliana. Plant Cell 16:S214–227

Otegui M, Lima C, Maldonado S, De Lederkremer RM (1998) Histological and chemical characterization of *Myrsine laetevirens* seed. Int J Plant Sci 159:762–772

Otegui M, Lima C, Madonado S, de Lederkremer RM (1999) Development of the endosperm of *Myrsine laetevirens* (Myrsinaceae). I. Cellularization and deposition of cell-wall storage carbohydrates. Int J Plant Sci 160:491–500

Otegui MS, Staehelin LA (2000a) Cytokinesis in flowering plants: more than one way to divide a cell. Curr Opin Plant Biol 3:493–502

Otegui MS, Staehelin LA (2000b) Syncytial-type cell plates: a novel kind of cell plate involved in endosperm cellularization of Arabidopsis. Plant Cell 12:933–947

Otegui MS, Austin JR 2nd (2006) Visualization of membrane-cytoskeletal interactions during plant cytokinesis. In: McIntosh JR (ed) Cellular electron microscopy. Methods in cell biology, vol 79. Elsevier, San Diego, CA, pp 221–240

Otegui MS, Mastronarde DN, Kang BH, Bednarek SY, Staehelin LA (2001) Three-dimensional analysis of syncytial-type cell plates during endosperm cellularization visualized by high resolution electron tomography. Plant Cell 13:2033–2051.

Pena MJ, Ryden P, Madson M, Smith AC, Carpita NC (2004) The galactose residues of xyloglucan are essential to maintain mechanical strength of the primary cell Walls in Arabidopsis during growth. Plant Physiol 134:443–451

Perrin RM, DeRocher AE, Bar-Peled M, Zeng W, Norambuena L, Orellana A, Raikhel NV, Keegstra K (1999) Xyloglucan fucosyltransferase, an enzyme involved in plant cell wall biosynthesis. Science 284:1976–1979

Perrin RM, Jia Z, Wagner TA, O'Neill MA, Sarria R, York WS, Raikhel NV, Keegstra K (2003) Analysis of xyloglucan fucosylation in Arabidopsis. Plant Physiol 132:768–778

Petkowicz CLD, Reicher F, Chanzy H, Taravel FR, Vuong R (2001) Linear mannan in the endosperm of *Schizolobium amazonicum*. Carbohydr Polym 44:107–112

Philippe S, Robert P, Barron C, Saulnier L, Guillon F (2006) Deposition of cell wall polysaccharides in wheat endosperm during grain development: Fourier transform-infrared microspectroscopy study. J Agric Food Chem 54:2303–2308

Pickett-Heaps JD, Gunning BES, Brown RC, Lemmon BE, Cleary AL (1999) The cytoplast concept in dividing plant cells: cytoplasmic domains and the evolution of spatially organized cell division. Am J Bot 86:153–172

Reid JSG (1985) Cell wall storage carbohydrates in seeds: Biochemistry of the seed "gums" and "hemicelluloses". Adv Bot Res 11:125–155

Reid JSG, Bewley DJ (1979) A dual role for the endosperm and its galactomannan reserves in the germinative physiology of fenugreek (*Trigonella foenum-graecum* L.), an endospermic leguminous seed. Planta 147:145–150

Reid JSG, Edwards ME, Dickson CA, Scott C, Gidley MJ (2003) Tobacco transgenic lines that express fenugreek galactomannan galactosyltransferase constitutively have structurally altered galactomannans in their seed endosperm cell walls. Plant Physiol 131:1487–1495

Ren YL, Picout DR, Ellis PR, Ross-Murphy SB, Reid JSG (2005) A novel xyloglucan from seeds of *Afzelia africana* Se. Pers. – extraction, characterization, and conformational properties. Carbohydr Res 340:997–1005

Richmond TA, Somerville CR (2000) The cellulose synthase superfamily. Plant Physiol 124:495–498

Richmond TA, Somerville CR (2001) Integrative approaches to determining Csl function. Plant Mol Biol 47:131–143

Samuels LA, Giddings TH, Staehelin LA (1995) Cytokinesis in tobacco BY-2 and root tip cells: a new model of cell plate formation in higher plants. J Cell Biol 130:1345–1357

Sanchez RA, deMiguel L (1997) Phytochrome promotion of mannan-degrading enzyme activities in the micropylar endosperm of *Datura ferox* seeds requires the presence of the embryo and gibberellin synthesis. Seed Sci Res 7:27–33

Scheible W-R, Pauly M (2004) Glycosyltransferases and cell wall biosynthesis: novel players and insights. Curr Opin Plant Biol 7:285–295

Segui-Simarro JM, Austin JR 2nd, White EA, Staehelin LA (2004) Electron tomographic analysis of somatic cell plate formation in meristematic cells of Arabidopsis preserved by high-pressure freezing. Plant Cell 16:836–856

Skendi A, Biliaderis CG, Lazaridou A, Izydorczyk MS (2003) Structure and rheological properties of water soluble β-glucans from oat cultivars of *Avena sativa* and *Avena bysantina*. J Cereal Sci 38:15–31

Somerville C, Bauer S, Brininstool G, Facette M, Hamann T, Milne J, Osborne E, Paredez A, Persson S, Raab T, Vorwerk S, Youngs H (2004) Toward a systems approach to understanding plant cell walls. Science 306:2206–2211

Sorensen MB, Mayer U, Lukowitz W, Robert H, Chambrier P, Jurgens G, Somerville C, Lepiniec L, Berger F (2002) Cellularisation in the endosperm of *Arabidopsis thaliana* is coupled to mitosis and shares multiple components with cytokinesis. Development 129:5567–5576

Tosh SM, Wood PJ, Wang Q, Weisz J (2004) Structural characteristics and rheological properties of partially hydrolyzed oat β-glucan: the effects of molecular weight and hydrolysis method. Carbohydr Polym 55:425–436

Vanzin GF, Madson M, Carpita NC, Raikhel NV, Keegstra K, Reiter W-D (2002) The *mur2* mutant of *Arabidopsis thaliana* lacks fucosylated xyloglucan because of a lesion in fucosyltransferase AtFUT1. Proc Natl Acad Sci USA 99:3340–3345

Vargas-Rechia C, Reicher F, Rita Sierakowski M, Heyraud A, Driguez H, Lienart Y (1998) Xyloglucan octasaccharide XXLGol derived from the seeds of *Hymenaea courbaril* acts as a signaling molecule. Plant Physiol 116:1013–1021

Verma DPS, Hong Z (2001) Plant callose synthase complexes. Plant Mol Biol 47:693–701

Wang Q, Ellis PR, RossMurphy SB, Reid JSG (1996) A new polysaccharide from a traditional Nigerian plant food: *Detarium senegalense* Gmelin. Carbohydr Res 284:229–239

Watanabe T, Takahashi K, Matsuda K (1980) Carbohydrates of jojoba seed 2. Isolation and characterization of oligosaccharides from purified cellulase digest of jojoba (*Simmondsia chinensis*) seed xyloglucan. Agric Biol Chem 44:791–797

Williams HA, Bewley D, Greenwood JS, Bourgault R, Mo B (2001) The storage cell walls in the endosperm of *Asparagus officinalis* L. seeds during development and following germination. Seed Sci Res 11:305–315

Woodward JR, Fincher GB, Stone BA (1983) Water-soluble (1 → 3), (1 → 4)-β-D-glucans from barley (*Hordeum vulgare*) endosperm. II. Fine structure. Carbohydr Polym 3:207–225

Yamagaki T, Mitsuishi Y, Nakanishi H (1998) Determination of structural isomers of xyloglucan octasaccharides using post-source decay fragment analysis in MALDI-TOF mass spectrometry. Tetrahedron Lett 39:4051–4054

Yang T, Davies PJ, Reid JB (1996) Genetic dissection of the relative roles of auxin and gibberellin in the regulation of stem elongation in intact light-grown peas. Plant Physiol 110:1029–1034

York WS, Harvey LK, Guillen R, Albersheim P, Darvill AG (1993) Structural analysis of tamarind seed xyloglucan oligosaccharides using beta-galactosidase digestion and spectroscopic methods. Carbohydr Res 248:285–301

Plant Cell Monogr (8)
O.-A. Olsen: Endosperm
DOI 10.1007/7089_2007_116/Published online: 24 March 2007

Starch Formation in the Cereal Endosperm

L. Curtis Hannah

Program in Plant Molecular and Cellular Biology & Department of Horticultural Sciences, University of Florida, Gainesville, FL 32611, USA
Hannah@mail.ifas.ufl.edu

Abstract Starch comprises ~ 70% of the dry weight of a cereal seed. Because the vast majority of this starch is in the cereal endosperm and cereals are the major agricultural plants grown world-wide, endosperm starch is the largest single contributor to the world's calorie supply. Starch biosynthesis in the endosperm is described herein. Emphasis is placed on classical "biochemical genetics", with the central theme of deciphering the biochemical lesions conditioned by mutants. These types of analyses have uncovered complexities in the pathway that were/are undetectable by other methods. Unexpected conclusions are highlighted. Starch synthesis in the endosperm is contrasted with its biogenesis in other tissues. Possible application of principles derived from endosperm starch synthesis to other biochemical pathways is discussed.

1 Introduction

Typical endosperm starch is composed of two major glucose polymers, amylose and amylopectin. The latter comprises usually ~ 75% of the total starch. Amylose is a rarely branched polymer of glucose with the sugar moieties linked through α 1,4 bonds. Amylopectin contains the same backbone as amylose; however this molecule is branched with α 1,6 bonds at the junction sites. These branches constitute approximately 5% of all glycosidic bonds. Another branched polymer, phytoglycogen is present in some plant mutants lacking a starch debranching enzyme. In contrast to amylopectin, phytoglycogen contains 10% α 1,6 bonds and is water soluble.

Cereal yields would not be nearly as high as they are today were it not for a peculiar property of amylopectin structure. The α 1,6 branch points are not random but rather are clustered. These clusters are believed important because the glucan side-chains are juxtaposed such that they allow formation of double-helical structures. These, in turn, allow dense and efficient packaging of glucose residues in an insoluble state and the synthesis of extraordinarily large molecules. A model accounting for the clustered nature of branches is described below.

The endosperm is the archetypical "sink" tissue for carbon storage. Sugars, primarily sucrose, flow from the leaf source tissue to the endosperm and, through a series of reactions, the carbon of sugars ends up in starch.

Biochemical studies of two classical mutants, *Sh1* (*shrunken1)* and *Mn1* (*miniature seed1*) (reviewed in Koch 2004) revealed that these affect early steps in the sucrose to starch conversion, namely the breakdown of sucrose. *Sh1* encodes the major sucrose synthase (sucrose + UDP $\Longleftrightarrow$ UDP-glucose and fructose) whereas *Mn1* affects cell wall invertase (sucrose $\Rightarrow$ fructose + glucose) activity. The enzymes do not function in a redundant or parallel fashion since a mutant kernel phenotype occurs when either gene (not both) is homozygous for a loss-of-function mutant allele. The sucrose synthase reaction is freely reversible; however, sucrose synthase functions physiologically as a sucrose degradase.

While developing maize kernels can synthesize sucrose as shown by the detection of sucrose in kernels grown on only glucose and/or fructose (Cobb and Hannah 1988), the ability to synthesize sucrose is not diminished in a *sh1* mutant. This shows that the physiological role of sucrose synthase is, in fact, the breakdown of sucrose. Hence, the most parsimonious explanation for all data is that sucrose appears at least twice in the starch biosynthetic pathway, an interpretation consistent with classical studies of Shannon (for example, Shannon 1968).

In contrast to earlier ideas, it is now clear that there is extensive intermediary metabolism in the sucrose to starch conversion. One of the clearest demonstrations of this came from labeling studies. Glawischnig et al. (2002) performed incorporation studies whereby every carbon of the incorporated and labeled glucose or sucrose contained ^{13}C. Analysis of the synthesized starch showed that the most highly abundant labeling patterns in the glucose molecules were 1,2,3-$^{13}C_3$ and 4,5,6-$^{13}C_3$. These patterns are best explained by widespread degradation and resynthesis of glucose. As suggested (Glawischnig et al. 2002) sucrose catabolism through glycolysis, glucogenesis and the pentose phosphate pathway most likely precedes starch biosynthesis.

Four enzymatic activities catalyze steps unique to starch biosynthesis (Table 1).

The first enzyme, adenosine diphosphate glucose pyrophosphorylase (AGPase) synthesizes the precursor for starch formation, ADP-glucose. This is used to elongate glucose polymers through formation of α 1,4 bonds by a class of enzymes termed starch synthases (SS). The α 1,6 bonds are formed by starch branching enzymes (SBE). Surprisingly, mutant analyses have definitively identified starch debranching enzymes (DBE) as necessary for the synthesis of wildtype starch levels. Evidence for the importance of these enzymes is given below. Also, because of the importance of starch and the active research in the area, a number of excellent recent reviews have been written. Readers are referred to them for additional information (Ball and Morell 2003; Hannah 1997, 2005; James et al. 2003; Martin and Smith 1995; Myers et al. 2000; Nelson and Pan 1995; Preiss and Levi 1980; Preiss et al. 1991; Tetlow et al. 2004).

Table 1 Enzymes of maize endosperm starch synthesis and the genes that encode them. Comparable genes have been reported in other cereals. A dichotomy exists in gene nomenclature. Names in parenthesis refer to the altered kernel phenotype detectable by eye. Other genes are named on the basis of the encoded enzyme.

Enzyme	Gene	Critical paper
Adenylate transporter	*Bt1 (brittle1)*	Sullivan et al. 1991
AGP large subunit	*Sh2 (shrunken2)*	Hannah & Nelson 1976
AGP small subunit	*Bt2 (brittle2)*	Hannah & Nelson 1976
Branching enzyme	*Ae (amylose-extender)*	Fisher et al. 1993
Branching enzyme	*Sbe1a*	Yao et al. 2004
Branching enzyme	*Sbe2a*	Blauth et al. 2001
GB starch synthase	*Wx (waxy)*	Nelson & Rines 1962
Isoamylase	*Su1 (sugary1)*	James et al. 1995
Pullulanase	*Zpu1*	Dinges et al. 2003
Starch synthase	*Du (Dull)*	Gao et al. 1998
Starch synthase	*Su2 (sugary2)*	Zhang et al. 2004

Starch synthesis in the endosperm differs fundamentally from starch synthesis in other tissues. Whereas the precursor of starch, adenosine diphosphate glucose is synthesized in the plastid in all non-endosperm tissues so far examined, this sugar is synthesized in the cytosol of endosperm cells. The biochemical genetics of the major enzymes unique to endosperm starch formation are described below.

2
Adenosine Diphosphate Glucose Pyrophosphorylase (AGPase)

AGPase synthesizes ADP-glucose and pyrophosphate from glucose-1-PO_4 and ATP. ADP-glucose is then used as the glucose donor for amylose, amylopectin and phytoglycogen biosynthesis. AGPase is a heterotetramer in all plant tissues and is a homotetramer in virtually all bacteria. In maize, the gene *Sh2 (shrunken2)* encodes the large subunit of the endosperm whereas the endosperm small subunit is encoded by *brittle-2 (Bt2)* (Hannah and Nelson 1976; Bhave et al. 1990; Bae et al. 1990). Loss-of-function *sh2* and *bt2* mutants accumulate high levels of sugars at the expense of starch synthesis.

That AGPase is uniquely situated in the cytosol of endosperm cells was mentioned above. Because earlier studies demonstrated a plastid localization for AGPases of the spinach leaf and of the potato tuber (Kim et al. 1989; Okita et al. 1979), it was expected that this would also hold true for endosperms. Early hints of a cytosolic positioning of AGPase came from the lack of a plastid transit peptide on the BT2 and SH2 proteins as well as the barley endosperm proteins (Giroux and Hannah 1994; Villand and Kleczkowski 1994). Also, the sizes of SH2 and BT2 proteins synthesized by in vitro transcription and translation were identical to those of the corresponding in vivo-synthesized proteins (Giroux and Hannah 1994). Definitive proof for a cytosolic location of AGPase came from metabolite analysis of a *brittle1* (*Bt1*) mutant. *Bt1* encodes a membrane-bound metabolite transporter (Sullivan et al. 1991) and its loss leads to a high sugar/low starch phenotype identical to that of a *sh2* or *bt2* mutant. Shannon and coworkers (Cao et al. 1995; Shannon et al. 1996, 1998) showed that ADP-glucose accumulated to high levels in the cytosol of a *bt1* mutant; however, this did not occur in a *sh2, bt1* double mutant. The build-up in the *bt1* mutant showed that the genetic lesion associated with this mutant must come after ADP-glucose synthesis but before its incorporation into starch. And the elevated ADP-glucose levels in a *bt1* mutant must come from the *Sh2*-encoded AGPase. Since *Bt1* encodes a membrane-bound transporter, the only logical conclusion was that ADP-glucose is synthesized in the cytosol and then transverses the membrane into the plastid. In addition, 85 to 95% of total AGPase activity was shown to be in the cytosol by use of enhanced cellular fractionation techniques (Denyer et al. 1996; Thorbjornsen et al. 1996a). Subsequently, from measurement of ADP-glucose and UDP-glucose levels, Beckles et al. (2001) concluded that a cytosolic location for AGPase is a feature of graminaceous endosperms but not of other starch-storing organs.

Evidence accumulated from a number of plants (and bacteria) leave no doubt that AGPase represents a rate limiting step in starch biosynthesis. It catalyzes the first committed step in this pathway and its allosteric properties are essential in controlling rates of starch biosynthesis (reviewed in Preiss et al. 1991; Preiss and Romeo 1989; Hannah 1997, 2005). Definitive genetic data proving the rate-limiting role played by this enzyme and the importance of its allosteric properties first came from characterization of the glycogen overproducing *E. coli* mutant, *glgC-16*. Importantly, expression of the allosterically enhanced *glgC-16* mutant gene in potato tubers led to greater than 35% increase in starch levels (Stark et al. 1992).

Most plant AGPases are activated by 3-phosphoglyceric acid (3PGA) and inhibited, or in at least one case deactivated, by inorganic phosphate (Pi). An exception to this is the barley endosperm AGPase which appears to be recalcitrant to allosteric effectors (Rudi et al. 1997; Doan et al. 1999). The importance of Pi inhibition came from analysis of a *Sh2* variant isolated through transposon mutagenesis. Giroux et al. (1996) isolated a *Sh2* variant via excision of the transposable element *dissociation (Ds)* termed *Rev6,* that conditioned a phosphate less-sensitive maize endosperm AGPase. This variant can cause up to a 40% increase in seed weight, depending on the genetic background and the growing conditions. Starch and other components were increased in this variant, suggesting that enhanced starch synthesis creates a stronger carbon sink within the endosperm. Likewise another allosterically enhanced *E. coli* AGPase variant was shown to increase seed weight up to 11% when expressed in rice (Sakulsingharoj et al. 2004).

Another important parameter affecting AGPase activity is the enzyme's lability to heat (reviewed in Greene and Hannah 1998a,b). Compared to other enzymes in the starch biosynthetic pathway, AGPase is one of the most if not the most heat labile enzyme. Accordingly, a more heat stable maize endosperm AGPase termed *HS33* was isolated by use of an *E. coli* expression system and by monitoring glycogen synthesis at elevated temperatures of mutagenized *Sh2* or *Bt2* genes (Greene and Hannah 1998b). This mutant functions by enhancing interactions between the SH2 and BT2 proteins.

The two variants, HS33 and Rev6 were placed in one gene and expressed in wheat (Smidansky et al. 2002), rice (Smidansky et al. 2003) and maize (Hannah and Greene, in preparation). Yield increases of 38% (wheat), 23% (rice) and 68% (maize) were recorded under standard field conditions. High temperatures amplified the yield difference observed in maize. These data point to the critical roles played by AGPase's allosteric and heat stability properties in conditioning grain yield. Surprisingly, yield increase in all three cereals was due to enhanced seed number rather than larger seeds. How alteration in the endosperm starch biosynthetic pathway affects seed number is under active research in a number of laboratories. Also, it would be interesting to determine whether the altered maize AGPase would enhance seed number in barley since the barley AGPase apparently is little effected by allosteric modifiers.

As is the case for many enzymes, different isoforms of AGPase are expressed in maize. The maize endosperm, embryo and leaf (Hannah et al. 2001) employ three different isoenzymes. A fourth form, a minor activity in the maize (Giroux and Hannah 1994) and wheat endosperm (Burton et al. 2002) has also been described.

Sequences of small and large subunits point to a series of gene duplications underlying the different isoforms. The small and large subunits share much sequence similarity with the lone bacterial subunit. This suggests that a gene duplication event occurred early in plant evolution. Subsequent duplications gave rise to the various forms of the small and of the large subunits.

It should be noted that while the small and large subunits share a common origin, mutant analysis has revealed that they have diverged to the point that they have complementary rather than duplicate functions. Loss of *Sh2* function or *Bt2* function leads to > 90% loss in maize endosperm AGPase activity (Hannah and Nelson 1976). If the two subunits retained duplicate function, a 90% activity loss would require mutation at both structural genes. Each subunit however, when expressed alone in an *E. coli* expression system, conditions 2 to 3% wildtype enzymes (Burger et al. 2003) showing that the ability to catalyze ADP-glucose synthesis is not the exclusive domain of either subunit.

While genes encoding each subunit have undergone duplications, the number and timing of these events differ dramatically for the two subunits.

Compared to the small subunit, duplication of the large subunit gene occurred early in evolution whereas the initial duplications of the small subunit did not occur until after the separation of monocots and eudicots (Hannah et al. 2001). Whether this timing in duplication events explains the more conserved nature of the small subunit is an open question. Another possibility was offered by Cross et al. (2004). They noted that the sequences of small subunit introns show greater conservation compared to introns of large subunit genes. This observation points to different intrinsic mutation rates of the two gene families.

A third feature that may explain the difference in sequence conservation is the fact that plants generally contain fewer small subunit genes than large subunit genes. For example, Arabidopsis contains only one small subunit but four large subunit genes (Crevillén et al. 2003) while rice has two small and four large subunit genes (Akihiro et al. 2005). In the case of barley, wheat, probably rice but not maize, one gene encodes the small subunit employed in the endosperm and leaf (Thorbjørnsen et al. 1996b; reviewed in Hannah 2005). Alternative RNA splicing facilitates the use of two different first exons, producing proteins with or without the chloroplast transit peptide. Hence, one small subunit must interact with more than one large subunit and must function in different cells types and in different subcellular locations. This may place constraints on the small subunits that do not affect the large subunit.

While it is clear that ADP-glucose is the primarily if not exclusive glucose donor for starch synthesis, how the remaining enzymes of the pathway, reviewed below, interact to give rise to the non-random placement of α 1,6 bonds remains the most important unanswered question of starch biosynthesis.

3
Starch Synthases (SS)

A series of different starch synthase isoenzymes elongates the linear chains of glucose residues. One form is exclusively bound to the starch granule and is termed the GBSS, while four others (SSI, SSII, SSIII and SSIV/V) are termed soluble starch synthases although some can also be found within the granule (reviewed in James et al. 2003; Ball and Morell 2003; Tetlow et al. 2004). Enzymes of the five classes bear much sequence similarity and the most parsimonious explanation is they were derived from a common progenitor through gene duplication events. In addition, there exist multiple forms of each isoenzyme. A recent expression analysis of rice seed development detected one functional gene for SSI, three genes for SSII, and two genes each for SSIII, SSIV and GBSS (Hirose and Herao 2004). While all are expressed in the developing seed, their developmental profiles do differ.

The first biochemical-genetic characterization of this enzyme system occurred over 40 years ago (Nelson and Rines 1962) with the enzymatic characterization of the maize *waxy (wx)* locus and mutants therein. It was known that endosperm starch lacking the *Wx* function is composed exclusively of amylopectin. Nelson and Rines showed that the enzymatic function associated with the *Wx* locus was the starch-bound, ADP-glucosyl transferase or starch synthase. This led to the surprising conclusion that branched amylopectin is not synthesized from straight-chained amylose.

With the advent of transposon tagging, the functions of two additional classic maize mutants with visible phenotypes were identified. Cloning and sequencing of the genes *dull1* (*Du1*) (Gao et al. 1998) and *sugary2* (*Su2*) (Zhang et al. 2004) showed that they encode the starch synthases, SSIII and SSIIa, respectively. Loss of these gene functions alters starch structure. Loss of SSIII in the *du1* mutation increases the relative amounts of amylose, decreases the number of long chains, and increases the frequency of branching (Wang et al. 1993). Loss of SSIIa in *su2* gives rise to an apparent increase in the amount of amylose and amylopectin composed of an increased proportion of short glucan chains (one to nine glucose residues) in length and a proportional decrease in chains of intermediate (10 to 23 glucose residues) length.

Recently, mutants of SS1 were isolated in rice via transposon mutagenesis (Fujita et al. 2006). An analysis of a series of allelic mutants showed that none conditioned a phenotype detectable by eye, nor was the size or shape of the seed or starch granules affected. Interestingly, glucans having 8 to 12 glucose residues were reduced in amount whereas chains with 6 to 7 and 16 to 19 glucose moieties were increased.

Mutant analyses are consistent with a model put forward by Commuri and Keeling (2001). This model is based on kinetic analysis of the various starch synthases and the lengths (in glucose residues) of likely substrates. Commuri and Keeling (2001) propose that SSI elongates short chain (< 10 glucose residues) glucans. Importantly, while increased chain length enhances binding, it drastically slows the catalytic reaction of elongation. Commuri and Keeling speculate that this phenomenon explains the fact that most SSI enzyme is found entrapped in the starch granule and that other starch synthases catalyze the longer chain elongation reactions. With the recent demonstration that only one SSI enzyme exists in rice (Hirose and Herao 2004) and knockout mutations in it do not destroy amylopectin synthesis, it seems reasonable to suggest that other starch synthases must have the ability to elongate short chains as well.

Some evidence points to the importance of phosphorylation and possible interaction with 14-3-3 regulatory proteins in controlling starch synthase activity. SSIIa activity is phosphorylated in the wheat endosperm (Tetlow et al. 2004) and this is associated with enhanced activity. The N-terminus of the *Du1*-encoded SSIII is much longer than that of other starch synthases and bears a binding motif for 14-3-3 protein. Interestingly, expression of an antisense 14-3-3 construct in Arabidopsis leaves doubles leaf starch content (Sehnke et al. 2001).

4 Starch Branching Enzymes (SBE)

Synthesis of α 1,6 linkages occurs via a series of starch branching enzymes. Acting perhaps in concert with starch synthases, these enzymes first cleave an α 1,4 bond generating a chain of glucose moieties. This is then attached to a glucose residue with formation of an α 1,6 bond.

Two major classes (I and II) of branching enzymes have been identified, based on sequence similarity and substrate preference. Longer glucans are transferred by SBEI which tends to branch polymers with fewer branches, compared to SBEII (reviewed in Satoh et al. 2003). Three distinguishable endosperm SBE isoforms are known; SBEIa, SBEIIa and SBEIIb (reviewed in Yao et al. 2004). In maize these are encoded by *Sbe1a*, *Sbe2a* and *Ae* (*amylose-extender*), respectively. The first two genes were defined via molecular analyses whereas *Ae* was first identified by its visible kernel phenotype and subsequently named for its increase in amylose content compared to amylopectin composition. Mutants and silencing constructs for branching enzymes have now been described in maize, rice and wheat. Analyses of these variants point to significant plant specific differences these endosperm enzymes play.

SBEI mutants have been reported in maize (Blauth et al. 2001; Yao et al. 2004) and in rice (Satoh et al. 2003). Whereas loss of SBEI in maize is reported to have no impact on amylopectin structure, Satoh and colleagues report that loss of this enzyme in rice changes the amylopectin structure. Chains containing 12 to 21 and more than 37 glucose residues were reduced in the mutant whereas chains having less than 10 and between 24 to 34 glucoses were enhanced relative to wildtype. Whether the maize/rice difference is caused by an additional maize SBEI gene has yet to be elucidated.

Loss of SBEIIa has been reported in maize (Blauth et al. 2001; Yao et al. 2004) and in wheat (Regina et al. 2006). Transposon mutagenesis was used in maize while a gene silencing construct was exploited in wheat. Whereas loss of SBEIIa in maize gave rise to leaf starch with little to no branching and, in turn, to non-lethal leaf senescence, endosperm starch was virtually unaltered in this mutant. In contrast, an SBEIIa gene silencing gene construct in wheat effectively doubled endosperm amylose content at the expense of amylopectin. Contrasting wheat/maize phenotypes have also been noted for loss of SBEIIb. The classical maize *amylose-extender* mutation is now known to encode SBEIIb. Loss of this gene product produces amylopectin with longer chains and fewer branches per cluster. In contrast, loss of SBEIIb function in wheat has no discernable effect. It is also of note that the relative levels of gene expression of SBEIIa and IIb differ in wheat and maize. This likely explains the difference in mutant phenotypes. Another surprising result is that removal of SBEIIb and SBEIa in the maize endosperm *increases* rather than decreases the number of branches within a cluster and *decreases* rather than increases chain length. Perhaps the presence of SBEIa activity somehow inhibits SBEIIa activity in the maize endosperm (Yao et al. 2004). Removal of SBEIa allows SBEIIa function.

Phosphorylation also appears to play a role in branching activity and perhaps other starch biosynthetic enzymes (Tetlow et al. 2004). The presence of ATP stimulates amylopectin but not amylose synthesis in the wheat endosperm. This led to studies showing that all three SBE enzymes require phosphorylation for maximal activity. Furthermore, phosphorylation is required to strengthen interactions involving SBEIIb, SBEI and starch phosphorylase. This latter observation provides an explanation for the partial loss of SBEI activity when SBEIIb activity is removed genetically (James et al. 2003).

5 Starch Debranching Enzymes (DBE)

Surprisingly, hydrolysis of α 1,6 linkages by starch debranching activity is required for synthesis of wildtype amylopectin levels. This unanticipated conclusion was first suggested by Pan and Nelson (1984) who reported that debranching activity was reduced in a *sugary1* (*Su1*) mutant. Mutant *su1* alleles are characterized by increased sucrose content, reduced starch, and, uniquely, a buildup of the highly branched polymer termed phytoglycogen. In contrast to amylopectin, phytoglycogen is water soluble and 10% of its glucose-glucose linkages are in an α 1,6 configuration (reviewed in Dinges et al. 2001; James et al. 2003). Amylopectin structure is altered in *su1* mutants.

DBE enzymes fall into two classes as judged by substrate specificity. The two isoforms are termed pullulanase and isoamylase. James et al. (1995) used transposon mutagenesis to clone *Su1* and showed that it encodes isoamylase. Pullulanase activity, however, is also reduced in *su1* mutants. Analysis of a series of *su1* mutants has highlighted the complexity of isoamylase function, its position in the starch pathway and its possible interaction with other starch metabolizing enzymes.

One *sugary1* mutant, termed *sugary1-Reference* (*su1-R*) has been particularly informative. This mutant genetically defined the *Su1* locus over 100 years ago. Sequencing of this allele (Dinges et al. 2001) revealed three amino acid changes, and from comparison to other *su1* alleles, *su1-R* appears to be a total loss-of-function allele. Likewise, a second total loss-of-function allele caused by insertion of the transposable element, *Mutator* and termed *su1-R4582::Mu1* produces no protein. A third allele, termed *su1-st (sugary-starchy)* was caused by insertion of a transposable element termed *Toad.* It exhibits alternative RNA splicing and produces drastically reduced amounts of the SU1 protein. While the protein is reduced in amount, it is sufficient to condition an intermediate phenotype in one, two or three doses in the endosperm, as judged by kernel phenotype and chemical composition. Crosses involving the null allele *su1-R4582::Mu1 and su1-st* gives rise to kernels with an intermediate phenotype. However, endosperms containing both *su1-R* and *su1-st* exhibit the extreme, total loss-of-function phenotype of *su1-R.*

Molecular investigations provide a possible explanation for these intriguing dominance relationships exhibited by the two heterozygotes. Isoamylase appears to be a large multimer of SU1 proteins (reviewed in James et al. 2003). Given this structure, the presence of the mutant SU1-R protein could "poison" or otherwise inactivate debranching activity of SU1 polymers. This inactivation would not occur in heterozygotes harboring *su1-R4582::Mu1* since this latter mutant does not produce a SU1 protein.

Another fascinating fact gleaned from characterization of *su1* mutants was the finding that the other debranching enzyme, pullulanase, is also reduced in *su1* mutants. This reduction appears to be related to elevated sugar levels since pullulanase is also reduced in *sh2* mutants (Wu et al. 2002). Another captivating feature of the *sugary1* locus is that although its enhancement of sugars has been exploited by the sweet corn industry for many years, the underlying reason for increased sugar is not known. It is possible that the lack of efficiently packaged glucose in starch backs up the pathway, giving rise to enhanced sugar levels.

A pullulanase maize mutant was recently isolated via transposon mutagenesis (Dinges et al. 2003). *Mu* insertion into the *maize pullulanase1* gene (*Zpu1*) and consequent mutant analysis showed that maize contains only one gene encoding this enzymatic activity. Furthermore, the enzyme is expressed into several tissues and is involved in both starch synthesis and degradation. Loss of *Zpu1* activity reduces the rate of leaf starch breakdown, increases the percentage of amylose in leaf starch and reduces germination and early seedling growth. Surprisingly, endosperm starch shows only minor modification. Total endosperm starch and sugars are unaffected, although phytoglycogen is moderately reduced. The lengths of phytoglycogen side branches in the mutant were noticeably shorter.

Double mutant analysis (Dinges et al. 2003) revealed that isoamylase and pullulanase likely have partially overlapping roles. Loss of both debranching enzymes conditions phytoglycogen content five times that of either single mutant and reduces starch content below that of either mutant. Interestingly, mutational obliteration of either pullulanase or isoamylase causes loss of starch branching enzyme IIa in endosperms. These data point to the complexity of starch synthesis and the likely involvement of complexes involving several enzymatic activities.

The physiologically relevant substrate(s) for starch debranching activity has been the topic of much debate and has been the focus of an immense amount of research. Elucidation of the true role of SBE will likely provide important insight into the actual mechanism of starch synthesis. No less than three roles have been proposed (reviewed in Myers et al. 2000; James et al. 2003; Ball and Morell 2003; Tetlow et al. 2004). These include the following: (1) creation of glucose chains to initiate starch synthesis, (2) synthesis of substrates for starch synthesis (the clearing model) through the degradation of phytoglycogen, and (3) removal of excess branches from amylopectin (the trimming model), thereby causing clustered branch points.

Note that of the three models, only the latter (the trimming model) accounts for altered amylopectin structure in a debranching mutant. As noted above, amylopectin structure is altered in *su1* mutants. Glucan side chains having 2 to 12 glucose residues are elevated relative to wildtype whereas chains with 15 to 25 glucose moieties are decreased in amount. In addition, the trimming model (Ball et al. 1996) provides an explanation for the clustering of α 1,6 linkages in amylopectin. As originally proposed the trimming model noted that the close spacing of α linkages within a cluster gives rise to parallel glucans. These can intertwine into double helices and, in turn, give rise to the crystalline structure of starch and to the growth of glucan side-chains that potentially are unlimited in length. Clustered α 1,6 linkages could come about if branching and debranching occur at random, but because of steric hindrance, debranching activity cannot cleave closely spaced branches (Myers et al. 2000).

6
Conclusion

The importance of starch biosynthesis to plant life and to mankind and agriculture as we know them can not be overstated. As pointed out by Myers et al. (2000), the evolutionary capture by plants of genes encoding enzymatic activities of bacterial glycogen biosynthesis and the selection of mutant forms of these enzymes that now synthesize very large, finely packaged, inert forms of glucose is a fabulous evolutionary success story. Key to this evolution was the ability to synthesize α-1,6 linkages in a non-random, clustered fashion. Debranching enzymes are essential to this process.

Acknowledgements The author thanks Martha James, Steven Ball, Mark Guiltinan, Matthew Morell, Karen Koch and Don McCarty for insightful discussions of the many details of starch biosynthesis and helpful directions to relevant literature. Errors are clearly the responsibility of the author. Research in this laboratory is supported by National Science Foundation Grants IOB-0444031, DBI-0077676, DBI-0606607, IOB-9982626 and USDA grant 2006-35100-17220.

References

Akihiro T, Mizuno K, Fujimura T (2005) Gene expression of ADP-glucose pyrophosphorylase and starch contents in rice cultured cells are cooperatively regulated by sucrose and ABA. Plant Cell Physiol 46:937–946

Bae JM, Giroux M, Hannah LC (1990) Cloning and molecular characterization of the *brittle-2* gene of maize. Maydica 35:317–322

Ball S, Guan H, James M, Myers A, Keeling A, Mouille G, Buleon A, Colonna P, Priess J (1996) From glycogen to amylopectin: a model for the biogenesis of the plant starch granule. Cell 86:349–352

Ball SG, Morell MK (2003) From bacterial glycogen to starch: Understanding the biogenesis of the plant starch granule. Ann Rev Plant Biol 54:207–233

Beckles DM, Smith AM, apRees T (2001) A cytosolic adp-glucose pyrophosphorylase is a feature of graminaceous endosperms, but not of other starch-storing organs. Plant Physiol 125:818–827

Bhave MR, Lawrence S, Barton C, Hannah LC (1990) Identification and molecular characterization of *shrunken-2* cDNA clones of maize. Plant Cell 2:581–588

Blauth SL, Yao Y, Klucinec JD, Shannon JC, Thompson DB, Guiltinan MJ (2001) Identification of *Mutator* insertional mutants of the starch-branching enzyme 2a in corn. Plant Physiol 125:1396–1405

Burger BT, Cross J, Shaw JR, Caren J, Greene TW, Okita TW, Hannah LC (2003) Relative turnover numbers of maize endosperm and potato tuber ADP-glucose pyrophosphorylases in the absence and presence of 3PGA. Planta 217:449–456

Burton RA, Johnson PE, Beckles DM, Fincher GB, Jenner HL, Naldrett MJ, Denyer K (2002) Characterization of the Genes Encoding the Cytosolic and Plastidial Forms of ADP-Glucose Pyrophosphorylase in Wheat Endosperm. Plant Physiol 130:1464–1475

Cao H, Sullivan TD, Boyer CD, Shannon JC (1995) *Bt1*, a structural gene for the major 39–44 kD amyloplast membrane polypeptides. Physiol Plant 95:176–186

Cobb BG, Hannah LC (1988) Shrunken-1 encoded sucrose synthase is not required for sucrose synthesis in the maize endosperm. Plant Physiol 88:1219–1221

Commuri P, Keeling PL (2001) Chain-length specificities of maize starch synthase I enzyme: studies of glucan affinity and catalytic properties. Plant J 25:475–486

Crevillén P, Ballicora MA, Mérida A, Preiss J, Romero J (2003) The different large subunit isoforms of Arabidopsis thaliana ADP-glucose pyrophosphorylase conferdistinct kinetic and regulatory properties to the heterotetrameric enzyme. J Biol Chem 278:28508–28515

Cross JM, Clancy M, Shaw JR, Greene TW, Schmidt RR, Okita TW, Hannah LC (2004) Both subunits of ADP-glucose pyrophosphorylase are regulatory. Plant Physiol 135:137–140

Denyer K, Dunlap F, Thorbjornsen T, Keeling P, Smith AM (1996) The major form of ADP-glucose pyrophosphorylase in maize endosperm is extra-plastidal. Plant Physiol 112:779–785

Dinges JR, Colleoni C, Myers AM, James MG (2001) Molecular structure of three mutations at the maize sugary1 locus and their allele-specific phenotypic effects. Plant Physiol 125:1406–1418

Doan DNP, Rudi H, Olsen OA (1999) The allosterically unregulated isoform of ADP-glucose pyrophosphorylase from barley endosperm is the most likely source of ADP-glucose incorporated into endosperm starch. Plant Physiol 121:965–975

Dinges JR, Colleoni C, James MG, Myers AM (2003) Mutational analysis of the pullulanase-type debranching enzyme in maize indicates multiple functions in starch metabolism. Plant Cell 15:666–680

Fujita N, Yoshida M, Asakura N, Ohdan T, Miyao A, Hirochika H, Nakamura Y (2006) Functin and characterization of starch synthase I using mutants in rice. Plant Physiol 140:1070–1084

Gao M, Wanat J, Stinard PS, James MG, Myers AM (1998) Characterization of dull1, a maize gene coding for a novel starch synthase. Plant Cell 10:399–412

Giroux MJ, Hannah LC (1994) ADP-glucose pyrophosphorylase in shrunken2 and brittle2 mutants of maize. Mol Gen Genet 243:400–408

Giroux MJ, Shaw J, Barry G, Cobb BG, Greene T, Okita T, Hannah LC (1996) A single gene mutation that increases maize seed weight. Proc Natl Acad Sci USA 93:5824–5829

Glawischnig E, Gierl A, Tomas A, Bacher A, Eisenreich W (2002) Starch biosynthesis and intermediary metabolism in maize kernels. Quantitative analysis of metabolite flux by nuclear magnetic resonance. Plant Physiol 130:1717–1727

Greene TW, Hannah LC (1998a) Assembly of maize endosperm ADP-glucose pyrophosphorylase requires motifs located throughout the large and small subunit units. Plant Cell 10:1295–1306

Greene TW, Hannah LC (1998b) Enhanced stability of maize endosperm ADP-glucose pyrophosphorylase is gained through mutants that alter subunit interactions. Proc Natl Acad Sci USA 95:13342–13347

Hannah LC (1997) Starch synthesis in the maize endosperm. In: Larkins BA, Vasil IK (eds) Advances in Cellular and Molecular Biology of Plants. Cellular and Molecular Biology of Plant Seed Development. Kluwer Academic Publishers, Dordrecht, The Netherlands 4:375–405

Hannah LC (2005) Starch synthesis in the maize endosperm. Maydica 50:497–506

Hannah LC, Nelson OE (1976) Characterization of ADP-glucose pyrophosphorylase from shrunken-2 and brittle-2 mutants of maize. Biochem Genet 14:547–560

Hannah LC, Shaw JR, Giroux M, Reyss A, Prioul JL, Bae JM, Lee JY (2001) Maize genes encoding the small subunit of ADP-glucose pyrophosphorylase. Plant Physiol 127:173–183

Hirose T, Herao T (2004) A comprehensive expression analysis of the starch synthase gene family in rice (Oryza sativa L.). Planta 220:9–16

James MG, Robertson DS, Myers AM (1995) Characterization of the maize gene sugary1, a determinant of starch composition in kernels. Plant Cell 7:417–429

James MG, Denyer K, Myers A (2003) Starch synthesis in the cereal endosperm. Curr Opin Plant Biol 6:215–222

Kim WT, Franceschi VR, Okita TW, Robinson NL, Morell M, Preiss J (1989) Immunocytochemical localization of ADPglucose pyrophosphorylase in developing potato tuber cells. Plant Physiol 91:217–220

Koch KE (2004) Sucrose metabolism: regulatory mechanisms and pivotal roles in sugar sensing and plant development. Curr Opin Plant Biol 7:235–246

Martin C, Smith AM (1995) Starch biosynthesis. Plant Cell 7:971–985

Myers AM, Morell MK, James MG, Ball SG (2000) Recent progress toward understanding biosynthesis of the amylopectin crystal. Plant Physiol 122:989–997

Nelson OE, Pan D (1995) Starch synthesis in maize endosperms. Ann Rev Plant Physiol Plant Mol Biol 46:475–496

Nelson OE, Rines HW (1962) The enzymatic deficiency in the waxy mutant of maize. Biochem Biophys Res Commun 9:297–300

Okita TW, Greenberg E, Kuhn DN, Preiss J (1979) Subcellular localization of the starch degradative and biosynthetic enzymes of spinach leaves. Plant Physiol 64:187–192

Pan D, Nelson OE (1984) A debranching enzyme deficiency in endosperms of the sugary-1 mutants of maize. Plant Physiol 74:324–328

Preiss J, Levi C (1980) Starch biosynthesis and degradation. In: Preiss J (ed) The Biochemistry of Plants, vol 3. Academic Press, New York, pp 371–423

Preiss J, Romeo T (1989) Physiology, biochemistry and genetics of bacterial glycogen synthesis. Adv Microbiol Phys 30:183–238

Preiss J, Ball K, Hutney J, Smith-White B, Li L, Okita TW (1991) Regulatory mechanisms involved in the biosynthesis of starch. Pur Appl Chem 63:535–544

Regina A, Bird A, Topping D, Bowden S, Freeman J, Barsby T, Kosar-Hashemi B, Li Z, Rahman S, Morell M (2006) High-amylose wheat generated by RNA interference improves indices of large-bowel health in rats. PNAS 103:3546–3551

Rudi H, Doan DN, Olsen OA (1997) A (his)6-tagged recombinant barley (Hordeum vulgare L.) endosperm ADP- glucose pyrophosphorylase expressed in the baculovirus-insect cell system is insensitive to allosteric regulation by 3-phosphoglycerate and inorganic phosphate. FEBS Lett 419:124–130

Sakulsingharoj C, Choi S-B, Hwang S-K, Edwards GE, Bork J, Meyer CR, Preiss J, Okita TW (2004) Engineering starch biosynthesis for increasing rice seed weight: the role of the cytoplasmic ADP-glucose pyrophosphorylase. Plant Sci 167:1323–1333

Satoh H, Nishi A, Yamashita K, Takemoto Y, Tanaka Y, Hosaka Y, Sakurai A, Fujita N, Nakamura Y (2003) Starch-branching enzyme I-deficient mutation specifically affects the structure and properties of starch in rice endosperm. Plant Physiol 133:1111–1121

Sehnke PC, Chung HJ, Wu K, Ferl RJ (2001) Regulation of starch accumulation by granule associated plant 14-3-3 proteins. Proc Natl Acad Sci USA 98:765–770

Shannon JC (1968) Carbon-14 distribution in carbohydrates of immature Zea mays kernels following 14C CO2 treatment of intact plants. Plant Physiol 43:1215–1220

Shannon JC, Pien FM, Liu KC (1996) Nucleotides and Nucleotide Sugars in Developing Maize Endosperms: Synthesis of ADP-Glucose in brittle-1. Plant Physiol 110:835–843

Shannon JC, Pien FM, Cao H, Liu KC (1998) Brittle-1, an Adenylate Translocator, Facilitates Transfer of Extraplastidial Synthesized ADP-Glucose into Amyloplasts of Maize Endosperms. Plant Physiol 117:1235–1252

Smidansky ED, Clancy M, Meyer FD, Lanning SP, Blake NK, Talbert LE, Giroux MJ (2002) Enhanced ADP-glucose pyrophosphorylase activity in wheat endosperm increases seed yield. Proc Natl Acad Sci 99:1724–1729

Smidansky ED, Martin JM, Hannah LC, Fischer AM, Giroux MJ (2003) Seed yield and plant biomass increases in rice are conferred by deregulation of endosperm ADP-glucose pyrophosphorylase. Planta 216:656–664

Stark DM, Timmerman KP, Barry G, Preiss J, Kishore GM (1992) Regulation of the amount of starch in plant tissues by ADP-glucose pyrophosphorylase. Science 258:287–292

Sullivan TD, Strelow LI, Illingworth CA, Phillips RL, Nelson OE Jr (1991) The maize brittle-1 locus: molecular characterization based on DNA clones isolated using the dSpm-tagged brittle-1-mutable allele. Plant Cell 3:1337–1348

Tetlow IJ, Morell MK, Emes MJ (2004) Recent developments in understanding the regulation of starch metabolism in higher plants. J Exp Bot 55:2131–2145

Thorbjornsen T, Villand P, Leszek A, Kleczkowski L, Olsen O-A (1996a) A single gene encodes two different transcripts for the ADP-glucose pyrophos-phorylase small subunit from barley (*Hordeum vulgare*). Biochem J 313:149–154

Thorbjornsen T, Villand P, Denyer K, Olsen O, Smith AM (1996b) Distinct isoforms of ADPglucose pyrophosphorylase occur inside and outside the amyloplast in barley endosperm. Plant J 10:243–250

Villand P, Kleczkowski LA (1994) Is there an alternative pathway for starch biosynthesis in cereal seeds? Z Naturforsch 49:215–219

Wang YJ, White P, Pollak L, Jane J (1993) Characterization of starch structures of 17 maize endosperm mutant genotypes with oh43 inbred line background. Cereal Chem 70:171–179

Wu C, Colleoni C, Myers AM, James MG (2002) Enzymatic properties and regulation of ZPU1, the maize pullulanase-type starch debranching enzyme. Arch Biochem Biophys 406:21–32

Yao Y, Thompson D, Guiltinan M (2004) Maize starch-branching enzymeisoforms and amylopectin structure. In the absence of starch-branching enzyme IIb, the further absence of starch-branching enzyme Ia leads to increased branching. Plant Physiol 136:3515–3523

Zhang X, Colleoni C, Ratushna V, Sirghie-Colleoni M, James MG, Myers AM (2004) Molecular characterization demonstrates that the Zea mays gene sugary2 codes for starch synthase isoform SSIIa. Plant Mol Biol 54:865–879

Plant Cell Monogr (8)
O.-A. Olsen: Endosperm
DOI 10.1007/7089_2007_114/Published online: 24 March 2007

The Structure and Expression of Cereal Storage Protein Genes

N. G. Halford · P. R. Shewry (✉)

Rothamsted Research, Harpenden AL5 2JQ, UK
peter.shewry@bbsrc.ac.uk

Abstract The cereal endosperm stores several types of protein. All cereals appear to store 7S globulins in their aleurone cells and prolamins in their starchy endosperm cells. In addition, the major storage proteins in the starchy endosperm of rice and oats are related to the 11S globulins of dicotyledonous seeds. Smaller amounts of proteins related to 11S globulins are also present in wheat while rice, maize and wheat also contain α-globulins. There are also clear gradients in protein amount and composition within the starchy endosperm, with higher protein contents in the subaleurone cells and in the vitreous parts of the maize endosperm. A number of mutant genes have been identified in maize, sorghum and barley which result in reductions in the accumulation of either all prolamins or specific prolamin groups leading to a high lysine phenotype. The prolamins are encoded by multigene families whose expression is regulated temporally and spatially. This regulation is determined by promoter elements such as the prolamin box which comprises endosperm (E) and nitrogen (N) elements, the opaque-2 (O2) binding site and the HMW glutenin subunit gene enhancer. Transcription factors that bind to these sites include prolamin box binding factor (PBF) (a Dof class Cys_2Cys_2 zinc finger DNA binding protein), opaque-2 (O2) and OHP1 (basic leucine zipper transcription factors) from maize, and ESBF-I and -II, SPA and BLZ1 from wheat. Prolamin genes may be silenced by several mechanisms including the presence of in-frame stop codons within coding sequences, promoter inactivity and reduced RNA stability.

1 Introduction

The cereal endosperm typically comprises over 70% starch and hence is frequently regarded primarily as a source of calories for human and animal nutrition. However, it also contains significant amounts of protein (typically about 8–15% dry wt) and it can be calculated that the amount of cereal protein consumed in food and feed for animals greatly exceeds that consumed in more protein-rich seeds such as legumes and oilseeds. For example, Shewry (2000) calculated that the total protein yield from cereal grains exceeded 250 million tonnes a year compared with about 110 million tonnes of protein from pulses, soybean, groundnut and oilseeds combined [based on yield data reported by FAO (1997)]. Hence, cereal seeds are the major source of protein in many livestock diets as well as for human communities in some parts of the world. Finally, it must be remembered that the grain proteins of wheat form the gluten fraction which confers the cohesive and viscoelastic properties that are essential for processing wheat flour into bread, other baked goods, pasta and noodles. Hence, understanding the control of storage protein gene expression in the developing endosperm is important in relation to improving the protein content and composition for nutrition and for food processing.

In the present work we review our current knowledge of the structure and expression of storage protein genes focusing on wheat, barley and maize. However, before doing this it is necessary to briefly introduce the major types of storage proteins present in these species and to describe their distributions within the grain.

2 Storage Protein Families

It can be argued that many of the proteins present in the mature cereal endosperm are present in sufficient quantities to act as storage proteins (ie., are digested during germination and provide nutrition to the seedling). However, the discussion in this work will be restricted to two groups of proteins which are the quantitatively major components and appear to have no biological role except storage: the prolamins and globulins.

2.1 Prolamins

Prolamins were initially defined on the basis of their solubility properties, being insoluble in water but soluble in alcohol-water mixtures (Osborne 1924). Furthermore, the name "prolamin" was coined to reflect their high contents of proline and amide nitrogen (now known to be derived from glutamine). This definition has subsequently been refined to reflect our increased knowledge, particularly the fact that a proportion of the prolamins may be present in disulphide-stabilised polymers which are only extracted in the presence of a reducing agent. Nevertheless, the prolamins still form a clearly defined group of proteins which are present only in the starchy endosperm cells of cereal seeds.

The prolamins of wheat, barley and rye (the tribe Triticeae) have been most intensively studied. They are called gliadins (monomeric components) and glutenins (polymeric) in wheat, hordeins in barley and secalins in rye. The prolamins in the three species have a high degree of structural relatedness and form part of a larger group of proteins called the "prolamin superfamily" (Kreis et al. 1985). In addition to prolamins from other species this superfamily also includes a range of small sulphur-rich proteins, many but not all of which are restricted to seeds (including puroindolines, cereal inhibitors of trypsin or α-amylase, 2S storage albumins, non-specific lipid transfer proteins), and a group of hydroxyproline-rich cell wall proteins (see Shewry et al. 2004).

The prolamins of the Triticeae are classified into three families based on their amino acid sequences (and hence evolutionary relationships). The sulphur-rich prolamins comprise the α-gliadins, γ-gliadins (monomeric) and low molecular weight (LMW) subunits of glutenin (polymeric) of wheat, the sulphur-poor prolamins comprise the ω-gliadins (monomeric) and the high molecular weight (HMW) prolamins comprise the high molecular weight subunits of glutenin (also polymeric). The characteristics of these groups are summarised in Table 1. It should be noted that all contain repeated sequences but these vary in their sequence motifs and extent of re-iteration. These repeated sequences are responsible for the unusual amino acid compositions of the prolamins and also appear to adopt unusual extended conformations, which may be similar to those formed by structural proteins from animal tissues (see Tatham and Shewry 2000; Shewry et al. 2003). Homologous groups of proteins related to γ-gliadins (γ-hordeins), ω-gliadins (C hordeins), LMW subunits (B hordeins) and HMW subunits (D hordeins) are also present in barley (see Shewry and Darlington 2002).

The prolamins of the Triticeae are also related to three groups of prolamins present in maize, called the β-, γ- and δ-zeins. These proteins have limited sequence similarity to the S-rich and HMW prolamins, notably in the number and distributions of cysteine residues. However, only the M_r 27 000 and M_r 16 000 γ-zeins contain repeated sequences, which are either two or eight tandem repeats of Pro.Pro.Pro.Val.His.Leu. The β-zeins and δ-zeins are rich in methionine residues which in the β-zeins are clustered in a region close to the C-terminus.

In contrast, the major α-zeins of maize appear to be unrelated to the prolamins of the Triticeae or indeed to any other protein family with the exception of α-type prolamins in related cereals (sorghum and millets) (Shewry and Halford 2002). The α-zeins comprise two sub-classes of proteins which are often called Z19 and Z22 zeins based on their relative mobilities on SDS-PAGE, although their true masses are higher (about 23 000–24 000 and 26 500–27 000, respectively). They both contain repeated blocks of about 20 amino acids (nine in the Z19 and 10 in the Z22 proteins) but these are poorly conserved in sequence and rich in Leu, Val and Ala (ie., amino acids with aliphatic side chains) as well as in Gln.

The β-, γ- and δ-zeins are also relatively rich in cysteine and present only in polymers whereas the α-zeins contain only one or two cysteine residues and are present as monomers, dimers and oligomers.

Table 1 Summary of the types and characteristics of wheat grain prolamins (gluten proteins) and zeins of maize. Based on Shewry and Halford (2002) incorporating data on zeins from Coleman and Larkins (1999) and other sources

Components	M_r (% total prolamins)	Polymers or monomers	Partial amino acid composition (mol%)
Gluten protein			
HMW prolamins			
HMW subunits of glutenin	65–90 000 (6–10%)	polymers	30–35% Gly, 10–16% Pro, 15–20% Gly, 0.5–1.5% Cys, 0.7–1.4% Lys
S-rich prolamins			
γ-gliadins	30–45 000	monomers	30–40% Gln, 15–20% Pro,
α-gliadins	(70–80%)	monomers	2–3% Cys, < 1.0% Lys
B- and C-type LMW subunits of glutenin [a]		polymers	
S-poor prolamins			
ω-gliadins	30–75 000	monomers	40–50% Gln, 20–30% pro,
D-type LMW subunits of glutenin [a]	(10–20%)	polymers	8–9% Phe, 0–0.5% Lys, 0–0.5% Cys [b]
Zeins			
α-zeins			
Z19 [c]	23–24 000	monomers,	≈ 20% Gln, 13–14% Ala,
Z22 [c]	26 500–27 000 (75–85%)	dimers and oligomers	17–19% Leu, 3–7% Val, 9–10% Pro, < 1% Cys, 0% Trp
β-zein			
	17 500 (10–15%) [d]	Polymers	16.5% Gln, 9% Pro, 14% Ala, 4.4% Cys, 11.4% Met, 0% Trp
γ-zeins			
27K [c]	22 000	polymers	15–19% Gln, 15–25% Pro,
16K [c]	17 800 (5–10%) [d]	polymers	≈ 7% Cys, < 2% Met, 0–0.6% Trp
δ-zeins			
10K [c]	14 400	polymers	12% Gln, 16% Pro,
18K [c]	21 100 (< 5%) [d]	polymers	2–4% Cys, 23–27% Met, 0–1% Trp

[a] C-type LMW subunits are essentially polymeric forms of γ-and α-gliadins and D-type subunits polymeric ω-gliadins. The B-type LMW subunits constitute a discrete group of S-rich prolamins;

[b] Cys is present in D-type LMW subunits but not ω-gliadins;

[c] Designations based on relative molecular mass determined by SDS-PAGE;

[d] Quantitative data on zeins are from Esen (1987) but Coleman and Larkins (1999) describe γ-zeins as the second most abundant group.

2.2 Globulins

Globulins are classically defined on their solubility in dilute salt solutions, typically 0.5–1.0 M NaCl. Maize, barley and wheat all contain 7S storage globulins which are related to the "vicilin-like" globulins of dicotyledonous seeds (notably legumes) (see Burgess and Shewry 1986; Yupsanis et al. 1990; Kriz 1999). However, in all three species they appear to be restricted in distribution to the embryo and aleurone layer, where they form the major storage protein fraction. They have been studied in most detail from maize embryos, where two components called GLB1 and GLB2 together account for about 10–20% of the total protein (Kriz 1999). They resemble the 7S globulins of legumes in having low contents of cysteine and methionine residues and probably have a similar structure with three subunits of M_r 50–70 000 which undergo post-translational processing (Yupsanis et al. 1990; Kriz 1999).

The major storage protein type present in most dicoyledonous seeds is a 11-12S globulin ("legumin"), with related proteins forming the major storage protein fractions in oats and rice (see Casey 1999; Shotwell 1999; Takaiwa et al. 1999). A related protein is also present in wheat where it is called triticin and has been estimated to account for up to 5% of the total grain protein (Singh et al. 1991a,b). Triticin resembles typical 11S globulins in comprising subunits which are post-translationally cleaved to give heavy (M_r 52 000, M_r 58 000) and light (M_r 23 000, M_r 22 000) chains but has not been established to have the typical hexameric structure of 11S globulins. Triticins are present in protein bodies in the starchy endosperm, confirming that they are storage proteins (Singh and Shepherd 1987; Bechtel et al. 1991).

Related 11S globulin proteins are presumably also present in barley but have not been characterised but Woo et al. (2001) have demonstrated the presence of transcripts encoding legumin-like proteins in developing endosperms of maize.

It appears, therefore, that most, if not all, cereals contain 7S and 11S storage proteins which are related to those present in dicotyledonous seeds and belong to the "cupin" superfamily of plant proteins as defined by Dunwell and colleagues (Dunwell and Gane 1998; Dunwell et al. 2004).

However, recent work has shown that a third class of globulin storage proteins may also be present, the α-globulins. Woo et al. (2001) demonstrated the presence of transcripts encoding an M_r 18 000 protein in developing endosperms of maize and called it α-globulin based on a previously characterised protein from rice seeds (Nakase et al. 1996), although the protein was not characterised and there was no evidence that it had typical globulin solubility properties. They subsequently used antibodies raised against recombinant protein to locate the protein in protein bodies of starchy endosperm cells. We have since used a similar approach to demonstrate the presence of a related storage protein in wheat endosperm (unpublished results of PR Shewry and co-workers). The α-globulins have not been quantified but the fact that they were initially detected at the transcript level rather than as proteins indicates that they are not major components. It should also be noted that they are members of the "prolamin superfamily", being related in sequence to prolamins, 2S storage albumins and other small S-rich seed proteins.

3 Location of Storage Proteins in the Endosperm

The cereal endosperm comprises two distinct tissues, the aleurone and starchy endosperm (Becraft, in this volume; Brown and Lemmon, in this volume), which differ in their protein composition. In particular, the aleurone cells contain 7S storage globulins which are not present in the starchy endosperm. Despite its separate ontogeny, 7S globulins are also stored in the embryo, being present in the scutellum and axis of maize embryos (Kriz 1999). This may reflect the functional similarity between the embryo and aleurone during seed germination. The starchy endosperm is also non-homogeneous, exhibiting clear gradients in protein content and composition.

3.1 Maize

It is well established that the cells in the floury parts of the maize endosperm contain less protein than those in the vitreous (horny) parts (see Watson 1987), with one or more layers of small sub-aleurone cells being particularly rich in protein (up to almost 30%) (Hinton 1953). Larkins and co-workers have shown that there are also gradients in protein composition, using specific antibodies and in situ hybridisation to study protein deposition and gene expression, respectively, during grain development. Immunolocalisation using light and electron microscopy showed clear differences in zein distribution at 14 and 18 days after pollination, with α-zeins being concentrated in the interior part of the starchy endosperm tissue and β- and γ-zeins in the outer (ie., sub-aleurone) layers of cells (Lending and Larkins 1989). In situ analyses (Woo et al. 2001) carried out between 10 and 25 days after pollination gave results which were consistent with those from immunolocalisation but failed to show significant levels of any storage protein transcripts in the crown and central regions which were not studied by Lending and Larkins (1989). Woo et al. (2001) also showed that transcripts for the α-globulins were not as abundant as those for most zeins and were particularly concentrated in the outer cells on the opposite side of the grain to the embryo.

3.2
Wheat and Barley

There is also considerable variation in the protein content of wheat and barley starchy endosperm cells with the sub-aleurone cells being rich in protein but containing little starch. In fact, Kent (1966) was able to prepare flour fractions corresponding to the sub-aleurone and inner endosperm cells containing about 45% and 11% protein, respectively, compared with about 13% protein for whole flour.

A different approach was used by Millet et al. (1991) to determine the distribution of protein in the barley grain. They used a pearling mill to remove sequential fractions from the outside of the grain by abrasion. This showed differences in the distribution of hordeins and of hordein polymers between the outer and inner layers of the starchy endosperm of two cultivars differing in their malting quality. In particular, the central part of the starchy endosperm was enriched in D hordein (16–23% compared with 14–16%) and the peripheral starchy endosperm in polymeric B hordeins. Shewry et al. (1996) used a similar approach to show that whereas the sub-aleurone layer of barley was richer in hordein this comprised mainly the S-rich and S-poor components (B and C hordeins, respectively) with the HMW D hordein being concentrated in fractions obtained from between the sub-aleurone and inner starchy endosperm. Complementary studies were reported by Davies et al. (1993) who used in situ hybridisation of serial sections cut from developing barley grain to show that mRNA for B hordein was most highly expressed in the sub-aleurone cells.

Although detailed studies of the distribution of gluten proteins in wheat have not been reported, evidence currently available indicates that similar differences in distribution of S-rich, S-poor and HMW components may exist to those reported in barley. In particular, the use of the GUS reporter gene to study promoter activity in the developing grain of transgenic lines shows clear differences in the patterns of expression conferred by promoters from LMW subunit (Stöger et al. 2001) and HMW subunit (Lamacchia et al. 2001) genes. This clearly shows that the LMW subunit promoter is most strongly expressed in the sub-aleurone cells and outer part of the starchy endosperm and the HMW subunit promoter in the central part of the starchy endosperm —these distributions are consistent with the analyses of barley discussed above.

3.3
Implications of Differences in Protein Distribution

The differences in protein distribution discussed above imply that the storage protein gene promoters have elements which specify their spatial as well as their temporal expression within the starchy endosperm. The differences also imply that flour fractions derived from different parts of the wheat endosperm will differ in their nutritional quality and functional properties as well as in their total protein content. This means that endosperm fractionation could be used to prepare flour fractions for specific end uses. In the longer term it may also be possible to use mutagenesis or transgenesis to alter the distributions of individual proteins within the endosperm to produce flours with improved or modified end use properties.

4 Mutant Genes that Affect Storage Protein Synthesis

A number of mutant genes have been identified which affect the accumulation or processing of storage proteins. In maize these are readily identified due to their effect on the phenotype of the grain, giving kernels which are soft and opaque to transmitted light. The demonstration that two well-characterised maize mutants, *opaque 2*, and *floury 2*, are enriched in the essential amino acid lysine (Mertz et al. 1964; Nelson et al. 1965) led to more systematic screening of maize mutants for genetic sources of the high lysine phenotype (see Bright and Shewry 1983) and to attempts to identify similar mutants in barley and sorghum. Similar phenotypic screening was used to identify two "high lysine" genes in sorghum, the spontaneous mutant *hl* gene (Singh and Axtell 1973) and the induced *P721opaque* gene (Axtell et al. 1979).

However, screening for a visual phenotype cannot be carried out in barley and it was necessary to use a direct chemical screen for total basic amino acid content. As in sorghum this led to the identification of one spontaneous mutant gene, the *lys* gene, present in the Ethiopian line Hiproly (Munck et al. 1970). In addition a number of other mutant genes were identified after treatment with a range of chemical and physical mutragens (Doll et al. 1974; Bright and Shewry 1983).

No systematic searches have been carried out for high lysine genes in wheat. Furthermore, to have an effect in the hexaploid or tetraploid background they would need to be dominant in their effect. Most high lysine genes in other cereals are recessive, although two dominant genes have been discovered in maize (as discussed below).

Many, but not all, of these high lysine mutations are characterised by low contents of prolamins (zein in maize, kafirins in sorghum and hordeins in barley) and in some cases also in the compositions of their prolamin fractions. Only a small number of these effects are understood at the level of gene regulation and these are discussed here.

4.1
The *lys3a* Gene of Barley

The mutant *lys3a* gene was generated by treating barley cv Bomi with the chemical mutagen ethylinimine (Doll et al. 1974). It is characterised by a reduction in the proportion of the total grain N present in hordein from about 45% to 15%, with compensatory increases in other proteins and free amino acids (Shewry et al. 1980). There are also dramatic changes in the composition of the hordein fraction, with an absence of C hordein, changes in the pattern of the S-rich B and γ-hordeins but no apparent effects on D hordein (Shewry et al. 1980).

4.2
The *opaque2* and *opaque15* Genes of Maize

opaque2 was the first high lysine cereal gene to be discovered (Mertz et al. 1964) and has been the most intensively studied. It results in an overall decrease in zeins, but a specific decrease in the Z22 α-zeins. This effect results from the wild-type 02 gene encoding a transcriptional activator of the Z22 zein genes (Ueda et al. 1992).

The *opaque2* mutation was initially identified by the presence of a soft floury phenotype, which may result from the presence of smaller protein bodies which affect the packing of the starch granules (Coleman and Larkins 1999; Kumamaru et al., in this volume). This floury texture also results in a number of negative impacts which limited exploitation of the trait by breeders, in particular greater damage during harvesting and handling and increased susceptibility to pests and pathogens (Coleman and Larkins 1999). This soft phenotype can be converted to the wild-type state by "genetic modifiers" allowing the production of "Quality Protein Maize" (QPM) with a hard phenotype (Villegas et al. 1992; Geevers and Lake, 1992). One impact of these genetic modifiers is an increase in the amount of γ-zein (Wallace et al. 1990) and this increase appears to be correlated with the degree of modification of the phenotype (Lopes and Larkins 1991, 1995). The effects of the *opaque2* mutation on zein gene expression are discussed in detail below.

The *opaque15* gene was identified by screening a mutagenised population of maize for reduced levels of γ-zein and appears to map to the same locus as one of the *opaque2* modifier genes (Dannenhoffer et al. 1995). The gene has not been isolated and its mechanism of action is not known but it clearly causes reductions in the amounts of γ-zein mRNA and protein.

4.3 The *floury2, De*B30* and *Mucronate* (*Mu*) Mutations of Maize

floury2 was the second high lysine gene to be discovered in maize (Nelson et al. 1965) while *De*B30* and *Mu* were discovered more recently by Salamini et al. (1979, 1983). Although the three mutant genes result in similar phenotypes, they differ in that *fl2* is recessive while *De*B30* and *Mu* are dominant. They also all result in broad decreases in all zeins accompanied by increases in the ER-located chaperone BiP (Boston et al. 1991). The *fl2* and *De*B30* genes also result in the synthesis of apparently novel zein components with slightly slower mobilities on SDS-PAGE than the major Z22 and Z19 α-zeins, respectively (Gillikin et al. 1997; Kim et al. 2004). These proteins appear to be encoded by mutant genes in which amino acid substitutions prevent cleavage of the signal peptide (Gillikin et al. 1997; Kim et al. 2004). This results in binding of the malfolded protein to BiP and tethering to the ER, disrupting zein synthesis and protein body formation (both mutations are associated with small misshapen protein bodies). It is possible that the *Mu* gene has a similar effect but this has not been established.

5 Endosperm-Specific Gene Expression

5.1 Prolamin Gene Families

Prolamin genes are organized into a number of multigenic loci. In barley, these loci are all present on chromosome 1H, with the *Hor1*, *Hor2* and *Hor5* loci, at which the C, B and γ hordein genes, respectively, are located, on the short arm and the *Hor3* locus, at which the D hordein genes are located, on the long arm.

The pattern is more complicated in wheat and analysis is made more difficult by the presence of multiple genomes. The *Glu-1* loci encoding the high molecular weight subunits of glutenin are present on the long arms of the group 1 chromosomes and each locus comprises two genes, designated *x* and *y*. Durum wheat cultivars therefore have four HMW subunit genes (Margiotta et al. 1993), while hexaploid wheat contains six: *Glu-A1x*, *Glu-A1y*, *Glu-B1x*, *Glu-B1y*, *Glu-D1x* and *Glu-D1y* (Forde J. et al. 1985). However, the *Glu-A1y* gene is always silent, while the *Glu-A1x* and *Glu-B1y* genes are silent in some cultivars.

The *Gli-2* loci encoding α-type gliadins are present on the group 6 chromosomes. Estimates of the number of genes at these loci range from 25-35 copies per haploid genome (Harberd et al. 1985) to 100 or more (Okita et al. 1985; Anderson and Greene et al. 1997). This far exceeds the number of proteins that can be separated by 2D electrophoresis, suggesting again that gene silencing has occurred.

The *Gli-1/Glu3* loci encoding the γ-gliadins and ω-gliadins (including the D group of the LMW subunits) on the group 1 chromosomes have proved particularly difficult to study, with an estimated 39 γ gliadin genes, 16 ω gliadin genes and 39 LMW subunit genes (Sabelli and Shewry 1991). As with the other gene families, the presence of some silent genes has been demonstrated (Rafalski 1986; Hsia and Anderson 2001).

The α-zein gene family of maize is even more complex, comprising an estimated 75–100 genes (Hagen and Rubenstein 1981; Wilson and Larkins 1984). The number that are expressed remains unclear, but evidence from the analysis of cDNA libraries suggests that many are silent (Marks et al. 1985a,b). The genes encoding α-zeins can be subdivided into three groups with the Z22 zeins being encoded by genes at loci on the short arm of chromosome 4 and Z19 zeins encoded by genes at loci on the long arms of chromosome 4 (B group) and chromosome 1 (D group) (reviewed by Woo et al. 2001).

In contrast, only one or two genes encoding β-, γ- and δ-zeins are present per genome (Pederson et al. 1986; Prat et al. 1987; Kirihara et al. 1988; Chui and Falco 1995). The β-zein genes are located on the short arm of chromosome 6, the γ-zein genes on the long arm of chromosome 7 and the δ-zein genes on the long arm of chromosome 9 (reviewed by Woo et al. 2001).

There is evidence for differential temporal expression of endosperm storage protein genes. For example, γ-zeins are expressed in maize endosperm before the α- and δ-zeins and show a more even distribution of expression (Woo et al. 2001). Similarly, although evidence from electrophoretic analyses of gluten proteins is inconclusive, expression profiling has shown that α/β gliadin genes can be divided into two groups showing early and late expression (Kawaura et al. 2005). However, broadly speaking prolamin genes are subject to similar tissue-specific and developmental regulation, the genes being expressed exclusively in the starchy endosperm during mid- and late-development, and nutritional regulation, responding sensitively to the availability of nitrogen and sulphur in the grain (Duffus and Cochrane 1992; Giese and Hopp 1984; DuPont et al. 2006). The expression of prolamin genes is controlled primarily at the transcriptional level (Bartels and Thompson 1986; Sørensen et al. 1989).

5.2 The Prolamin Box

A comparison of the nucleotide sequences of promoters of several wheat gliadin and barley hordein genes identified a conserved sequence approximately 30 bp long around 300 bp upstream of the ATG translation start site (Forde B.G. et al. 1985). This was first called the – 300 element and subsequently the prolamin box or endosperm element. It contains two conserved motifs, the endosperm or E motif, TGTAAAG, and the GCN4-like motif (GLM), nitrogen element or N motif, ATGAGTCAT (Hammond-Kosack et al. 1993; Müller and Knudsen 1993).

"Complete" prolamin boxes are present in the promoters of a wide range of prolamin genes, from wheat, barley, rye and oats. α-Zeins also contain a prolamin box, although the N motif is not perfectly conserved, while β- and γ-zeins contain an E motif in the expected position but the N motif is further upstream, in other words the two elements of the prolamin box are separate (the maize community often refers to the E motif as the prolamin or P element).

The α-gliadin and LMW subunit gene promoters also contain additional complete or partial boxes further upstream of the primary prolamin box, but these do not appear to be required for promoter activity and are not present in the γ- or ω- gliadin promoters. There is no evidence of a prolamin box or either of the E or N elements in the δ-zein promoters that have been characterised to date.

The E motif was shown to bind one or more maize endosperm proteins as long ago as 1987 (Maier et al. 1987) but it was not until 1997/1998 that maize and barley transcription factors that recognised the motif were cloned (Vicente-Carbajosa et al. 1997; Mena et al. 1998) (Fig. 1a). These transcription factors, called prolamin box binding factor (PBF) are members of the Dof class of Cys_2-Cys_2 zinc finger DNA binding proteins. In the meantime, another transcription factor that regulates zein gene expression had been the subject of much study. This was opaque-2 (*O2*), a basic leucine zipper transcription factor (Schmidt et al. 1990).

Mutations in O2 principally affect the Z22 class of α-zeins (Pedersen et al. 1980; Burr and Burr 1982; Langridge et al. 1982), causing a severe reduction in gene expression. The binding site for O2 was identified by Schmidt and coworkers (1992) and is adjacent to the prolamin box (Fig. 1b, Fig. 2b). PBF and O2 have been shown to interact in vitro (Vicente-Carbajosa et al. 1997), and to act additively to promote prolamin promoter activity in transient expression experiments (Hwang et al. 2004). There is also evidence that O2 activity is regulated diurnally through phosphorylation (Ciceri et al. 1997). O2 homologues have been identified in both sorghum and *Coix* (Pirovano et al. 1994; Vettore et al. 1998).

A second basic leucine zipper transcription factor that binds the O2 target site has been identified; this factor was given the name OHP1 (Pysh et al. 1993) (Fig. 2) and can bind to the site either as a homodimer or as a heterodimer with O2. However, OHP1 does not interact with PBF (Vicente-Carbajosa et al. 1997).

The primary O2 binding site that is present in Z22 α-zein genes is not present in other prolamin genes and even single base pair changes in the site have been shown to reduce O2 binding significantly (Ueda et al. 1992). There is evidence that O2 will also recognise the N motif (Yunes et al. 1994) and co-

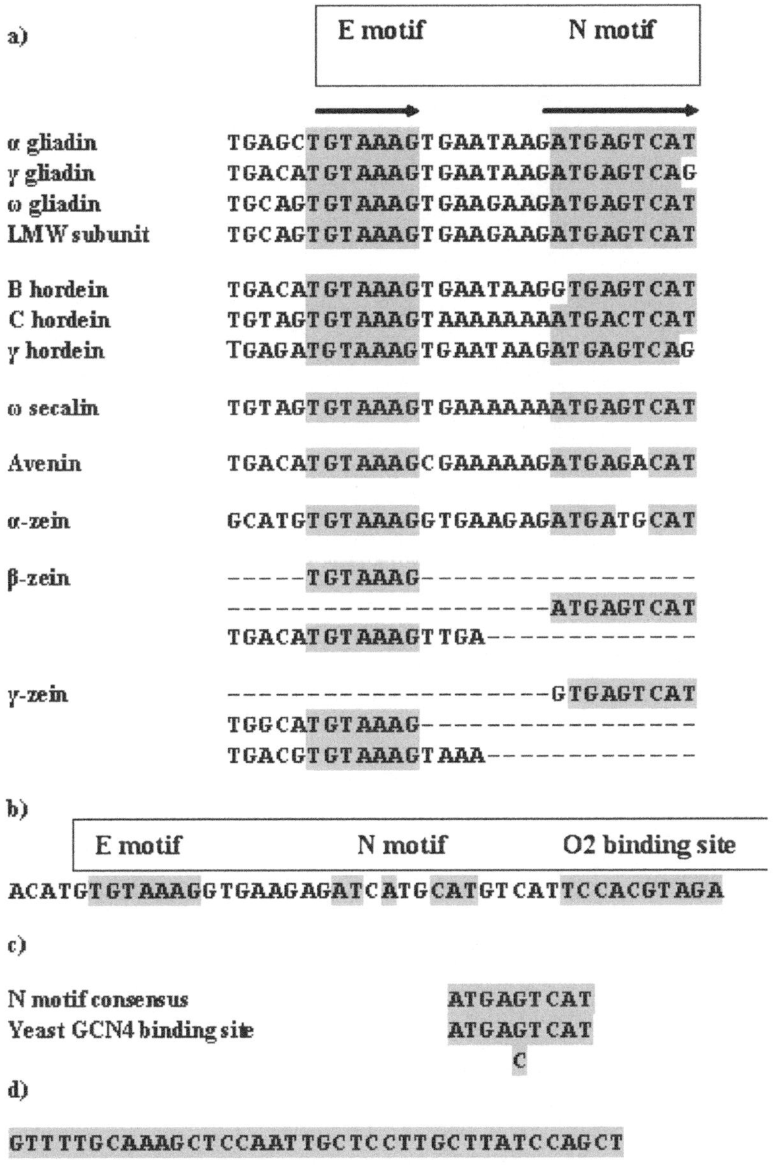

Fig. 1 Regulatory elements in prolamin promoters. (**a**) The prolamin box or – 300 element in the promoters of α, γ and ω gliadins (Genbank accession numbers X00627, AF234649 and AF280605), B hordein (X87232), C hordein (M36941), γ hordein (M36378), ω secalin (X60295), avenin (AY294282), α-zein (X61085), β-zein (M13507) and γ-zein (X58197). (**b**) The prolamin box and Opaque-2 binding site in a 22 kDa α-zein (X55722). (**c**) Comparison of the yeast GCN4 binding site with the N motif consensus from the prolamin box. (**d**) The HMW prolamin enhancer. Nucleotides matching the consensus sequence for the E and N motifs, the O2 and GCN4 binding sites and the HMW prolamin enhancer are highlighted.

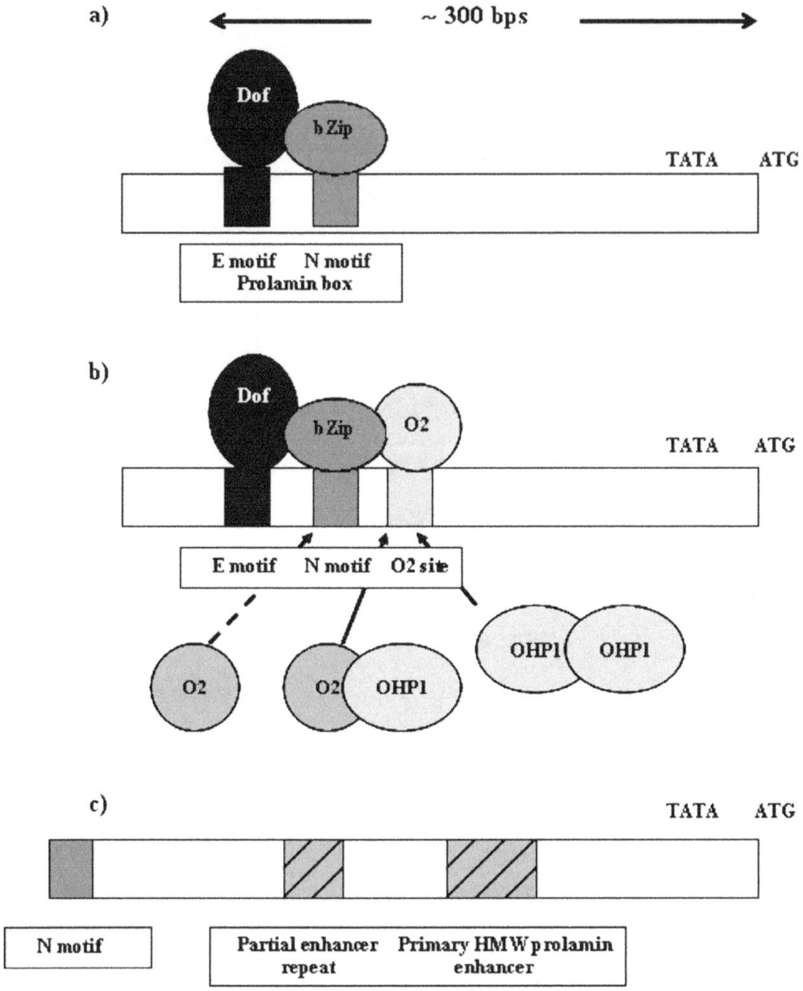

Fig. 2 (**a**) Schematic diagram showing the prolamin box of genes encoding α, γ and ω gliadins, B, C and γ hordeins, ω secalins, avenins, α-, β- and γ-zeins. The prolamin box contains two distinct elements, the E and N motifs that bind transcription factors of the Dof and bZip families, respectively. Note that the promoters of some but not all of these genes contain additional E and N motifs elsewhere in the promoter. (**b**) Schematic diagram of a 22 kDa α-zein gene promoter. The prolamin box is adjacent to the primary binding site for the Opaque-2 (O2) transcription factor. This site is also recognised by another transcription factor, OHP1, which can bind as a homodimer or as a heterodimer with O2. O2 also recognises the N motif and interacts with the Dof-class transcription factor that binds to the E motif, whereas OHP1 does not. (**c**) Schematic diagram of a HMW prolamin promoter. The primary enhancer is unique to the HMW prolamin genes, although it contains a sequence that is similar to the E motif. Part of it is repeated further upstream in the promoter. Additional E and N motif sequences are also present further upstream in the promoter but the primary enhancer on its own drives high levels of endosperm-specific expression

expression of O2 has been shown to enhance the expression of a reporter gene driven by a wheat LMW glutenin subunit gene promoter in plant protoplasts and yeast (Holdsworth et al. 1995); this promoter contains an N motif but not a primary O2 binding site. However, it should be remembered that O2 appears to regulate the Z22 class of α-zeins specifically, despite the prolamin box being present in other groups of zein genes.

The first functional analyses of wheat prolamin promoters involved the stable transformation of tobacco with LMW subunit gene promoter/chloramphenicol acetyl transferase (CAT) reporter gene constructs (Colot et al. 1987). Perhaps surprisingly, the promoter conferred the same tissue-specific and developmental regulation of gene expression in tobacco as it did in wheat. The shortest promoter length tested in the study and found to be active was 326 bp, long enough to include the prolamin box. More detailed analyses were carried out by Müller and Knudsen (1993), establishing conclusively a regulatory role for the prolamin box by using a transient expression system involving particle bombardment of cultured barley endosperms with C hordein promoter/β-glucuronidase (GUS) constructs. These experiments also confirmed that the E and N motifs were separate elements, with the N motif acting as a negative element at low nitrogen levels and interacting with the E motif to give high expression when nitrogen levels were adequate.

In the same year, Hammond-Kosack and co-workers used in vivo footprinting and gel retardation assays to show that the E motif within the prolamin box of a LMW subunit gene bound a putative transcription factor, ESBF-1, during early grain development (Hammond-Kosack et al. 1993). The N motif bound a second putative transcription factor, ESBF-II, prior to maximum expression of the gene. A functional analysis of the prolamin box of this gene in transgenic tobacco showed that both motifs were required for seed-specific expression. Transcription factors of the BZIP class that recognize the N motif of the prolamin box have since been cloned from wheat and barley and given the names SPA and BLZ1, respectively (Albani et al. 1997; Vicente-Carbajosa et al. 1998). They show some similarity to O2 and OHP1 from maize.

Full activation of the LMW subunit gene promoter, when introduced into yeast, requires a functional yeast GCN4; much lower levels of activation are obtained in yeast containing a defective GCN4 gene (Holdsworth et al. 1995). This is intriguing given that the N motif matches the yeast GCN4 binding site (Fig. 1c) (Hill et al. 1986). GCN4 is a transcriptional activator, the expression of which is controlled at the translational level in response to amino acid levels through the action of a protein kinase, GCN2 (Hinnebusch and Natarajan 2002). GCN2 phosphorylates eukaryotic translation initiation factor-2α (eIF-2α), bringing about a general reduction in protein synthesis at the same time as an increase in the synthesis of GCN4. The primary action of GCN4 is to induce the expression of amino acid biosynthesis genes, but it regulates a plethora of other genes affecting numerous areas of metabolism microarray analysis identified 539 yeast genes that are induced by GCN4 (Natarajan 2001). It is tempting to suggest that a GCN4-like transcription factor could be involved in regulating prolamin gene expression in response to nitrogen availability, since nitrogen deprivation could lead to a reduction in amino acid levels. If there is a functional homologue in plants, it has not been possible to identify it based solely on sequence similarity despite the extensive plant genome information that is now available. However, plants do contain a homologue of the protein kinase, GCN2 (Zhang et al. 2003).

5.3 The HMW Prolamin Enhancer

HMW prolamin genes show the same patterns of gene expression as gliadin and LMW subunit genes. However, the HMW prolamin enhancer shows only very limited similarity with the prolamin box and was identified independently. As with other prolamin genes, HMW prolamin gene promoters are active in tobacco and drive high levels of endosperm-specific expression. This has been demonstrated using a complete HMW glutenin subunit gene from wheat (Robert et al. 1989) and chimaeric constructs of HMW subunit gene promoters attached to CAT and GUS reporter genes (Colot et al. 1987; Halford et al. 1989).

Halford et al. (1989) showed that the region of the promoter of a *Glu-D1x* gene from – 277 to + 39 with respect to the transcription start site was sufficient to drive high levels of GUS expression, and the major enhancer was finally located in a 38 bp sequence from position – 186 to – 148 of a *Glu-D1y* gene by Thomas and Flavell (1990) (Figs. 1d and 2c). Simple BLAST searches with this sequence using the tools and databases available at the National Center for Biotechnology Information (http://www.ncbi.nlm.nih.gov) reveal that it is extremely tightly conserved in HMW prolamins from a variety of species, with identical or near-identical sequences being present in D hordein genes from barley and HMW secalin genes from rye, as well as HMW prolamin genes from a range of wild and cultivated *Triticum* and *Aegilops* species, and ornamental grasses of the genus *Leymus*. However, no similar sequences have been identified to date in maize or rice.

In HMW glutenin subunit genes from wheat, the palindromic sequence TTTGCAAA within the primary enhancer is repeated further upstream. Part of this sequence resembles the E motif of the prolamin box (above) and it has been referred to as a prolamin box or P element by some workers (Norre et al. 2002). Sequences corresponding to parts of the E and N motifs are also present further upstream. Both the primary enhancer and the upstream N motif have been shown to bind maize nuclear proteins, one of which could be O2 (Norre et al. 2002). However, it is not clear how relevant this is to other species, since maize does not make HMW prolamins and no maize genes contain the HMW prolamin enhancer.

These "additional" E and N motifs upstream of the primary enhancer are not required for activity of the HMW subunit promoter in transgenic tobacco. In the experiments of Halford and co-workers, higher levels of expression were obtained with a promoter that included the N-box than with shorter promoters that contained the HMW enhancer but none of the other motifs, but a statistically significant difference could not be demonstrated (Halford et al. 1989).

The high level of endosperm-specific expression driven by HMW prolamin promoters [each active HMW subunit gene in bread wheat encodes a protein that accounts for approximately 2% of the total seed protein at maturity (Halford et al. 1992)] make them potentially very useful for the seed-specific expression of transgenes. High levels of expression have been obtained from additional copies of HMW subunit genes introduced into transgenic wheat (Altpeter et al. 1996; Barro et al. 1997), and a HMW subunit gene promoter has been shown to drive tight, readily-detectable, endosperm-specific expression of a GUS reporter gene in transgenic wheat (Lamacchia et al. 2001). A word of caution, however: there is some evidence that the wheat HMW subunit gene promoter is active in pollen when introduced into maize or barley (Scott et al. 2002; Zhang et al. 2001). There are reports from confidential sources that the minimal HMW subunit gene promoter (ie the region including the primary enhancer but no more) does not give this problem but these reports remain anecdotal to date.

5.4 Gene Silencing

As we describe above, many prolamin genes contain in-frame stop codons. Examples include the *Glu-D1x* gene from *A. cylindrica*, which contains a CAA (Gln) → TAA (Stop) mutation in the 343rd codon (Wan et al. 2002), and the *Glu-A1y* and *Glu-A1x* genes of *Triticum aestivum* (Forde J. et al. 1985; Bustos et al. 2000), resulting from similar CAG (Gln) → TAG (Stop) and CAA (Gln) → TAA (Stop) mutations. Stop codons are present in silent gliadin and LMW subunit genes (Anderson and Greene 1997; Rafalski 1986), and in zein genes, including for example the Z22 zein genes *pML1, pML2, zA1* and Z7 (Langridge and Feix 1983; Wandelt and Feix 1989; Spena et al. 1982; Kridl et al. 1984). However, mutation is not the only way in which the coding regions of prolamins have been disrupted, one wheat *Glu-A1y* allele, for example, having been shown to contain a large transposon-like insertion (Harberd et al. 1987).

The presence of a stop codon would not necessarily prevent a gene from being expressed, but a prolamin gene that has been truncated in this way has never been shown to be active. Where the presence of a protein on an SDS-PAGE gel can readily be attributed to a particular gene as in the case of the HMW prolamins, no protein has ever been shown to be encoded by a gene containing an internal, in-frame stop codon. Western blot analysis of total proteins extracted from single seeds of *A. cylindrica*, for example, failed to detect any protein of the size predicted for a truncated HMW subunit protein encoded by the *Glu-D1x* gene (Wan et al. 2002).

The obvious conclusion to be drawn from this is that the promoters of prolamin genes that contain internal stop codons or are disrupted in another way have become silent and this has been demonstrated for the promoter of a silent *Glu-A1y* gene (Halford et al. 1989). However, the basis for the inactivity has never been explained satisfactorily because the silent *Glu-A1y* gene contains an apparently intact enhancer element. Similarly, the silent Z22 α-zein genes listed above contain either a perfect O2 binding site (*pML2* and *zA1*) or a site with only a single nucleotide substitution (*pML1* and Z7) (Schmidt et al. 1992).

There is evidence that prolamin gene silencing could be caused in part by RNA instability. *Escherichia coli* cells containing a plasmid designed to express the truncated *Glu-D1x* gene from *A. cylindrica* failed to accumulate any RNA transcribed from the heterologous gene (Wan et al. 2002), while four other HMW subunit genes expressed well in the same system. However, RNA instability would not explain why a promoter from a silent HMW subunit gene would fail to drive expression of a reporter gene.

6
Conclusions

We now have a reasonably complete understanding of the various regulatory elements present in the promoter regions of the genes encoding major types of cereal prolamins and of the factors that bind to these. However, little is known about how gene expression is actually triggered by developmental and environmental cues. Similarly, we have no explanation for the wide differences (10-fold or greater) in expression levels which occur within multigene families or for the subtle differences in temporal and spatial patterns of expression which may exist both within and between gene families (as discussed above). Furthermore, once we extend beyond the major groups of prolamins our level of understanding falls to practically zero. A more detailed understanding of the control of endosperm gene expression is clearly required if we are to exploit the potential to use the cereal endosperm as a factory for novel and modified products, for example to produce high value pharmaceuticals or high volume biomaterials. We therefore hope that the current interest in such applications will stimulate more research on fundamental aspects of gene expression in the developing cereal endosperm.

Acknowledgements Rothamsted Research receives grant-aided support from the Biotechnology and Biological Sciences Research Council of the UK.

References

Albani D, Hammond-Kosack MCU, Smith C, Conlan S, Colot V, Holdsworth M, Bevan MW (1997) The wheat transcriptional activator SPA: A seed-specific bZIP protein that recognizes the GCN4-like motif in the bifactorial endosperm box of prolamin genes. Plant Cell 9:171–184

Altpeter F, Vasil V, Srivastava V, Vasil IK (1996) Integration and expression of the high-molecular-weight glutenin subunit 1Ax1 gene into wheat. Nat Biotechnol 14:1155–1159

Anderson OD, Greene FC (1997) The α-gliadin gene family. II. DNA and protein sequence variation, subfamily structure, and origins of pseudogenes. Theor Appl Genet 95:59–65

Axtell JD, Van Scoyoc SW, Christensen PJ, Ejeta G (1979) Current status of protein quality improvement in grain sorghum. In: Seed Protein Improvement in Cereals and Grain Legumes, vol II. Int Atomic Energy Agency, Vienna, pp 357–365

Barro F, Rooke L, Békés F, Gras P, Tatham AS, Fido R, Lazzeri PA, Shewry PR, Barceló P (1997) Transformation of wheat with high molecular weight subunit genes results in improved functional properties. Nat Biotechnol 15:1295–1299

Bartels D, Thompson RD (1986) Synthesis of messenger-RNAs coding for abundant endosperm proteins during wheat grain development. Plant Sci 46:117–125

Bechtel DB, Wilson JD, Shewry PR (1991) Immunocytochemical localization of the wheat storage protein triticin in developing endosperm tissue. Cereal Chem 68:573–577

Boston RS, Fontes EBP, Shank BB, Wrobel RL (1991) Increased expression of the maize immunoglobulin binding protein homolog b-70 in three zein regulatory mutants. Plant Cell 3:497–505

Bright SWJ, Shewry PR (1983) Improvement of protein quality in cereals. CRC Crit Rev Plant Sci 1:49–93

Burgess SR, Shewry PR (1986) Identification of homologous globulins from embryos of wheat, barley, rye and oats. J Exp Bot 37:1863–1871

Burr FA, Burr B (1982) Three mutations in *Zea mays* affecting zein accumulation: a comparison of zein polypeptides, in vitro synthesis and processing, mRNA levels, and genomic organization. J Cell Biol 94:201–206

Bustos AD, Rubio P, Jouve N (2000) Molecular characterisation of the inactive allele of the gene *Glu-A1* and the development of a set of AS-PCR markers for HMW glutenins of wheat. Theor Appl Genet 100:1085–1094

Casey R (1999) Distribution of Some Properties of Seed Globulins. In: Shewry PR, Casey R (eds) Seed Proteins. Kluwer Academic Publishers, Dordrecht, pp 4159–4169

Chui C-F, Falco SC (1995) A new methionine-rich seed storage protein from maize. Plant Physiol 107:291

Ciceri P, Gianazza E, Lazzari B, Lippoli G, Genga A, Hoschek G, Schmidt RJ, Viotti A (1997) Phosphorylation of Opaque2 changes diurnally and impacts its DNA binding activity. Plant Cell 9:97–108

Coleman CE, Larkins BA (1999) The Prolamins of Maize. In Shewry PR, Case R (eds) Seed Proteins. Kluwer Academic Publishers, Dordrecht, pp 109–139

Colot V, Robert LS, Kavanagh TA, Bevan MW, Thompson RD (1987) Localization of sequences in wheat endosperm protein genes which confer tissue-specific expression in tobacco. EMBO J 6:3559–3564

Dannenhoffer JM, Bostwick DE, Or E, Larkins BA (1995) opaque-15, a maize mutation with properties of a defective opaque-2 modifier. Proc Natl Acad Sci USA 92:1931–1935

Davies JT, Shewry PR, Harris N (1993) Spatial and temporal patterns of B hordein synthesis in developing barley (*Hordeum vulgare* L.) caryopses. Cell Biol Int Rep 17:195–203

Doll H, Køie B, Eggum BO (1974) Induced high lysine mutants in barley. Rad Bot 14:73–80
Duffus CM, Cochrane MP (1992) Grain structure and composition. In: Shewry PR (ed) Barley: Genetics, Biochemistry, Molecular Biology and Biotechnology. CAB International, Wallingford, pp 291–317
Dunwell JM, Gane PJ (1998) Microbial relatives of seed storage proteins: conservation of motifs in a functionally diverse superfamily of enzymes. J Mol Evol 46:147–154
Dunwell JM, Purvis A, Khuri S (2004) Cupins: the most functionally diverse protein superfamily? Phytochem 65:7–17
DuPont FM, Hurkman WJ, Vensel WH, Chan R, Lopez R, Tanaka CK, Altenbach SB (2006) Differential accumulation of sulfur-rich and sulfur-poor wheat flour proteins is affected by temperature and mineral nutrition during grain development. J Cereal Sci 44:101–112
Esen A (1987) A proposed nomenclature for the alcohol-soluble proteins (zeins) of maize (*Zea mays* L). J Cereal Sci 5:117–128
FAO (1997) Production Volume 51, FAO, Rome
Forde BG, Heyworth A, Pywell J, Kreis M (1985) Nucleotide sequence of a B1 hordein gene and the identification of possible upstream regulatory elements in endosperm storage protein genes from barley, wheat and maize. Nucl Acids Res 13:7327–7339
Forde J, Malpica J-M, Halford NG, Shewry PR, Anderson OD, Greene FC, Miflin BJ (1985) The nucleotide sequence of a HMW glutenin subunit gene located on chromosome 1A of wheat (*Triticum aestivum* L.). Nucl Acids Res 13:6817–6831
Geevers HO, Lake JK (1992) Development of modified *opaque-2* maize in South Africa. In: Mertz E (ed) Quality Protein Maize. Am Soc Cereal Chem, St Paul, MN, pp 49–78
Giese H, Hopp E (1984) Influence of nitrogen nutrition on the amount of hordein, protein Z and β-amylase messenger RNA in developing endosperms of barley. Carlsberg Res Commun 49:365–383
Gillikin JW, Zhang F, Coleman CE, Bass HW, Larkins BA, Boston RS (1997) A defective signal peptide tethers the *floury-2* zein to the endoplasmic reticulum membrane. Plant Physiol 114:345–352
Hagen G, Rubenstein I (1981) Complex organisation of zein genes in maize. Gene 13:239–249
Halford NG, Forde J, Shewry PR, Kreis M (1989) Functional analysis of the upstream regions of a silent and an expressed member of a family of wheat seed protein genes in transgenic tobacco. Plant Sci 62:207–216
Halford NG, Field JM, Blair H, Urwin P, Moore K, Robert L, Thompson R, Flavell RB, Tatham AS, Shewry PR (1992). Analysis of HMW glutenin subunits encoded by chromosome 1A of bread wheat (*Triticum aestivum* L.) indicates quantitative effects on grain quality. Theor Appl Genet 83:373–378
Hammond-Kosack MCU, Holdsworth MJ, Bevan MW (1993) in vivo footprinting of a low molecular weight glutenin gene (*LMWG-1D1*) in wheat endosperm. EMBO J 12:545–554
Harberd NP, Bartels D, Thompson RD (1985) Analysis of the gliadin multigene loci in bread wheat using nullisomic-tetrasomic lines. Mol Gen Genet 198:234–242
Harberd NP, Flavell RB, Thompson RD (1987) Identification of a transposon-like insertion in a *Glu-1* allele of wheat. Mol Gen Genet 209:326–332
Hill DE, Hope IA, Macke JP, Struhl K (1986) Saturation mutagenesis of the yeast *his3* regulatory site: requirements for transcriptional induction and for binding by GCN4 activator protein. Science 234:451–457
Hinnebusch AG, Natarajan K (2002) Gcn4p, a master regulator of gene expression, is controlled at multiple levels by diverse signals of starvation and stress. Eukaryot Cell 1:22–32

Hinton JJC (1953) The distribution of protein in the maize kernel in comparison with that of wheat. Cereal Chem 30:441–445

Holdsworth MJ, Munoz-Blanco J, Hammond-Kosack M, Colot V, Schuch W, Bevan MW (1995) The maize transcription factor Opaque-2 activates a wheat glutenin promoter in plant and yeast cells. Plant Mol Biol 29:711–720

Hsia CC, Anderson OD (2001) Isolation, characterization of wheat ω-gliadin genes. Theor Appl Genet 103:37–44

Hwang Y-S, Ciceri P, Parsons RL, Moose SP, Schmidt RJ, Huang N (2004) The maize O2 and PBF proteins act additively to promote transcription from storage protein gene promoters in rice endosperm cells. Plant Cell Physiol 45:1509–1518

Kawaura K, Mochida K, Ogihara Y (2005) Expression profile of two storage protein gene families in hexaploid wheat revealed by large-scale analysis of expressed sequence tags. Plant Physiol 139:1870–1880

Kent NL (1966) Subaleurone endosperm cells of high protein content. Cereal Chem 43:585–601

Kim CS, Gunter BG, Kraft J, Boston RS, Yans S, Jung R, Larkins BA (2004) A defective signal peptide in a 19-kD α-zein protein causes the unfolded protein response and an opaque endosperm phenotype in the maize *De*-B30* mutant. Plant Physiol 134:380–387

Kirihara JA, Petri JB, Messing JW (1988) Isolation and sequence of a gene encoding a methionine-rich 10-kDA zein protein from maize. Gene 71:359–370

Kreis M, Forde BG, Rahman S, Miflin BJ, Shewry PR (1985) Molecular evolution of the seed storage proteins of barley, rye and wheat. J Mol Biol 183:499–502

Kridl JC, Vieira J, Rubenstein I, Messing J (1984) Nucleotide sequence analysis of a zein genomic clone with a short open reading frame. Gene 28:113–118

Kriz AL (1999) 7S Globulins of Cereals. In: Shewry PR, Casey R (eds) Seed Proteins. Kluwer Academic Publishers, Dordrecht, pp 477–498

Lamacchia C, Shewry PR, Di Fonzo N, Forsyth J, Harris N, Lazzeri PA, Napier JA, Halford NG, Barcelo P (2001) Endosperm-specific activity of a storage protein gene promoter in transgenic wheat seed. J Exp Bot 355:243–250

Langridge P, Feix G (1983) A zein gene of maize is transcribed from two widely separated promoter regions. Cell 34:1015–1022

Langridge P, Pintor-Toro JA, Feix G (1982) Direction of zein gene transcription in maize genomic clones. Biochem Biophys Res Commun 107:1236–1242

Lending CR, Larkins BA (1989) Changes in the zein composition of protein bodies during maize endosperm development. Plant Cell 1:1011–1023

Lopes MA, Larkins BA (1991) Gamma-zein content is related to endosperm modification in quality protein maize. Crop Sci 31:1655–1662

Lopes MA, Larkins BA (1995) Genetic analysis of opaque-2 modifier gene activity in maize endosperm. Theor Appl Genet 19:274–281

Maier U-G, Browm JWS, Toloczyki C, Feix G (1987) Binding of a nuclear factor to a consensus sequence in the 5' flanking region of zein genes from maize. EMBO J 6:17–22

Margiotta B, Colaprico G, D'Ovidio R, Lafiandra D (1993) Characterization of high M_r subunits of glutenin by combined chromatographic (RP-HPLC) and electrophoretic separations and restriction fragment length polymorphism (RFLP) analyses of their encoding genes. J Cereal Sci 17:221–236

Marks MD, Lindell JS, Larkins BA (1985a) Quantitative analysis of the accumulation of zein mRNA during maize endosperm development. J Biol Chem 260:16445–16450

Marks MD, Lindell JS, Larkins BA (1985b) Nucleotide sequence analysis of zein mRNAs from maize endosperm. J Biol Chem 260:16451–16459

Mena M, Vicente-Carbajosa J, Schmidt RJ, Carbonero P (1998) An endosperm-specific DOF protein from barley, highly conserved in wheat, binds to and activates transcription from the prolamin-box of a native β-hordein promoter in barley endosperm. Plant J 16:53–62

Mertz ET, Bates LS, Nelson OE (1964) Mutant gene that changes protein composition and increases lysine content of maize endosperm. Science 145:279–280

Millet M-O, Montembault A, Autran J-C (1991) Hordein compositional differences in various anatomical regions of the kernel between two different barley types. Sciences des Alimentations 11:155–161

Müller M, Knudsen S (1993) The nitrogen response of a barley C-hordein promoter is controlled by positive and negative regulation of the GCN4 and endosperm box. Plant J 4:343–355

Munck L, Karlsson KE, Hagberg A, Eggum BO (1970) Gene for improved nutritional value in barley seed protein. Science 168:985–987

Nakase M, Hotta H, Adachi T, Aoki N, Nakamura R, Masumura T, Tanaka K, Matsuda T (1996) Cloning of the rice seed α-globulin-encoding gene: sequence similarity of the 5'-flanking region to those of the genes encoding wheat high-molecular-weight glutenin and barley D hordein. Gene 170:223–226

Natarajan K, Meyer MR, Jackson BM, Slade D, Roberts C, Hinnebusch AG, Marton MJ (2001) Transcriptional profiling shows that Gcn4p is a master regulator of gene expression during amino acid starvation in yeast. Mol Cell Biol 21:4347–4368

Nelson OE, Mertz ET, Bates LS (1965) Second mutant gene affecting the amino acid pattern of maize endosperm proteins. Science 150:1469–1470

Norre F, Peyrot C, Garcia C, Rancé I, Drevet J, Theisen M, Gruber V (2002) Powerful effect of an atypical bifactorial endosperm box from wheat HMWG-Dx5 promoter in maize endosperm. Plant Mol Biol 50:699–712

Okita TW, Cheesbrough V, Reeves CD (1985) Evolution and heterogeneity of the alpha-/beta-type and gamma-type gliadin DNA sequences. J Biol Chem 260:8203–8213.

Osborne TB (1924) The Vegetable Proteins, 2nd edn. Longmans, Green & Co, London, p 154

Pedersen K, Bloom KS, Anderson JN, Glover DV, Larkins BA (1980) Analysis of the complexity and frequency of zein genes in the maize genome. Biochem 19:1644–1650

Pedersen K, Argos P, Naravanna SVL, Larkins BA (1986) Sequence analysis and characterisation of a maize gene encoding a high-sulfur zein protein of M_r 15 000. J Biol Chem 261:6279–6284

Pirovano L, Lanzini S, Hartings H, Lazzaroni N, Rossi V, Joshi R, Thompson RD (1994) Structural and functional analysis of an Opaque-2 related gene from sorghum. Plant Mol Biol 24:515–523

Prat S, Perez-Grau L, Puigdomenech P (1987) Multiple variability in the sequence of a family of maize endosperm proteins. Gene 52:41–49

Pysh LD, Aukerman MJ, Schmidt RJ (1993) OHP1: a maize basic domain/leucine zipper protein that interacts with opaque2. Plant Cell 5:227–236

Rafalski JA (1986) Structure of wheat gamma-gliadin genes. Gene 43:221–229

Robert LS, Thompson RD, Flavell RB (1989) Tissue-specific expression of a wheat high molecular weight glutenin gene in transgenic tobacco. Plant Cell 1:569–578

Sabelli P, Shewry PR (1991) Characterization and organisation of gene families at the *Gli-1* loci of bread and durum wheats by restriction fragment analysis. Theor Appl Genet 83:209–216

Salamini F, Di Fonzo N, Gentinetta E, Soave C (1979) A dominant mutation interferring with protein accumulation in maize seeds. In: Seed Protein Improvement in Cereals and Grain Legumes. Proceedings of Food and Agriculture Organisation—Int Atomic Energy Agency Symp. Int Atomic Energy Agency, Vienna, pp 97–108

Salamini F, Foranasani E, Gentinetta E (1983) Mucuronate, Mc, a dominant gene of maize which interacts with opaque-2 to suppress zein synthesis. Theor Appl Genet 65:123–128

Schmidt RJ, Burr FA, Aukerman MJ, Burr B (1990) Maize regulatory gene opaque-2 encodes a protein with a "leucine-zipper" motif that binds to zein DNA. Proc Natl Acad Sci USA 87:46–50

Schmidt RJ, Ketudat M, Aukerman M, Hoschek G (1992) *Opaque-2* is a transcriptional activator that recognizes a specific target site in 22-kDa zein genes. Plant Cell 4:689–700

Scott MP, Sangtong V, Moran D (2002) A Wheat DNA Fragment Exhibits Reduced Pollen Transmission in Transgenic Maize. Maize Genetics Conference Abstracts 44:130

Shewry PR (2000) Seed Proteins. In: Black M, Bewley JD (eds) Seed Technology and its Biological Basis. Sheffield Academic Press, Sheffield, pp 42–84

Shewry PR, Darlington H (2002) The proteins of the mature barley grain and their role in determining malting performance. In: Slafer GA, Molina-Cano JL, Savin R, Araus JL, Romagosa I (eds) Barley Science: Recent Advances from Molecular Biology to Agronomy of Yield and Quality. Food Products Press, NY, pp 503–521

Shewry PR, Halford NG (2002) Cereal seed storage proteins: structures, properties and role in grain utilization. J Exp Bot 53:947–958

Shewry PR, Faulks AR, Miflin BJ (1980) The effect of high-lysine mutations on the protein fractions of barley grain. Biochem Genet 18:133–151

Shewry PR, Brennan C, Tatham AS, Warburton T, Fido R, Smith D, Griggs D, Cantrell I, Harris N (1996) The development, structure and composition of the barley grain in relation to its end use properties. In: Cereals 96—Proc 46th Australian Cereal Chemistry Conf, Sydney, September 1996 158–162

Shewry PR, Halford NG, Belton PS, Tatham AS (2003) Gluten, the elastomeric protein of wheat seeds. In: Shewry PR, Tatham AS, Bailey A (eds) Elastomeric Proteins. Cambridge University Press, Cambridge, pp 279–301

Shewry PR, Jenkins J, Beaudoin F, Mills ENC (2004) The classification, functions and evolutionary relationships of plant proteins in relation to food allergens. In: Mills ENC, Shewry PR (eds) Plant Food Allergens. Blackwell Science, Oxford, pp 24–41

Shotwell MA (1999) Oat Globulins. In: Shewry PR, Casey R (eds) Seed Proteins. Kluwer Academic Publishers, Dordrecht, pp 389–400

Singh NK, Axtell JD (1973) High lysine mutant gene (*hl*) that improves protein quality and biological value of grain sorghum. Crop Sci 3:535–539

Singh NK, Shepherd KW (1987) Solubility behaviour, synthesis, degradation and subcellular location of a new class of disulphide-linked proteins in wheat endosperm. Aust J Plant Physiol 14:245–252

Singh NK, Shepherd KW, Langridge P, Green LC (1991a) Purification and biochemical characterization of triticin, a legume-like protein in wheat endosperm. J Cereal Sci 3:207–219

Singh NK, Shepherd KW, Langridge P, Gruen LC (1991b) Purification and biochemical characterization of triticin, a legumin-like protein in wheat endosperm. J Cereal Sci 13:207–219

Sørensen MB, Cameron-Mills V, Brandt A (1989) Transcriptional and post-transcriptional regulation of gene expression in developing barley endosperm. Mol Gen Genet 217:195–201

Spena A, Viotti A, Pirrotta V (1982) A homologous repetitive block structure underlies the heterogeneity of heavy and light chain zein genes. EMBO J 1:1589–1594

Stöger E, Parker M, Christou P, Casey R (2001) Pea legumin overexpressed in wheat endosperm assembles into an ordered paracrystalline matrix. Plant Physiol 125:1732–1742

Takaiwa F, Ogawa M, Okita TW (1999) Rice Glutelins. In: Shewry PR, Casey R (eds) Seed Proteins. Kluwer Academic Publishers, Dordrecht, pp 401–425

Tatham AS, Shewry PR (2000) Elastomeric proteins—biological roles, structures and mechanisms. Trends Biochem Sci 25:567–571

Thomas MS, Flavell RB (1990) Identification of an enhancer element for the endopserm-specific expression of high molecular weight glutenin. Plant Cell 2:1171–1180

Ueda T, Waverczak W, Ward K, Sher N, Ketudat M, Schmidt RJ, Messing J (1992) Mutations of the 22- and 27-kD zein promoters affect transactivation by the opaque-2 protein. Plant Cell 4:701–709

Vettore AL, Yunes JA, Cord Neto G, Da Silva M, Arruda P, Leite A (1998) The molecular and functional characterization of an opaque-2 homologue gene from Coix and a new classification of plant bZIP proteins. Plant Mol Biol 36:249–263

Villegas E, Vasal SK, Bjarnarson M (1992) Quality Protein Maize—What is it and how was it developed. In: Mertz ET (ed) Quality Protein Maize. Am Soc Cereal Chem, St Paul, MN, pp 27–48

Vicente-Carbajosa J, Carbonero P (2005) Seed maturation: developing an intrusive phase to accomplish a quiescent state. Int J Dev Biol 49:645–651

Vicente-Carbajosa J, Moose SP, Parsons R, Schmidt RJ (1997) A maize zinc-finger protein binds the prolamin box in zein gene promoters and interacts with the basic leucine zipper transcriptional activator Opaque2. Proc Natl Acad Sci USA 94:7685–7690

Wallace JC, Lopes MA, Paiva E, Larkins BA (1990) New methods for extraction and quantitation of zeins reveal a high content of γ-zein in modified opaque-2 maize. Plant Physiol 92:191–196

Wan Y, Wang D, Shewry PR, Halford NG (2002) Isolation, characterization of five novel high molecular weight subunits of glutenin genes from *Triticum timopheevi* and *Aegilops cylindrica*. Theor Appl Genet 104:828–839

Wandelt C, Feix G (1989) Sequence of a 21 kD zein gene from maize containing an in-frame stop codon. Nucl Acids Res 17:2354

Watson SA (1987) Structure and Composition. In: Watson SA, Ramstad PE (eds) Corn: Chemistry and Technology. Am Assoc Cereal Chem, St Paul, MN, pp 53–82

Wilson DR, Larkins BA (1984) Zein gene organization in maize and related grasses. J Mol Biol 20:330–340

Woo Y-M, Hu DW-N, Larkins BA, Jung R (2001) Genomics analysis of genes expressed in maize endosperm identifies novel seed proteins and clarifies patterns of zein gene expression. Plant Cell 13:2297–2317

Yupsanis T, Burgess SR, Jackson PJ, Shewry PR (1990) Characterization of the major component from aleurone cells of barley (*Hordeum vulgare* L). J Exp Bot 41:385–392

Yunes JA, Cord Neto G, Silva G, Leite A, Ottoboni LMM, Aruda P (1994) The transcriptional activator Opaque-2 recognizes two different target sequences in the 22 kDa-like prolamin genes. Plant Cell 6:237–249

Zhang Y, Shewry PR, Jones H, Barcelo P, Lazzeri PA, Halford NG (2001) Expression of antisense SnRK1 protein kinase sequence causes abnormal pollen development and male sterility in transgenic barley. Plant J 28:431–442

Zhang Y, Dickinson JR, Paul MJ, Halford NG (2003) Molecular cloning of an Arabidopsis homologue of GCN2, a protein kinase involved in co-ordinated response to amino acid starvation. Planta 217:668–675

Plant Cell Monogr (8)
O.-A. Olsen: Endosperm
DOI 10.1007/7089_2007_112/Published online: 19 May 2007

Genomic Imprinting in *Arabidopsis thaliana* and *Zea mays*

Jon Penterman · Jin Hoe Huh · Tzung-Fu Hsieh · Robert L. Fischer (✉)

Department of Plant and Microbial Biology, University of California,
Berkeley, California 94720, USA
rfischer@calmail.berkeley.edu

Abstract Genomic imprinting is the differential expression of paternal and maternal alleles. In plants, gene imprinting occurs in the endosperm and has not been found in the embryo or adult plants. Imprinting can affect every allele of a locus (locus-dependent imprinting) or be specific to a particular allele (allele-dependent imprinting). Allele-dependent imprinting was the first type of gene imprinting discovered and has only been documented in maize. Locus-specific imprinting is found in both maize and Arabidopsis. Recent studies have revealed the integral role of female and male gametophytes in gene imprinting and mechanisms by which the parental alleles are distinguished from one another. Herein we will focus on the mechanisms of locus-specific gene imprinting in Arabidopsis and maize.

1 Introduction

Gene imprinting is the differential expression of parental alleles in the same nucleus and has independently evolved in both angiosperm and mammal lineages. Gene imprinting occurs primarily in the placenta of mammals (Fowden et al. 2006) and in the endosperm of angiosperms (Gehring et al. 2004). Both of these structures nourish the next generation and are integral to the success of their respective progenies. According to the parental conflict theory, it is this vital function that led to the evolution of imprinting in angiosperms and mammals (Haig and Westoby 1989, 1991).

The parental conflict theory states that both paternal and maternal parents compete over the amount of resources allocated to their progeny (Haig and Westoby 1989, 1991). The maternal parent, which may have progeny with multiple fathers, equally divides her resources among all of her progeny. The paternal parent might sire only a subset of the total progeny nourished by the mother. It is therefore advantageous for him to direct the most maternal resources to his progeny. During evolution this conflict would lead to a situation where the alleles of genes that repress endosperm growth are largely maternally expressed, whereas the alleles of genes that promote endosperm growth are largely paternally expressed. There are four known genes imprinted in the endosperm of Arabidopsis: *MEDEA* (*MEA*), *FERTILIZATION INDEPENDENT SEED2* (*FIS2*), *FWA*, and *PHERES1* (*PHE1*) (Kinoshita et al.

1999, 2004; Kohler et al. 2005; Jullien et al. 2006a). *MEA*, *FIS2*, and *FWA* are maternally expressed and paternally silent in the endosperm whereas *PHE1* is paternally expressed and maternally repressed. In maize, locus-specific imprinting occurs at *fertilization independent endosperm1* (*fie1*), *fie2*, *maternally expressed gene1* (*meg1*), and *no-apical-meristem related protein1* (*nrp1*) (Danilevskaya et al. 2003; Guo et al. 2003; Gutierrez-Marcos et al. 2003, 2004), all of which are maternally expressed and paternally silenced during endosperm development.

For genes to be imprinted, the alleles must be treated differently from one another in the same nucleus. In a highly inbred population like Arabidopsis, these alleles often have identical DNA sequences. The differential regulation of imprinted genes is therefore not encoded within the DNA sequence but is determined by inherited epigenetic modifications at the locus. These epigenetic modifications include cytosine methylation and histone modifications, both of which can repress gene transcription. Epigenetic modifications can be established, stably maintained during development, and reversed. Most importantly, the epigenetic modifications of one allele can co-exist with different epigenetic modifications of another allele, the end result of which is differential expression of alleles.

Chromatin is the environment in which genes function. Chromatin is composed of non-histone proteins and nucleosomes. A nucleosome is DNA wound around a histone octamer, which consists of two subunits each of histones H2A, H2B, H3, and H4. Nucleosomes are linked by histone H1. Covalent modifications of histones and associated chromatin remodeling can have a profound effect on gene activity (Hsieh and Fischer 2005; Martin and Zhang 2005). Polycomb group (PcG) proteins, which modify histones and chromatin structure, are directly involved in Arabidopsis gene imprinting (Kohler et al. 2005; Baroux et al. 2006; Gehring et al. 2006; Jullien et al. 2006b). PcG proteins form a multimeric complex that covalently methylates lysine 27 of H3 (H3K27) and compacts chromatin (Guitton and Berger 2005; Grimaud et al. 2006). Chromatin compaction can form heterochromatin, which makes genes physically inaccessible to transcription factors and transcriptional complexes (Hsieh and Fischer 2005; Martin and Zhang 2005).

Cytosine DNA methylation can also lead to heterochromatin formation and transcriptional repression. Cytosine methylation is found in CG, CNG, and CNN contexts (N = A, C, or T). CG methylation is maintained by METHYLTRANSFERASE 1 (MET1) and non-CG methylation is maintained by CHROMOMETHYLTRANSFERASE3 and DOMAINS REARRANGED DNA METHYLTRANSFERASE2 (DRM2) (Bender 2004). At many loci, there is an intimate connection between DNA methylation and histone modifications (Jackson et al. 2002; Lippman et al. 2004; Ebbs et al. 2005). These epigenetic modifications are often interdependent; loss of DNA methylation coincides with a loss of histone modifications and vice versa (Lippman and Martienssen 2004; Chan et al. 2005). DNA methylation and histone modifications are integral to heterochromatin formation and maintenance (Lippman and Martienssen 2004; Chan et al. 2005).

CG methylation regulates the imprinting of several Arabidopsis genes (Xiao et al. 2003; Kinoshita et al. 2004; Jullien et al. 2006a) and has been implicated in maize genomic imprinting, too (Gutierrez-Marcos et al. 2006). Methylation at CG sites is symmetrical with cytosines methylated on both DNA strands. CG methylation is established by the de novo methyltransferase DRM2 and can be maintained indefinitely by MET1, which methylates CG sites of the daughter strand immediately after DNA replication (Bender 2004). Establishment and maintenance of Arabidopsis CG methylation are independent processes; the majority of CG methylation is maintained by MET1 with little or no contribution by DMR2 (Bender 2004; Chan et al. 2005).

In this work, we discuss the functions of Arabidopsis and maize imprinted genes and focus on the molecular mechanisms that operate to establish and maintain their imprinting in the endosperm. Readers also might wish to read the following reviews on the relationship of ploidy and genomic imprinting (Dilkes and Comai 2004), the history of plant genomic imprinting (Gehring et al. 2004), mechanisms of mouse genomic imprinting (Lewis and Reik 2006), and the comparison of mammalian and plant genomic imprinting (Scott and Spielman, 2006).

2 Functions of Imprinted Genes

MEA, *FIS2*, *FWA*, and *PHE1* imprinted genes are expressed in the endosperm. In the sections below, we review the functions of *MEA*, *FIS2*, *FWA*, and *PHE1* and the evidence showing that these genes are imprinted.

2.1 Arabidopsis Imprinted Genes

2.1.1 *MEA* and *FIS2* Functions and Imprinting

Genetic experiments indicate that *MEA* and *FIS2* have several functions during Arabidopsis reproduction. Unfertilized female gametophytes that inherit a mutant *mea* or *fis2* allele develop endosperm-like tissue because of precocious nuclear replication of the diploid central cell (Chaudhury et al. 1997; Kiyosue et al. 1999). These seed-like structures cause siliques to elongate in the absence of fertilization. Thus, one function of *FIS2* and *MEA* is to repress nuclear division of the central cell prior to fertilization.

mea and *fis2* mutations also have effects when ovules are fertilized. Siliques from self-pollinated *MEA*/*mea* and *FIS2*/*fis2* heterozygotes contain a 1 : 1 ratio of aborted to viable seed. Reciprocal crosses to wild type showed that *mea* and *fis2* seed abortion is caused by the maternal inheritance of these mutations and that paternally inherited *fis2* and *mea* mutations do not affect seed development (Grossniklaus et al. 1998; Kiyosue et al. 1999). Thus, *mea* and *fis2* have a parent-of-origin effect on seed development. The endosperm of *mea* and *fis2* aborting seeds fails to cellularize, lacks a morphologically defined anterior-posterior polarity, and is enlarged due to proliferation of nuclei (Grossniklaus et al. 1998; Kiyosue et al. 1999; Luo et al. 1999; Sorensen et al. 2001; Gehring et al. 2004) (See Berger et al. in this volume). Thus, one function of the maternal MEA and FIS2 proteins is to repress endosperm growth. Imprinting of the *MEA* and *FIS2* genes is therefore in accord with the predictions of the parental conflict theory because *MEA* and *FIS2*, which repress endosperm growth, are only expressed from the maternal allele.

MEA encodes a SET-domain PcG protein that is homologous to *Drosophila Enhancer of Zeste* [E(z)] (Grossniklaus et al. 1998; Kiyosue et al. 1999), and *FIS2* encodes a zinc-finger transcription factor homologous to *Drosophila Suppressor of Zeste 12* [Su(z)12](Luo et al. 1999). Su(z)12, E(z), and other PcG proteins in *Drosophila* form a large multimeric complex that can remodel chromatin and methylate histone 3 on lysine 27 (H3K27) (Guitton and Berger 2005; Grimaud et al. 2006). H3K27 is methylated by the histone methyltransferase activity of the E(z) SET domain. MEA has also been found to function in a large, multimeric PcG complex (Kohler et al. 2003b), and plants with mutations in other components of the MEA/FIS2 PcG complex, such as *FERTILIZATION INDEPENDENT ENDOSPERM* (*FIE*), have phenotypes similar to *mea* and *fis2* mutants (Hsieh et al. 2003; Guitton and Berger, 2005) (See Berger et al., in this volume). It is not known whether the SET domain of MEA has histone methyltransferase activity, however, H3K27 methylation at a MEA/FIS2 polycomb target requires MEA (Gehring et al. 2006; Jullien et al. 2006b).

Imprinting of *FIS2* and *MEA* has been shown using a number of techniques. One method is to cross two ecotypes polymorphic for *FIS2* or *MEA* coding sequence and then amplify *MEA* and *FIS2* transcripts from seeds or tissues of dissected seeds (i.e. embryo, endosperm) using reverse transcriptase-polymerase chain reaction (RT-PCR) (Kinoshita et al. 1999; Jullien et al. 2006a). The RT-PCR products are then incubated with a restriction enzyme that specifically cuts transcripts from one ecotype, making it possible to distinguish transcripts from maternal and paternal alleles. For the imprinted *MEA* and *FIS2* genes, only the transcripts from the maternal *MEA* and *FIS2* alleles are detected in the endosperm (Kinoshita et al. 1999; Gehring et al. 2006; Jullien et al. 2006a).

Another technique is to monitor the expression of a reporter gene in reciprocal crosses. In this method, the promoter and coding sequence of the gene of interest are fused to the *GREEN FLUORESCENT PROTEIN* (*GFP*) or *β-GLUCURONIDASE* (*GUS*). The expression of a *FIS2::GUS* reporter gene mirrors the expression of the endogenous gene. Maternally inherited *FIS2::GUS* is expressed in the central cell and endosperm whereas paternally inherited *FIS2::GUS* is silent in the endosperm (Luo et al. 2000; Jullien et al. 2006a).

Imprinting has also been shown using in situ hybridization to detect nascent transcripts from the primary endosperm nuclei (Vielle-Calzada et al. 1999). In this method, probes hybridizing to nascent transcripts appear as nuclear dots. If a gene is not imprinted, then there should be three nuclear dots, representing two maternal genomes and one paternal genome. In the primary endosperm cell only two nuclear dots were observed, suggesting that the two maternal *MEA* alleles are expressed and paternal allele is silenced. Besides showing *MEA* imprinting, this method indicates that the endosperm inherits a transcriptionally active maternal *MEA* allele.

2.1.2
FWA Function and Imprinting

FWA encodes a homeodomain transcription factor and is best known for its effect on Arabidopsis flowering time when it is aberrantly expressed in vegetative tissues (Soppe et al. 2000). The promoter and transcription start site of *FWA* reside within two pairs of near-perfect direct repeats. Repetitive sequences are often targets of DNA methylation and repressive histone modifications. The DNA methylation status of these repeats is correlated with the transcriptional state of *FWA* (Soppe et al. 2000). *FWA* is expressed when these sites are not methylated and transcriptionally silent when they are methylated. Mutations that cause genome-wide loss of CG methylation invariably lead to *FWA* expression and late flowering (Soppe et al. 2000; Kankel et al. 2003). *FWA* is normally expressed in the central cell and endosperm but nowhere else (Soppe et al. 2000; Kinoshita et al. 2004). The *FWA* gene was shown to be imprinted using RT-PCR followed by ecotype-specific restriction enzyme analysis and reporter genes (Kinoshita et al. 2004). Loss-of-function mutations in *fwa* do not affect flowering time and seed development and the function of *FWA* in the seed is not known (Soppe et al. 2000; Kinoshita et al. 2004).

2.1.3
PHE1 Function and Imprinting

PHE1 encodes a type 1 MADS box transcription factor that is expressed in the endosperm and embryo (Kohler et al. 2003a). *PHE1* imprinting was shown using RT-PCR followed by ecotype-specific restriction enzyme analysis and by analyzing a reporter gene (Kohler et al. 2005). In seeds, the paternal *PHE1* allele is transcribed significantly more than the maternal *PHE1* allele, and the expression of the *PHE1* reporter gene suggests that *PHE1* is imprinted in the endosperm. Mutations in *PHE1* have no overt effect on seed development. However, *PHE1* is a direct target of the MEA/FIS2 PcG complex, and down-regulation of *PHE1* partially rescues *mea* seed abortion (Kohler et al. 2003a). This indicates that one of the reasons *mea* seeds abort is because *PHE1* is not properly repressed.

2.2 Maize Imprinted Genes

In *Zea mays* there are a number of imprinted genes that fall into two general groups. In one group are genes that exhibit locus-specific imprinting and in the other group are genes that exhibit allele-specific imprinting. In the section below we focus on the genes *fie1*, *fie2*, *meg1*, and *nrp1*, the imprinting of which is inherent to the locus. For a review of allele-specific imprinting we refer our readers to the review by Gehring et al. (2004).

2.2.1 *fie1* and *fie2* Function and Imprinting

fie1 and *fie2* were the first maize genes found to exhibit locus-specific imprinting (Danilevskaya et al. 2003; Gutierrez-Marcos et al. 2003). These genes encode WD-repeat PcG proteins homologous to the Arabidopsis *FIE* gene and *Drosophila Extra Sex Combs* (Springer et al. 2002). Genes encoding other components of the PcG complex have also been identified in maize, suggesting that there is a large multimeric PcG complex that regulates transcription (Springer et al. 2002). Mutations in *fie1* and *fie2* have not been found and their functions during maize seed development are not known. *fie1* expression is only detected in the endosperm (Danilevskaya et al. 2003; Gutierrez-Marcos et al. 2003) whereas *fie2* expression is detected in all reproductive and vegetative tissues tested so far (Springer et al. 2002; Danilevskaya et al. 2003).

The *fie1* and *fie2* genes were shown to be imprinted by reciprocally crossing maize inbred lines and using RT-PCR to follow expression of line-specific polymorphisms in the endosperm (Danilevskaya et al. 2003; Gutierrez-Marcos et al. 2003, 2006). For both genes only expression from the maternal alleles are detected in the endosperm of young developing seeds. *fie1* imprinting is stable throughout seed development (Danilevskaya et al. 2003; Gutierrez-Marcos et al. 2003). By contrast, *fie2* imprinting is less stable and expression from both maternal and paternal alleles is detected in the endosperm during later stages of seed development (Danilevskaya et al. 2003; Gutierrez-Marcos et al. 2003). Imprinting of the *pFIE1-GUS* and *pFIE2-GUS* reporter genes suggests that epigenetic modifications at the promoter are sufficient for *fie1* and *fie2* imprinting (Gutierrez-Marcos et al. 2006).

2.2.2
meg1 Function and Imprinting

meg1 is specifically expressed in the endosperm basal transfer region (Gutierrez-Marcos et al. 2004), a tissue whose primary function is to facilitate nutrient transfer from the maternal tissue to the developing seed (see Royo et al. in this volume). *meg1* encodes a novel protein that is glycosylated and localized in or adjacent to the cell wall ingrowths of basal transfer cells (Gutierrez-Marcos et al. 2004). The role of MEG1 in seed development is not known.

Locus-specific imprinting of *meg1* was discovered using allelic message display (AMD) (Gutierrez-Marcos et al. 2004). AMD is a differential display PCR technique that allows the detection and visualization of multiple transcripts and was used to monitor the transcriptome of endosperm from seeds produced by selfing and reciprocally crossing Mo17 and F2 inbred maize lines (Gutierrez-Marcos et al. 2003). These inbred lines are highly polymorphic and produce differently sized transcripts for many genes. Gutierrez-Marcos et al. (2004) followed the maternal and paternal expression of transcripts and found that *meg1* was maternally expressed and paternally silenced during early endosperm development. At later stages of endosperm development biallelic expression of *meg1* was observed (Gutierrez-Marcos et al. 2004), indicating that *meg1* imprinting is not stable throughout endosperm development. A *pMEG1-GUS* reporter gene was also imprinted, suggesting that the regulation of *meg1* imprinting lies in the promoter region (Gutierrez-Marcos et al. 2004). The epigenetic mechanism operating at *meg1* is not known.

2.2.3
nrp1 Function and Imprinting

nrp1 encodes a putative transcription factor of the NAM gene family. The *nrp1* gene is expressed exclusively in the endosperm; there is no expression detected in the gametophyte generations or in any other sporophytic tissue (Guo et al. 2003). The function of *nrp1* in seeds is not known.

Imprinting of *nrp1* was discovered in a global gene expression study of endosperm tissues of multiple maize hybrids and their inbred parents (Guo et al. 2003). Gene expression was analyzed using GeneCalling (Shimkets et al. 1999), which is a differential display PCR technique. Restriction enzyme sites are essential to the GeneCalling technique and therefore polymorphisms between inbred lines allow the origin of transcripts to be inferred. In this analysis *nrp1* transcripts in the hybrid endosperm were almost totally derived from the female parent (Guo et al. 2003). After discovering line specific polymorphisms, Guo et al. (2003) confirmed *nrp1* imprinting using RT-PCR followed by high-performance liquid chromatography, which allowed them to observe and quantify the expression levels of maternal and paternal alleles. The epigenetic mechanism underlying *nrp1* imprinting is not known.

3
Epigenetic Regulation of Arabidopsis and Maize Gene Imprinting

The mechanisms regulating gene imprinting span the generational divide between gametophyte and sporophyte. Recent research has revealed the integral roles of the female and male gametophytes and uncovered several epigenetic mechanisms used to regulate gene imprinting. In this section, we discuss the regulatory mechanisms that operate at maternal and paternal alleles of the imprinted genes *FWA*, *MEA*, *FIS2*, *fie1*, *fie2*, and *PHE1*.

3.1
Regulation of the Maternal *FWA*, *FIS2*, and *MEA* Alleles

The maternal *FWA*, *FIS2*, and *MEA* alleles are expressed in the central cell and endosperm. The epigenetic regulation of the maternal *MEA* allele is a paradigm that likely applies to *FWA*, *FIS2*, and other unknown imprinted genes. In the paragraphs that follow, we introduce two genes, *DEMETER* (*DME*) and *MET1*, and explain how these genes interact at and transcriptionally regulate the *MEA* locus. We then discuss how the paradigm of *MEA* regulation applies to the regulation of the maternal *FWA* and *FIS2* alleles.

Like *mea* and *fis2*, mutations in *dme* have a parent-of-origin effect on seed viability (Choi et al. 2002). Seeds that inherit a maternal *dme* allele abort regardless of the paternal allele. *DME* is primarily expressed in the central cell before fertilization and decreases substantially afterward (Choi et al. 2002). *DME* expression was not detected in stamens and *DME* reporter genes were not expressed in the egg cell, embryo, and endosperm (Choi et al. 2002). One of the reasons *dme* seeds abort is because *MEA* expression is absent in unfertilized female gametophytes and during early endosperm development (Choi et al. 2002). Thus, DME is required for expression of *MEA* in the central cell and during the early stages of endosperm development.

DME is a bifunctional helix-hairpin-helix DNA glycosylase. DNA glycosylases function in the base excision repair (BER) pathway that replaces damaged, mismatched, and methylated bases from DNA (Scharer and Jiricny 2001). The DNA glycosylase activity of DME is required for Arabidopsis seed viability (Choi et al. 2004). Bifunctional helix-hairpin-helix DNA glycosylases have both DNA glycosylase and apurinic/apyrimidinic (AP) lyase activities. DME DNA glycosylase activity initiates the BER pathway by excising a 5-methylcytosine and the phosphate backbone is then nicked by DME AP lyase activity. An AP endonuclease subsequently generates a 3'-hydroxyl to which a repair DNA polymerase adds a non-methylated cytosine. A DNA ligase seals the nick, completing the BER process.

The discovery that *met1* methyltransferase mutations suppress *dme* seed abortion provided clues about the function of *DME* at the *MEA* locus. Mutations in *met1* suppressed *dme* only when inherited maternally, indicating that *met1* suppression of *dme* occurred in the female gametophyte (Xiao et al. 2003). Furthermore, *met1* suppression of *dme* required a functional maternal *MEA* allele and mutations in *met1* also fully restored *MEA* expression in *dme* mutants (Xiao et al. 2003). This indicates that *met1* suppression of *dme* is upstream of or at the *MEA* locus. The *MEA* locus is methylated at four regions: three distinct 5' promoter regions and a repetitive region flanking the 3' end of *MEA* (*MEA-ISR*) (Cao and Jacobsen 2002; Xiao et al. 2003). Mutations in *met1* eliminate most CG methylation at the *MEA* promoter (Xiao et al. 2003). These observations led to the hypothesis that MET1 methylation directly represses maternal *MEA* allele expression and 5-methylcytosine excision (i.e. demethylation) by DME DNA glycosylase leads to maternal *MEA* allele transcription in the central cell.

This hypothesis was tested by reciprocally crossing ecotypes with polymorphisms in *MEA*'s methylated regions and extracting and bisulfite sequencing DNA from both endosperm and embryo (Gehring et al. 2006). Bisulfite sequencing quantitatively measures the methylation of each cytosine, and because the methylated regions have polymorphisms, the methylation of both paternal and maternal alleles could be distinguished. The maternal *MEA* allele in the endosperm was found to be hypomethylated relative to the paternal endosperm allele at the 3' repeats and a promoter region about – 500 base pairs upstream of the transcription start site (Gehring et al. 2006). In the embryo, both the paternal and maternal alleles were as highly methylated as the paternal *MEA* endosperm allele. Thus, maternal *MEA* allele expression in the endosperm is correlated with hypomethylation.

The maternal endosperm allele is derived from the central cell, which expresses DME DNA glycosylase. Hypomethylation of the maternal *MEA* allele in the endosperm might be due to DME DNA glycosylase excision of methylated cytosine. Indeed, the maternal *MEA* allele is not hypomethylated in endosperm derived from *dme* mutant females crossed to wild-type males (Gehring et al. 2006). DME excises 5-methylcytosine in *E. coli* (Gehring et al. 2006), and purified DME DNA glycosylase protein excises 5-methylcytosine in CG, CNG, and CNN contexts from double-stranded oligonucleotides (Gehring et al. 2006; Morales-Ruiz et al. 2006). Collectively, these data indicate that DME excises 5-methylcytosine from the *MEA* locus, which leads to transcriptional activation of *MEA* in the central cell.

DME is also required for the expression of the maternal *FWA* allele and *FWA-GFP* reporter gene in the central cell and endosperm (Kinoshita et al. 2004). In the sporophyte, *FWA* is transcriptionally silenced by MET1-maintained CG methylation (Soppe et al. 2000; Kankel et al. 2003). Therefore, in the gametophyte, maternal *FWA* expression might be the result of hypomethylation of *FWA*. Kinoshita et al. used bisulfite sequencing to show that the *FWA* loci of the endosperm were indeed hypomethylated (Kinoshita et al. 2004). Lack of *FWA* expression in embryo and seed coat correlated with high levels of *FWA* methylation as observed in the sporophyte (Kinoshita et al.

2004). It is likely that endosperm *FWA* hypomethylation and expression is due to DME-dependent demethylation in the central cell.

The maternal *FIS2* allele and *FIS2::GUS* reporter genes are also positively regulated by DME (Jullien et al. 2006a). There is MET1-dependent methylation at the promoter of *FIS2*, and the paternal *FIS2* endosperm allele is silenced by DNA methylation. It seems likely that DME demethylates the *FIS2* locus in the central cell, which leads to upregulation of the maternal *FIS2* allele in the central cell and early endosperm.

3.2
Regulation of the Paternal *FWA, FIS2*, and *MEA* Alleles

In contrast to maternal allele regulation, the epigenetic mechanisms that regulate the paternal *MEA, FIS2*, and *FWA* alleles are distinct. The paternal *FWA* and *FIS2* alleles are silenced by DNA methylation in the endosperm whereas PcG proteins silence the paternal *MEA* endosperm allele. Identification of the epigenetic mechanisms that silence the paternal *FWA*, *FIS2*, and *MEA* alleles has revealed the critical role of the male gametophyte in gene imprinting. In the following paragraphs, we first discuss the paternal regulation of the *FWA* and *FIS2* alleles and then the paternal regulation of the *MEA* allele.

3.2.1
Paternal *FWA* and *FIS2* Allele Regulation

During vegetative development *FWA* expression is repressed by a combination of DNA and histone methylation. In *met1* mutants CG methylation at the *FWA* promoter is lost and *FWA* is expressed (Bender, 2004). Thus, CG methylation is one factor silencing *FWA*. Another epigenetic modification involved in repressing *FWA* is methylation of histone 3 at lysine 9 (H3K9). Mutations in *DECREASE IN DNA METHYLATION1* (*DDM1*), which is a SWI/SNF chromatin remodeling factor, cause loss of both DNA and H3K9 methylation at the *FWA* promoter (Soppe et al. 2000; Lippman et al. 2004). The combination of DNA and H3K9 methylation is a hallmark of silent heterochromatin, which makes promoters inaccessible to transcription factors (Lippman et al. 2004; Lippman and Martienssen 2004).

The paternal *FWA* allele is silent in both the endosperm and embryo (Kinoshita et al. 2004). Because *FWA* is silenced in the sporophyte, it was hypothesized that silencing of the paternal *FWA* allele might be an inherited condition from the paternal parent. To test this hypothesis, Kinoshita *et al.* examined *FWA* imprinting in the endosperm of seeds from WT females crossed to *met1* mutant males (Kinoshita et al. 2004). In this cross they found that the paternal *FWA* allele was expressed in the endosperm. Crosses to *cmt3* and *drm2* methyltransferase mutant males, which lack *FWA* CNG and CNN methylation, did not affect paternal *FWA* allele silencing (Kinoshita et al. 2004). Thus, the silencing of the paternal *FWA* allele requires MET1-maintained CG methylation.

FIS2 expression is restricted to the central cell and endosperm. Unlike *FWA*, *FIS2* is not expressed in vegetative tissues of *met1* mutant plants and thus, the restricted pattern of *FIS2* expression is not due solely to MET1 methylation. However, silencing of the paternal *FIS2* endosperm allele depends on MET1-mediated CG methylation (Jullien et al. 2006a). In crosses to *met1* homozygous and *met1/MET1* heterozygous males, expression of the paternal *FIS2* allele is detected in the endosperm (Jullien et al. 2006a).

FIS2 and *FWA* are methylated and silenced in WT anthers, but are expressed in anthers of *MET1/met1* heterozygotes (Jullien et al. 2006a). *FWA* and *FIS2* expression in *MET/met1* heterozygotes likely reflects allele hypomethylation and expression in male gametophytes because in *MET1/ met1* heterozygotes methylation is lost during the mitotic divisions of haploid gametophytes that inherit a *met1* mutation (Saze et al. 2003). Indeed, *pFWA::FWA-GFP* gene fusion, which is normally silenced in pollen, is expressed in pollen produced by a *MET1/met1* plant (Jullien et al. 2006a). Thus, MET1-mediated methylation is required in the male gametophyte to maintain a transcriptionally silent allele of both *FWA* and *FIS2*. The transcriptional silence of *FWA* and *FIS2* is then transmitted to and presumably maintained by *MET1* in the endosperm.

3.2.2 Paternal *MEA* Allele Regulation

Paternal allele silencing of *MEA* is distinct from *FWA* and *FIS2*. In crosses to *met1*, *cmt3*, and *drm2* males, the paternal *MEA* allele is transcriptionally silent in the endosperm, indicating that methylation is not required for paternal *MEA* allele silencing (Gehring et al. 2006; Jullien et al. 2006b). Thus, while the maternal allele is repressed by DNA methylation, an epigenetic mechanism other than DNA methylation represses the paternal *MEA* allele.

Clues as to what regulates the paternal *MEA* allele arose from studies on other PcG proteins. In rosette leaves, *MEA* expression was found to be down-regulated by the PcG proteins *FIE* and *CURLY LEAF* (*CLF*), which encode a WD40 protein homologous with *Drosophila* extra sex combs and a SET domain PcG protein, respectively (Katz et al. 2004; Jullien et al. 2006b). Mutations that disrupt other PcG proteins, such as VERNALIZATION 2 and EMBRYONIC FLOWER 2, also lead to *MEA* up-regulation (Jullien et al. 2006b). PcG-dependent H3K27 methylation is observed at the 5' and 3' ends of the *MEA* locus, indicating that PcG proteins directly regulate *MEA* expression (Gehring et al. 2006; Jullien et al. 2006b). Imprinting of the *MEA* gene in the endosperm depends on PcG proteins. The paternal *MEA* allele is expressed in the endosperm of seeds derived from crosses with heterozygous *FIE/fie* males (Jullien et al. 2006b). This observation indicates that paternal *MEA* allele silencing in the endosperm requires functional PcG complexes during male gametophyte development. In seeds that inherit a maternal *fie* or *mea* allele, the paternal *MEA* allele is also expressed (i.e. derepressed) in the endosperm (Baroux et al. 2006; Gehring et al. 2006; Jullien et al. 2006b). Derepression of the paternal *MEA* allele in endosperm correlates with loss of H3K27 methylation at the 5' region of the paternal *MEA* locus in seeds (Gehring et al. 2006). Thus, maternal and paternal PcG proteins, including maternal MEA, maintain the epigenetic modifications and transcriptional silence of the paternal *MEA* endosperm allele.

3.3 Regulation of *fie1* and *fie2* Imprinting

An impressive study by Gutierrez-Marcos et al. (2006) has identified a potential epigenetic mechanism governing maize *fie1* and *fie2* imprinting. Both *fie1* and *fie2* are maternally expressed and paternally silent during endosperm development (Danilevskaya et al. 2003). Because *pFIE1::GUS* and *pFIE2::GUS* reporter genes recapitulate the expression of *fie1* and *fie2* genes, it was hypothesized that the epigenetic mechanism controlling *fie1* and *fie2* imprinting lies in the promoter regions (Gutierrez-Marcos et al. 2006). Bisulfite sequencing uncovered two methylated regions at the *fie1* promoter and one methylated region at the *fie2* promoter (Gutierrez-Marcos et al. 2006). Methylated promoter regions of *fie1* contain CG and CNG methylation and the methylated promoter region of *fie2* contains CG, CNG, and CNN methylation. In early stage endosperm tissue, the promoters of the transcriptionally active maternal *fie1* and *fie2* alleles were hypomethylated and the promoters of the silent paternal *fie1* and *fie2* alleles were hypermethylated (Gutierrez-Marcos et al. 2006). Thus, in maize endosperm tissue the methylation level of parental *fie1* and *fie2* alleles correlates with their transcriptional state. These data suggest that differential DNA methylation is one epigenetic mechanism regulating maize *fie1* and *fie2* imprinting.

Further bisulfite sequencing showed that the methylation pattern of *fie1* alleles in the endosperm is the result of alleles inherited from the gametes. The *fie1* locus is hypermethylated in the egg cell and sperm cells, however, *fie1* is significantly hypomethylated in the central cell (Gutierrez-Marcos et al. 2006). Thus, in the endosperm hypomethylation of maternal *fie1* alleles and hypermethylation of the paternal *fie1* allele is the product of epigenetic alleles inherited from the central cell and sperm cell, respectively. Interestingly, hypomethylated *fie1* alleles in the central cell are not expressed in the ovule (Danilevskaya et al. 2003), indicating that factors other than DNA methylation regulate *fie1* expression. Collectively, these data show that differential DNA methylation of endosperm *fie1* alleles is established in the gametophyte generation.

Unlike *fie1*, the differential DNA methylation of endosperm *fie2* alleles is not the product of inheritance. Bisulfite sequencing of egg, central, and sperm cells showed that the *fie2* promoter is not methylated in any gamete, indicating that the endosperm does not inherit a hypermethylated paternal *fie2* allele (Gutierrez-Marcos et al. 2006). Instead, hypermethylation at the paternal *fie2* allele appears to be established at fertilization or during early endosperm development. Paternal *fie2* hypermethylation might be established by the de novo methyltransferase DRM2 (Gutierrez-Marcos et al. 2006). The mechanism behind preferential hypermethylation of the paternal *fie2* allele is not known. It is also not known if the gametophyte generation has a role in *fie2* imprinting.

3.4 Regulation of *PHE1* Imprinting

PHE1 imprinting is not an on/off situation, but instead, the paternal *PHE1* allele is preferentially expressed while the maternal allele is expressed at a significantly lower level (Kohler et al. 2005). The epigenetic modification regulating maternal *PHE1* allele repression has been identified.

PHE1 is negatively regulated by a PcG complex that includes MEA, FIE, and FIS2 proteins. *PHE1* expression is significantly up-regulated in siliques of selfed *MEA/mea*, *FIE/fie*, or *FIS2/fis2* plants, and chromatin immunoprecipitation assays show that MEA and FIE bind to the *PHE1* promoter (Kohler et al. 2003a). PcG proteins regulate *PHE1* imprinting by repressing the maternal endosperm allele. The maternal *PHE1* allele and a maternally derived *PHE1::GUS* reporter gene are significantly up-regulated in seeds that inherit a maternal *mea* mutation (Kohler et al. 2003a, 2005). At present, the role of the gametophyte generation in *PHE1* imprinting is not known.

4 Paradigms of Plant Gene Imprinting

At least three paradigms of gene imprinting exist in plants: DNA methylation and demethylation regulate *FWA* and *FIS2* imprinting; DNA methylation, demethylation and PcG proteins regulate *MEA* imprinting; and PcG proteins regulate *PHE1* imprinting. In the following paragraphs, we use the models of *FWA* and *MEA* imprinting to highlight some intriguing and important aspects of plant gene imprinting (Fig. 1a, b). We also discuss maize *fie1* and *fie2* imprinting, the latter of which might represent another distinct paradigm of gene imprinting.

4.1 *FWA* and *FIS2* Imprinting

The gametophyte generations are integral to *FIS2* and *FWA* imprinting. Activation of the maternal *FIS2* and *FWA* alleles is likely mediated by DME-

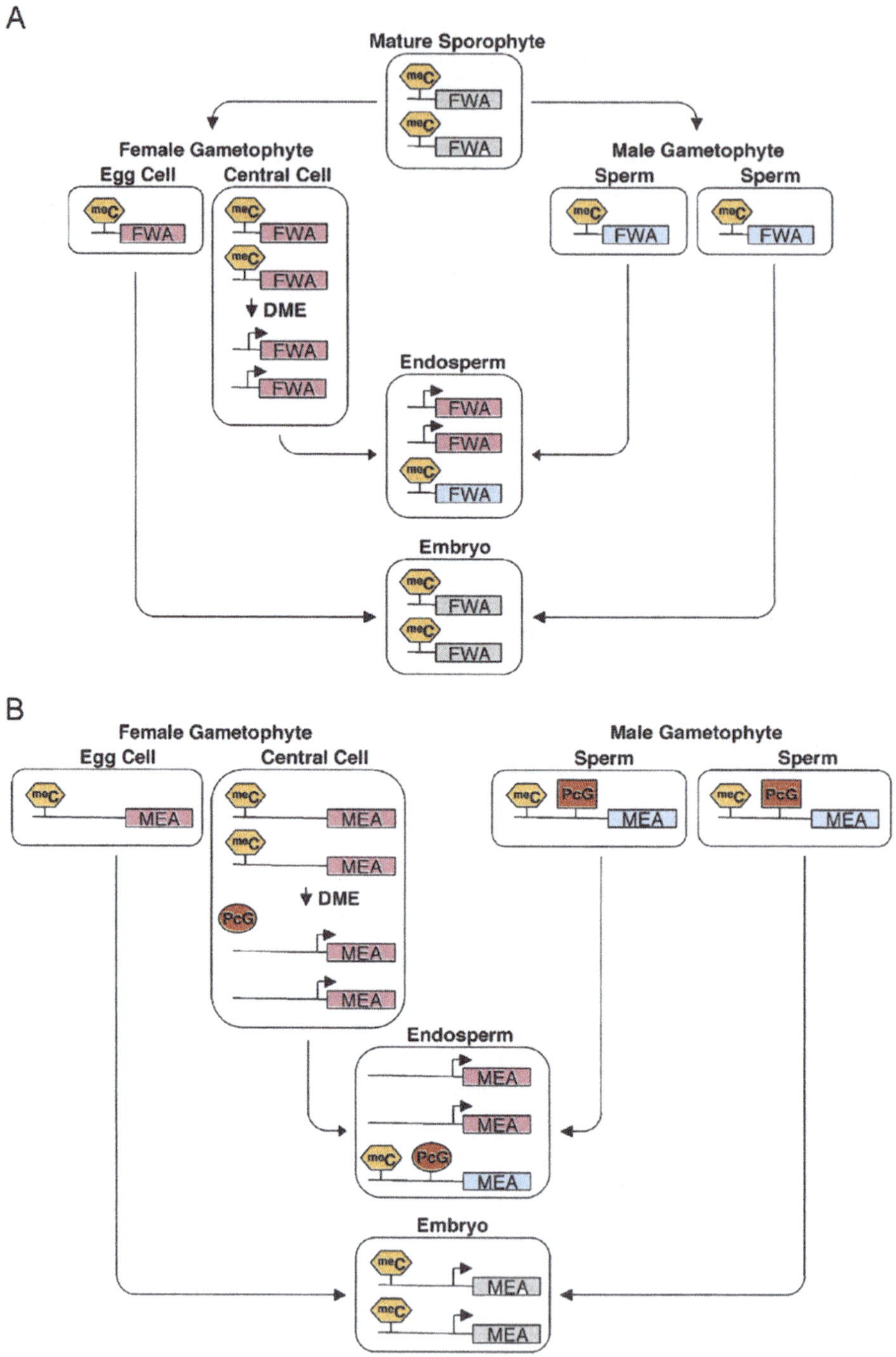
A
Mature Sporophyte
meC
FWA
meC
FWA
Female Gametophyte
Egg Cell
Central Cell
meC
FWA
meC
FWA
meC
FWA
DME
FWA
FWA
Male Gametophyte
Sperm
Sperm
meC
FWA
meC
FWA
Endosperm
FWA
FWA
meC
FWA
Embryo
meC
FWA
meC
FWA
B
Female Gametophyte
Egg Cell
Central Cell
meC
MEA
meC
MEA
meC
MEA
DME
PcG
MEA
MEA
Male Gametophyte
Sperm
Sperm
meC
PcG
MEA
meC
PcG
MEA
Endosperm
MEA
MEA
meC
PcG
MEA
Embryo
meC
MEA
meC
MEA

◀ **Fig. 1** Model of *FWA* and *MEA* imprinting. (**A**) In the sporophyte, *FWA* is transcriptionally silenced by MET1-maintained DNA methylation. A methylated and transcriptionally silent *FWA* allele is transmitted to the progenitors of the female and male gametophyte generation. During male gametophyte development, MET1 maintains the methylation at and transcriptional silence of *FWA*. During female gametophyte development, *FWA* methylation and transcriptional silence is maintained by MET1 in the cellular lineages leading to the egg cell. However, in the lineages leading to the central cell, the maternal *FWA* alleles are transcriptionally activated in a DME-dependent manner. The DME DNA glycosylase likely removes repressive DNA methylation. At fertilization, the primary endosperm nucleus inherits hypomethylated, transcriptionally active maternal *FWA* alleles and a methylated, transcriptionally silent paternal *FWA* allele. During endosperm development, the differential methylation of parental alleles is maintained by MET1 and the maternal *FWA* alleles are expressed while the paternal *FWA* allele is transcriptionally silent. The embryo, which gives rise to the sporophyte, inherits methylated, transcriptionally silent paternal and maternal *FWA* alleles. The silent state is maintained by MET1. (**B**) *MEA* imprinting is established in the gametophyte generation. In the female gametophyte, *MEA* transcription is repressed by DNA methylation. Activation of *MEA* in the central cell is due to DME-dependent demethylation of the *MEA* loci. *MEA, FIS2*, and other PcG genes are expressed and PcG protein complexes that contain MEA and FIS2 are assembled. Presumably *MEA* is not activated by DME in the egg. During male gametophyte development, PcG protein complexes transcriptionally silence *MEA* by methylating H3K27 and compacting chromatin. These PcG protein complexes do not contain MEA and FIS2 because *MEA* and *FIS2* are not expressed in the male gametophyte (Choi et al. 2004; Jullien et al. 2006a). At fertilization, the primary endosperm nucleus inherits hypomethylated, transcriptionally active *MEA* alleles and PcG protein complexes from the female gametophyte. A transcriptionally silent *MEA* allele is inherited from the male gametophyte. During endosperm development, PcG protein complexes that include MEA and FIS2 maintain the transcriptional silence of the paternal *MEA* allele and do not silence the expression of the maternal *MEA* allele. The embryo, which gives rise to the sporophyte, expresses both maternal and paternal *MEA* alleles even though it inherited a PcG-repressed paternal *MEA* allele and a highly DNA methylated maternal *MEA* allele. This may be due to the lack of maternally derived MEA and FIS2 PcG in the early embryo and the fact that DNA methylation does not repress *MEA* in the sporophyte

induced demethylation in the central cell (Fig. 1a) (Kinoshita et al. 2004; Jullien et al. 2006a). The endosperm inherits methylated, transcriptionally silent paternal alleles of *FIS2* and *FWA* from the male gametophyte (Fig. 1a) (Kinoshita et al. 2004; Jullien et al. 2006a). During endosperm development, the transcriptionally active maternal alleles of *FIS2* and *FWA* remain hypomethylated because CG methylation at *FWA* and *FIS2* is likely not re-established. The methylated paternal alleles, on the other hand, remain methylated during endosperm development because MET1 maintains the methylation during DNA replication (Jullien et al. 2006a). In wild-type plants, expression of *FWA* and *FIS2* genes is from the maternal allele and only occurs in the central cell and endosperm (Fig. 1a).

Throughout the plant the default state for *FWA* and *FIS2* is transcriptional silence, which for *FWA* is mediated by DNA methylation. Unlike the embryo, the endosperm does not genetically contribute to the next generation and therefore is a dead end for the hypomethylated and expressed maternal *FWA* and *FIS2* alleles. Silent *FWA* and *FIS2* alleles are inherited in the embryo every generation and maintained each generation by MET1 methylation (Fig. 1a) (Kinoshita et al. 2004; Jullien et al. 2006a; Scott and Spielman 2006).

4.2 *MEA* Imprinting

MEA imprinting is also established in the gametophyte generation (Fig. 1b). In the central cell, the *MEA* locus is demethylated by DME, which leads to its transcriptional activation (Fig. 1b) (Choi et al. 2002; Gehring et al. 2006). The transcriptionally active maternal *MEA* allele is transmitted to the endosperm. In the male gametophyte, PcG proteins repress the *MEA* allele and ensure that a silent paternal *MEA* allele is transmitted to the endosperm (Fig. 1b) (Jullien et al. 2006b). In the endosperm, PcG proteins, of which MEA and FIS2 are maternally derived, form a complex that maintains the transcriptionally repressed state of the paternal *MEA* allele (Fig. 1b) (Gehring et al. 2006; Jullien et al. 2006b). The maternal *MEA* allele of the endosperm remains transcriptionally active.

MEA is expressed outside of the central cell and endosperm (Kinoshita et al. 1999; Kiyosue et al. 1999; Vielle-Calzada et al. 1999; Gehring et al. 2006). Biallelic *MEA* expression is observed in the embryo, seedling, roots, and stems (Kinoshita et al. 1999). Thus, unlike *FIS2* and *FWA*, the default state of the *MEA* locus is not transcriptional silence. In the rosette leaf, male gametophyte and endosperm, PcG protein complexes repress *MEA* transcription, which suggests that PcG repression is a general mechanism that regulates *MEA* expression. PcG silencing of *MEA* in the male gametophyte may be a transcriptional state inherited from the sporophyte (Jullien et al. 2006b).

In the embryo, both *MEA* alleles are expressed despite being methylated (Gehring et al. 2006). Thus, DNA methylation is not repressing *MEA* transcription in the embryo. Yet, in the central cell, MET1 and DME functionally antagonize each other at the *MEA* locus (Xiao et al. 2003). The data suggest that DNA methylation at *MEA* has a repressive role in the central cell (Gehring et al. 2006).

4.3 Location and Timing of DNA Demethylation are Crucial to Gene Imprinting

DME regulates the imprinting of *FIS2*, *MEA*, and *FWA* directly and *PHE1* indirectly. *DME* expression is restricted to the central cell where the DME protein likely demethylates the maternal *FIS2*, *MEA*, and *FWA* alleles (Fig. 1a, b). After fertilization, *DME* expression is not detected in the primary endosperm nucleus, and thus, there is minimal exposure of the paternal *MEA*,

FIS2, and *FWA* to DME-dependent demethylation. The temporal regulation of DME demethylation is key to gene imprinting.

DME-dependent DNA demethylation occurs at the maternal *MEA* allele of the endosperm and not at the maternal *MEA* allele of the embryo (Fig. 1b). These alleles are derived from the central and egg cells, respectively. *DME* reporter genes are expressed in the central cell but not in egg cell, which suggests that egg cells lack DME-dependent demethylation because *DME* is not expressed there.

The restriction of DME-dependent DNA demethylation to the central cell is essential for *FWA* gene imprinting in the endosperm (Fig. 1a). *FWA* imprinting in the endosperm depends on the default state of *FWA* being transcriptionally silent. One unique characteristic of the *FWA* gene is that a hypomethylated *FWA* allele can be stably expressed for generations because remethylation is sporadic and slow (Soppe et al. 2000; Kankel et al. 2003). Thus, to maintain the transcriptionally silent default state of *FWA* and therefore *FWA* imprinting, the methylation at *FWA* must be maintained in the sporophyte and the lineages that give to the sporophyte. In wild-type plants, *FWA* is expressed only in the central cell and endosperm because DME demethylates the maternal *FWA* allele only in the central cell. If DME-dependent demethylation and transcriptional activation of *FWA* occurred in the egg cell, the embryo (i.e. sporophyte) would inherit a hypomethylated, transcriptionally active *FWA* allele and *FWA* imprinting would be disrupted in the next generation. Thus, the spatial regulation of DME-dependent demethylation in the female gametophyte is key to *FWA* imprinting.

4.4
fie1 and *fie2* Imprinting

It is tempting to speculate that *fie1* imprinting is similar to *FWA* imprinting in Arabidopsis. Like *FWA*, hypomethylated maternal *fie1* alleles are inherited from the central cell and a hypermethylated paternal *fie1* allele is inherited from the sperm cell (Kinoshita et al. 2004; Gutierrez-Marcos et al. 2006; Jullien et al. 2006a). Additionally, *fie1* expression is restricted to the maize endosperm and this resembles the pattern of *FWA* expression in Arabidopsis (Danilevskaya et al. 2003; Kinoshita et al. 2004). The default state for *fie1* might be hypermethylation and transcriptional silence, which is maintained in the male gametophyte and transmitted to the endosperm. DNA demethylation by a maize DME homolog could lead to *fie1* hypomethylation in the central cell, which is transmitted to and expressed in the endosperm. These events would lead to differential DNA methylation at and imprinting of *fie1* in maize endosperm tissue.

Imprinting of *fie2* might represent a new paradigm of plant gene imprinting. Paternal *fie2* allele hypermethylation in the endosperm is not inherited from the male gamete (Gutierrez-Marcos et al. 2006). Unlike *FWA*, *FIS2*, and

MEA, there is no DNA methylation at the *fie2* promoter in the gametes and embryo (Gutierrez-Marcos et al. 2006). These findings strongly suggest that hypomethylation is the default state for *fie2*, which is in agreement with the broad expression of *fie2* in maize (Danilevskaya et al. 2003; Kinoshita et al. 2004; Jullien et al. 2006a). A novel gene imprinting pathway might be regulating *fie2* (Gutierrez-Marcos et al. 2006). This pathway would involve specific de novo DNA methylation of the paternal *fie2* allele at fertilization or during early endosperm development.

5 Conclusion

In plants, two epigenetic mechanisms, DNA methylation and PcG proteins, regulate gene imprinting. Proper gene imprinting in Arabidopsis and possibly in maize requires DME-dependent DNA demethylation, which is spatially and temporally compartmentalized in the female gametophyte. Arabidopsis gene imprinting also requires pathways that maintain epigenetic marks, such as DNA and H3K27 methylation, during the male gametophyte generation. For genes imprinted in Arabidopsis, the transcriptional activity of maternal and paternal alleles often reflects their transcriptional state in the female and male gametophyte, respectively. Finally, the molecular events that establish gene imprinting are also necessary for endosperm development and seed viability.

References

Baroux C, Gagliardini V, Page DR, Grossniklaus U (2006) Dynamic regulatory interactions of Polycomb group genes: MEDEA autoregulation is required for imprinted gene expression in Arabidopsis. Genes Dev 20:1081–1086

Bender J (2004) DNA methylation and epigenetics. Annu Rev Plant Biol 55:41–68

Cao X, Jacobsen SE (2002). Locus-specific control of asymmetric and CpNpG methylation by the DRM and CMT3 methyltransferase genes. Proc Natl Acad Sci USA 99(4):16491–16498

Chan SW, Henderson IR, Jacobsen SE (2005) Gardening the genome: DNA methylation in *Arabidopsis thaliana*. Nat Rev Genet 6:351–360

Chaudhury AM, Ming L, Miller C, Craig S, Dennis ES, Peacock WJ (1997) Fertilization-independent seed development in *Arabidopsis thaliana*. Proc Natl Acad Sci USA 94:4223–4228

Choi Y, Harada JJ, Goldberg RB, Fischer RL (2004) An invariant aspartic acid in the DNA glycosylase domain of DEMETER is necessary for transcriptional activation of the imprinted MEDEA gene. Proc Natl Acad Sci USA 101:7481–7486

Choi Y, Gehring M, Johnson L, Hannon M, Harada JJ, Goldberg RB, Jacobsen SE, Fischer RL (2002) DEMETER, a DNA glycosylase domain protein, is required for endosperm gene imprinting and seed viability in *Arabidopsis*. Cell 110:33–42

Danilevskaya ON, Hermon P, Hantke S, Muszynski MG, Kollipara K, Ananiev EV (2003) Duplicated fie genes in maize: expression pattern and imprinting suggest distinct functions. Plant Cell 15:425–438

Dilkes BP, Comai L (2004) A differential dosage hypothesis for parental effects in seed development. Plant Cell 16:3174–3180

Ebbs ML, Bartee L, Bender J (2005) H3 lysine 9 methylation is maintained on a transcribed inverted repeat by combined action of SUVH6 and SUVH4 methyltransferases. Mol Cell Biol 25:10507–10515

Fowden AL, Sibley C, Reik W, Constancia M (2006) Imprinted genes, placental development and fetal growth. Horm Res 65(3):50–58

Gehring M, Choi Y, Fischer RL (2004) Imprinting and seed development. Plant Cell 16(1):S203–S213

Gehring M, Huh JH, Hsieh TF, Penterman J, Choi Y, Harada JJ, Goldberg RB, Fischer RL (2006) DEMETER DNA glycosylase establishes MEDEA polycomb gene self-imprinting by allele-specific demethylation. Cell 124:495–506

Grimaud C, Negre N, Cavalli G (2006) From genetics to epigenetics: the tale of Polycomb group and trithorax group genes. Chromosome Res 14:363–375

Grossniklaus U, Vielle-Calzada JP, Hoeppner MA, Gagliano WB (1998) Maternal control of embryogenesis by MEDEA, a polycomb group gene in *Arabidopsis*. Science 280:446–450

Guitton AE, Berger F (2005) Control of reproduction by Polycomb Group complexes in animals and plants. Int J Dev Biol 49:707–716

Guo M, Rupe MA, Danilevskaya ON, Yang X, Hu Z (2003) Genome-wide mRNA profiling reveals heterochronic allelic variation and a new imprinted gene in hybrid maize endosperm. Plant J 36:30–44

Gutierrez-Marcos JF, Pennington PD, Costa LM, Dickinson HG (2003) Imprinting in the endosperm: a possible role in preventing wide hybridisation. Philos Trans R Soc Lond 358:1105–1111

Gutierrez-Marcos JF, Costa LM, Dal Pra M, Scholten S, Kranz E, Perez P, Dickinson HG (2006) Epigenetic asymmetry of imprinted genes in plant gametes. Nat Genet 38:876–878

Gutierrez-Marcos JF, Costa LM, Biderre-Petit C, Khbaya B, O'Sullivan DM, Wormald M, Perez P, Dickinson HG (2004) maternally expressed gene1 Is a novel maize endosperm transfer cell-specific gene with a maternal parent-of-origin pattern of expression. Plant Cell 16:1288–1301

Haig D, Westoby M (1989) Parent-specific gene expression and the triploid endosperm. Am Nat 134:147–155

Haig D, Westoby M (1991) Genomic imprinting in endosperm: Its effect of seed development in crosses between species, and between different ploidies of the same species, and its implications for the evolution of apomixis. Philos Trans R Soc Lond 333:1–13

Hsieh TF, Fischer RL (2005) Biology of chromatin dynamics. Annu Rev Plant Biol 56:327–351

Hsieh TF, Hakim O, Ohad N, Fischer RL (2003) From flour to flower: how Polycomb group proteins influence multiple aspects of plant development. Trends Plant Sci 8:439–445

Jackson JP, Lindroth AM, Cao X, Jacobsen SE (2002) Control of CpNpG DNA methylation by the KRYPTONITE histone H3 methyltransferase. Nature 416:556–560

Jullien PE, Kinoshita T, Ohad N, Berger F (2006a) Maintenance of DNA methylation during the *Arabidopsis* life cycle is essential for parental imprinting. Plant Cell 18:1360–1372

Jullien PE, Katz A, Oliva M, Ohad N, Berger F (2006b) Polycomb group complexes self-regulate imprinting of the Polycomb group gene MEDEA in *Arabidopsis*. Curr Biol 16:486–492

Kankel MW, Ramsey DE, Stokes TL, Flowers SK, Haag JR, Jeddeloh JA, Riddle NC, Verbsky ML, Richards EJ (2003) *Arabidopsis* MET1 cytosine methyltransferase mutants. Genetics 163:1109–1122

Katz A, Oliva M, Mosquna A, Hakim O, Ohad N (2004) FIE and CURLY LEAF polycomb proteins interact in the regulation of homeobox gene expression during sporophyte development. Plant J 37:707–719

Kinoshita T, Yadegari R, Harada JJ, Goldberg RB, Fischer RL (1999) Imprinting of the MEDEA polycomb gene in the *Arabidopsis* endosperm. Plant Cell 11:1945–1952

Kinoshita T, Miura A, Choi Y, Kinoshita Y, Cao X, Jacobsen SE, Fischer RL, Kakutani T (2004) One-way control of FWA imprinting in *Arabidopsis* endosperm by DNA methylation. Science 303:521–523

Kiyosue T, Ohad N, Yadegari R, Hannon M, Dinneny J, Wells D, Katz A, Margossian L, Harada JJ, Goldberg RB, Fischer RL (1999) Control of fertilization-independent endosperm development by the MEDEA polycomb gene in *Arabidopsis*. Proc Natl Acad Sci USA 96:4186–4191

Kohler C, Page DR, Gagliardini V, Grossniklaus U (2005) The *Arabidopsis thaliana* MEDEA Polycomb group protein controls expression of PHERES1 by parental imprinting. Nat Genet 37:28–30

Kohler C, Hennig L, Spillane C, Pien S, Gruissem W, Grossniklaus U (2003a) The Polycomb-group protein MEDEA regulates seed development by controlling expression of the MADS-box gene PHERES1. Genes Dev 17:1540–1553

Kohler C, Hennig L, Bouveret R, Gheyselinck J, Grossniklaus U, Gruissem W (2003b) *Arabidopsis* MSI1 is a component of the MEA/FIE Polycomb group complex and required for seed development. EMBO J 22:4804–4814

Lewis A, Reik W (2006) How imprinting centres work. Cytogenet Genome Res 113:81–89

Lippman Z, Martienssen R (2004) The role of RNA interference in heterochromatic silencing. Nature 431:364–370

Lippman Z, Gendrel AV, Black M, Vaughn MW, Dedhia N, McCombie WR, Lavine K, Mittal V, May B, Kasschau KD, Carrington JC, Doerge RW, Colot V, Martienssen R (2004) Role of transposable elements in heterochromatin and epigenetic control. Nature 430:471–476

Luo M, Bilodeau P, Dennis ES, Peacock WJ, Chaudhury A (2000) Expression and parent-of-origin effects for FIS2, MEA, and FIE in the endosperm and embryo of developing *Arabidopsis* seeds. Proc Natl Acad Sci USA 97:10637–10642

Luo M, Bilodeau P, Koltunow A, Dennis ES, Peacock WJ, Chaudhury AM (1999) Genes controlling fertilization-independent seed development in *Arabidopsis thaliana*. Proc Natl Acad Sci USA 96:296–301

Martin C, Zhang Y (2005) The diverse functions of histone lysine methylation. Nat Rev Mol Cell Biol 6:838–849

Morales-Ruiz T, Ortega-Galisteo AP, Ponferrada-Marin MI, Martinez-Macias MI, Ariza RR, Roldan-Arjona T (2006) DEMETER and REPRESSOR OF SILENCING 1 encode 5-methylcytosine DNA glycosylases. Proc Natl Acad Sci USA 103:6853–6858

Saze H, Scheid OM, Paszkowski J (2003) Maintenance of CpG methylation is essential for epigenetic inheritance during plant gametogenesis. Nat Genet 34:65–69

Scharer OD, Jiricny J (2001) Recent progress in the biology, chemistry and structural biology of DNA glycosylases. Bioessays 23:270–281

Scott RJ, Spielman M (2006) Genomic imprinting in plants and mammals: how life history constrains convergence. Cytogenet Genome Res 113:53–67

Shimkets RA, Lowe DG, Tai JT, Sehl P, Jin H, Yang R, Predki PF, Rothberg BE, Murtha MT, Roth ME, Shenoy SG, Windemuth A, Simpson JW, Simons JF, Daley MP, Gold SA,

McKenna MP, Hillan K, Went GT, Rothberg JM (1999) Gene expression analysis by transcript profiling coupled to a gene database query. Nat Biotechnol 17:798–803
Soppe WJ, Jacobsen SE, Alonso-Blanco C, Jackson JP, Kakutani T, Koornneef M, Peeters AJ (2000) The late flowering phenotype of fwa mutants is caused by gain-of-function epigenetic alleles of a homeodomain gene. Mol Cell 6:791–802
Sorensen MB, Chaudhury AM, Robert H, Bancharel E, Berger F (2001) Polycomb group genes control pattern formation in plant seed. Curr Biol 11:277–281
Springer NM, Danilevskaya ON, Hermon P, Helentjaris TG, Phillips RL, Kaeppler HF, Kaeppler SM (2002) Sequence relationships, conserved domains, and expression patterns for maize homologs of the polycomb group genes E(z), esc, and E(Pc). Plant Physiol 128:1332–1345
Vielle-Calzada JP, Thomas J, Spillane C, Coluccio A, Hoeppner MA, Grossniklaus U (1999) Maintenance of genomic imprinting at the *Arabidopsis medea* locus requires zygotic DDM1 activity. Genes Dev 13:2971–2982
Xiao W, Gehring M, Choi Y, Margossian L, Pu H, Harada JJ, Goldberg RB, Pennell RI, Fischer RL (2003) Imprinting of the MEA Polycomb gene is controlled by antagonism between MET1 methyltransferase and DME glycosylase. Dev Cell 5:891–901

Subject Index

GPSR Compliance

The European Union's (EU) General Product Safety Regulation (GPSR) is a set of rules that requires consumer products to be safe and our obligations to ensure this.

If you have any concerns about our products, you can contact us on ProductSafety@springernature.com

In case Publisher is established outside the EU, the EU authorized representative is:

Springer Nature Customer Service Center GmbH
Europaplatz 3
69115 Heidelberg, Germany

Batch number: 10777861

Printed by Printforce, the Netherlands